THE HANDBOOK OF MANUFACTURING ENGINEERING
Second Edition

Product Design and Factory Development

THE HANDBOOK OF MANUFACTURING ENGINEERING
Second Edition

Product Design and Factory Development

EDITED BY

Richard Crowson

CRC Press
Taylor & Francis Group
Boca Raton   London   New York

CRC Press is an imprint of the
Taylor & Francis Group, an **informa** business

CRC Press
Taylor & Francis Group
6000 Broken Sound Parkway NW, Suite 300
Boca Raton, FL 33487-2742

First issued in paperback 2019

© 2006 by Taylor & Francis Group, LLC
CRC Press is an imprint of Taylor & Francis Group, an Informa business

No claim to original U.S. Government works

ISBN-13: 978-0-8493-5519-6 (hbk)
ISBN-13: 978-0-367-39140-9 (pbk)
ISBN 13: 978-0-8247-2341-5 (set)

Library of Congress Cataloging-in-Publication Data

Catalog record is available from the Library of Congress

Visit the Taylor & Francis Web site at
http://www.taylorandfrancis.com

and the CRC Press Web site at
http://www.crcpress.com

Preface

Handbooks are generally considered to be concise references for specific subjects. Today's fast-paced manufacturing culture demands that such reference books provide the reader with how-to information with no frills. Some use handbooks to impart buzzwords on a particular technical subject that will allow the uninitiated to gain credibility when discussing a technical situation with more experienced practitioners.

The second edition of *The Manufacturing Engineering Handbook* was written to equip executives, manufacturing professionals, and shop personnel with enough information to function at a certain level on a variety of subjects. The level of professional skill developed by reading this handbook is determined by the reader.

This series of books is written to serve as a textbook or supplemental material for classes in engineering schools for students pursuing a manufacturing engineering bachelor's or graduate degree.

Manufacturing engineering professionals are taking the lead in a number of jobs not traditionally thought to be covered in the discipline: decision-making executives of manufacturing businesses; product design engineers; process design engineers; and positions in tooling, facility planning, and production control are now filled by engineers that first learned the trade of manufacturing engineering.

The resurrection of an undesignated apprentice in our manufacturing culture has become necessary as the technology continues to grow at a phenomenal rate. Companies can no longer afford to allow the manufacturing engineer to learn on the job. As more universities graduate engineers with bachelors to doctorate degrees in manufacturing engineering, the challenges presented by the fiercely technical world we live and work in are met by the manufacturing engineer. Therefore, the manufacturing engineer must gain as much knowledge from as many credible sources as possible. That is why this series of books was developed: to be an aid to successfully applying manufacturing engineering skills in school or on the job.

Many manufacturing engineers choose to stay in the field of manufacturing engineering due largely to the sense of accomplishment gained from creatively seeing every step of the process by which raw material is turned into a finished product. However, an increasingly large number of engineers who started as manufacturing engineers are moving into positions of supervision in the design engineering, research and development, and administrative areas. This is true because the manufacturing engineer brings the quality of knowledge that is hands-on, commonsense, and practical.

The advent of smaller and faster computer processors has made some of the systematic approaches to manufacturing affordable to the mid-sized to small manufacturer. Therefore the quality information presented by Geoffrey Boothroyd's DFMA® (Design for Manufacture and Assembly), which has historically been used by automotive and aircraft manufacturers, is becoming available to the smaller, more flexible manufacturer.

As a manufacturing engineer uses this handbook to study history and apply principles to an existing manufacturing firm, new ideas will be spawned that will allow improvements in process flow and product flow. The successful efforts of many years' experience are captured in these chapters and can be used profitably by any reader willing to think out of the box when facing challenges on a daily basis.

Human factors greatly influence productivity in the workplace, as does workplace safety. The public awareness of hazardous practices requires the manufacturing engineer to put the ergonomic and safety practices that are described in this handbook into practice.

The historic data for anthropometry is based upon the U.S. military data and statistics compiled in the 1970s and 1980s. As the culture of manufacturing expands to include the world, these data are no longer relevant. It is up to the new manufacturing engineer to seek to add to these data and generate ethnicity-specific data to assist in the design of ergonomically based products.

The cost of a product is directly impacted by injuries, product liability, and low productivity. Thus, a manufacturing engineer must strive to become expert in the mitigation of these areas.

This second edition of *The Manufacturing Engineering Handbook* is dedicated to Jack M. Walker, who was unable to participate in the editing of this book but who contributed greatly in the last few months of his life. Much written by Jack still remains unchanged in this edition because of its value in the workplace and manufacturing environment today. Jack was a wonderful and inventive man who loved to look for ways to do things that had not been thought of before.

I will always remember with joy the hours he and I spent together while he gave example after example of inventions and designs that he had authored. Most of these were incredibly simple and ingenious. Jack delighted in swimming upstream by solving problems too quickly for others. He would recount examples of being made to prove over and over again the idea he had come up with quickly. This did not discourage him, and he delighted in the fact that the proving of his ideas took many times longer than the invention of that idea. His favorite saying was that waiting for others to catch up with him was "better than a sharp stick in the eye."

His smile and the gleam in his eye let you know he had experienced a full life and had a deep understanding of the human interaction in the manufacturing engineering process. We will miss Jack, but we will all enjoy his spirit of joy at the discovery of new ways to manufacture.

RICHARD D. CROWSON
SET, CMfgT, CMfgE

Editor

Richard D. Crowson

Richard Crowson is currently a mechanical engineer at Controlled Semiconductor, Inc., in Orlando, Florida. He has worked in the field of engineering, especially in the area of lasers and in the development of semiconductor manufacturing equipment, for over 25 years. He has experience leading multidisciplinary engineering product development groups for several Fortune 500 companies as well as small and start-up companies specializing in laser integration and semiconductor equipment manufacture.

Crowson's formal engineering training includes academic undergraduate and graduate studies at major universities including the University of Alabama at Birmingham, University of Alabama in Huntsville, and Florida Institute of Technology. He presented and published technical papers at Display Works and SemiCon in San Jose, California.

He has served on numerous SEMI task forces and committees as a voting member. His past achievements include participating in writing the SEMI S2 specification, consulting for the 9th Circuit Court as an expert in laser welding, and sitting on the ANSI Z136 main committee that regulates laser safety in the United States.

Contributors

Geoffrey Boothroyd
Boothroyd Dewhurst, Inc.
Wakefield, Rhode Island

Richard D. Crowson
Consultant
Melbourne, Florida

Vijay S. Sheth
McDonnell Douglas Corporation
Titusville, Florida

Timothy L. Murphy
McDonnell Douglas Corporation
Titusville, Florida

Allen E. Plogstedt
McDonnell Douglas Aerospace East
Titusville, Florida

Marc Plogstedt
ITEC
Orlando, Florida

Paul Riedel
Certified Industrial Hygienist
Rockledge, Florida

Jeffery W. Vincoli
CSP, REP
Titusville, Florida

Jack M. Walker
Consultant
Merritt Island, Florida

William L. Walker
National High Magnetic Field Laboratory
Florida State University
Tallahassee, Florida

Contents

1 Product Development

Jack M. Walker

and

Richard D. Crowson

with

Geoffrey Boothroyd

The greatest durability contest in the world is trying to get a new idea into a factory.
—Charles F. Kettering, CEO, General Motors
but

It's one heck of a lot easier to implement World Class Manufacturing principles than to go out of business.
—William P. Conlin, President, Cal Comp, Inc.
and

You can't achieve World Class Manufacturing without first having a World Class Product Design!
—Jack M. Walker

1.0 INTRODUCTION TO PRODUCT DEVELOPMENT

This chapter on product development discusses the impact of concurrent engineering on product development. In the first subchapter we discuss the need for teamwork among the various factory functions, starting with the customer's wants and his or her early involvement in the product development cycle. The role of the engineering manager, manufacturing engineer, project manager, or product engineer when guiding the product development process includes five basic steps in a systematic approach to product development. Pahl and Beitz, in their book *Engineering Design: A Systematic Approach* (1993), defined five conditions that must be satisfied when employing a systematic approach to guide the product development process. Those five steps have proven to be very crucial to the editor in creating the corporate atmosphere conducive to cooperation and concurrent engineering design in the past 30 years of product design for major corporations.

1

The steps are: (1) ensure the requisite motivation for the solution of the task; (2) clarify the boundary conditions; (3) dispel prejudice; (4) look for variants, i.e., a number of possible solutions that will best serve as the solution to the design problem; and (5) make decisions.

The motivation to introduce a new product is not always equally shared by each participant in the concurrent process. Some may feel the current product does not need to be changed. These feeling may stem from the insecurity change in a company brings. The uncertainty in the financial implications of the risk/reward ratio of moving into new territory that is yet unexplored may unsettle some who do not normally participate in the design process. Teams are made up of all disciplines within the company and equally share in the risks and rewards from the effort. Some team members may simply be satisfied with the status quo and do not want to learn new things. It is up to the leader of this effort to understand that the first three steps involve removing any "psychological inertia" that might deflect him or her from the objective of introducing a new innovative product to design teams. A key factor in the presentation order of these five steps is that the first three steps and about 60% of the effort are consumed in effectively communicating the product vision. The leader must present a roadmap to define what the product will be, how it will be developed, and what it will cost. The leader must also define what the product will not be, and he or she must attempt to answer any questions that may have negative connotations from those prejudiced against the product. The concurrent team is the leader's best source of information that will come from the customer and ultimately the end user. The leader must listen to team members and others and consider as many possible solutions as is practical before leading the team/company in any direction. The last step involves making decisions. These objective decisions will cause the success or failure of any product development program. Past history has shown us that products succeed for a number of reasons; unless there is cooperation in the team that develops the product, the company introducing the product to the market may simply spawn an idea others bring to profitable conclusions. The team with a vision for success is one that has agreed to rally around a winning idea. This team will succeed with a good leader at who gathers data, defines problems, and then defines solutions for those problems with the help of his or her team. This systematic approach will be discussed in more detail in later chapters dealing with machine design and product troubleshooting.

Other subchapters describe in some detail two of the commercial systems (tools) available that may be used to formalize the satisfaction of the customer and to design products that will make money for the company. These are design for assembly (DFA) and quality function deployment (QFD). The final subchapter introduces some of the rapid prototyping techniques that allow design concepts to be produced quickly to assist in product development and production start-up prior to availability of the planned "hard" tooling.

This chapter covers the first half of product development and design—the systems of determining who our real customers are and what they want, and the design of a product that meets these requirements at the lowest cost.

Chapter 2 introduces the product development of machines; the detailed design parameters for parts fabricated by machining operations, by casting and forging, and of sheet metal; and a brief introduction of the materials and processes involved in each.

Chapter 3 discusses the role of the professional manufacturing engineer in product development, debugging or troubleshooting the newly developed product, and the fundamentals of factory layout. Troubleshooting is one skill that all product development professionals must master if they are to be successful leaders. Many problems of the mechanical, electronic, and human kind follow very similar patterns. The process of understanding methods for designing solutions to problems is an important tool for all involved in the process. All of the remaining chapters in this second edition of *The Manufacturing Engineering Handbook* provide additional information that is needed in order for the manufacturing engineer to participate properly in the product design process. Both this Chapter and 2 are part of the same process, but they look at the product from different perspectives.

The entire book is a tool for use by the manufacturing engineer, engineering manager, and product development engineer, and will serve to lead the inquiring mind to other sources for additional information on various topics. This book seeks to provide enough information to give the beginning practitioner and the experienced professional useful information that will help in the execution of the product development task. Additionally, this book seeks to join others in the profession who have helped to define the role of the manufacturing engineer as he or she fills the position of design engineering manager, product development engineer, and research and development director, enriching the position with a great depth of understanding of the application of manufacturing while guiding the research and design tasks of product development.

1.0.1 World-Class Manufacturing

The world is shrinking fast. With the advent of Internet commerce, faster travel, and better communication, we have to understand manufacturing around the world and be equipped to compete in world markets. Some have termed the current century the "E-Century" after the e-commerce that is beginning to be a major factor in industry. Vast changes are taking place at a record pace. The most prominent language of the Internet is expected to be Mandarin Chinese very soon. What does the term "world-class manufacturing" mean today? Successful companies are very much involved in researching the cultures of other countries; they consider it a requirement for survival in the competitive business environment. Factors that were never part of the manufacturing world before are now very important to the successful businessman. Some of those factors include:

- Eye contact with a certain people, group, or gender, and what that means in terms of the relationship, both personal and professional
- Why should one be aware of the past history between two people or groups of people when attempting to broker a manufacturing deal?
- Is an insult if a Dutch executive with a prospective distributor of your product prepares a dossier on you?
- What economic risks are tied to the government instability of the region?

It is difficult to quantify the performance improvements required to become a world-class manufacturer. However, one significant study of several successful companies several years ago reveals the following features:

Costs to produce down 20–50%
Manufacturing lead time decreased by 50–90%
Overall cycle time decreased by 50%
Inventory down 50%+
Cost of quality reduced by 50%+ (less than 5% of sales)
Factory floor space reduced by 30–70%
Purchasing costs down 5–10% every year
A minimum of (25) inventory turns of raw material
Manufacturing cycle times with less than 25% queue time
On-time delivery 98%
Quality defects from any cause less than 200 per million

While we can be certain that no one would agree that all of the above performance improvements are required, there is a strong message here that no one can dispute. We must make a paradigm shift in the way we operate to achieve all of these "future state" conditions—and must make some major changes to achieve any one of them! There is also general agreement that world-class manufacturing cannot be achieved without first having a *world-class product design*.

The product development and design process really starts by listening to the customer (the one who ultimately pays the bill!) and understanding his or her needs (and wants). The customer's requirements should be listed, prioritized, and answered completely during the product development and design phase. The other customer we must learn to listen to is our corporation—and especially our factory. We have heard messages such as:

Use the facilities we have.
Don't spend any more capital dollars beyond those needed to support our programs.
Watch your cash requirements. (This translates to any expenditure in the chain leading to product delivery—inventory, work in process, etc.)
Do the job with fewer people.
Continuous improvement is necessary, but we also need a paradigm shift in the way we do business!
Listen to our customers.

These are all great thoughts, but it is difficult to actually run a manufacturing operation with only these high-level goals. The bottom line (dollars of profit) was the *only* focus of many companies during the 1980s and 1990s. The MBA mentality was micromanaging profits on a quarterly basis—to the detriment of the longer-term success of many companies. The top-level focus changed, and must trickle down to the working elements of the company, including the manufacturing engineering function. In the new millennium, financial uncertainty, terrorism, wars, and natural disasters constantly impact the manufacturing industry. The rush to keep the delicate balance of JIT, or just-in-time, manufacturing facilities running overshadows new innovations. A recent multiple-state power outage is expected to ripple throughout many sectors as the factories depending on electric power shut down and then restarted. This interruption caused several delays that will ripple throughout many products and ultimately affect the customer's pocket. While we must do some rather uncomfortable things to keep our companies operating in the short term, the basic change in the way companies will operate in the future is our most important focus. The trained, experienced professional in today's manufacturing world—the person responsible for getting a fantastic product idea or dream or vision out the door of the factory—is facing a tremendous challenge. He or she will find that his or her training and experience have a lot of holes in them. It has nothing to do with how bright or hard-working the person is—the *job* has changed! This manufacturing engineer will be uncomfortable trying to do all elements of the job today, and his or her management will be even more uncomfortable! Management today is facing a worldwide struggle to keep their companies afloat. Many years ago, the owner of a struggling company understood this "new, recently discovered" requirement. One of my favorite stories may illustrate this.

Henry Ford defied all his experts by insisting that the wood shipping container for the Model A battery (the battery was just coming into play as an addition to his latest model) be specified in a particular way. It must be made of a certain type of wood, be reinforced in places, and contain some rather peculiar vent holes in most of the pieces. He also insisted that this would not increase the cost of the battery from the supplier, since the quantities would be so large. (He was correct in this.) As the first Model A of the new series was coming down the line, he called all his department heads—engineers, buyers, accountants, and others—to the factory floor. The battery box was knocked down to get at the battery, and the battery was installed under the floor at the driver's feet. Henry then picked up the pieces of the box and fitted them above the battery, exactly forming the floorboards on the driver's side of the car. The holes in the boards were for the brake, starter switch, etc. The screws that held the box together were then used to bolt the floorboards down. Henry really understood the problems of total product design, cost of materials, just-in-time delivery, zero stock balance, low inventory cost, no material shortages, no waste-disposal costs, good quality, and a host of others. He was a one-man product design team and good manufacturing engineer!

In order to compete in today's marketplace, we must find new ways to increase productivity, reduce costs, shorten product cycles, improve quality, and be more responsive to customer needs. A good product design is probably the most important element to focus on.

Continuous Improvement is not a new idea. Jack Walker visited the IBM laptop computer assembly plant in Austin, Texas, in the late 1980s. He also visited the Pro Printer manufacturing plant at about this time. These were excellent examples of the concurrent engineering process required for IBM to outperform their Japanese competitors. After studying these examples of computer-integrated manufacturing (CIM), the McDonnell Douglas Missile Assembly Plant sponsored two

CIM application transfer studies with IBM's assistance: one concentrating on the production of low-cost, high-rate products, and the other on the production of high-cost, low-rate products.

Our strategy was for the studies to establish a road map for the transition to integrated manufacturing. This plan defined our existing state, what we wanted to become, projects and schedules that would be required, and the cost/benefits analyses. At that time, we had a somewhat fragmented approach to industrial automation, which included material requirements planning (MRP), manufacturing resource planning (MRPII), shop floor control, cost, process planning, bill of material, purchasing, stores, and so on. Several of these were in independent functional areas with their own hardware, software, communications, files, databases, and so on. This was the situation that we were determined to improve. Our goal was to provide a factory that would support today's changing program requirements and that would be even more productive in the future. We believe that manufacturing systems and processes that simply modernize today's operations are not adequate for tomorrow's business environment. Rather, we need greater control over product cost and quality. We need to outcompete our competition in both quality and cost—particularly when we compete oversees and find an average 17% tariff levied against U.S. products competing abroad.

Today, our company is well on the way to becoming the leading facility within the corporation in developing an architecture that ties our systems applications together and fulfills the requirements of a truly integrated set of databases. CIM has evolved into the computer-integrated enterprise (CIE), tying the various plants in the corporation together, and now into an "extended enterprise" that includes our key suppliers. The goal of supporting the complete life cycle of product systems from design concept to development, manufacturing, customer support, and eventually "disassembly" in a logical manner to perform maintenance or modifications is still the focus.

1.0.2   Cost Analysis and Constraints

Although it may appear that many of us will build different products and systems, we see a network of common elements and challenges. All products have requirements for lower costs and improved quality. All can expect competition from producers of similar products, as well as from competing different products—both U.S. and foreign. All must accommodate rapid change in both product design and manufacturing processes.

The term "low-cost" is difficult to define. If your cost is about as low as your chief competitor, whatever the product or production rate is, you are probably a low-cost producer. Benchmarking against your most successful competitors is almost a necessity. In the manufacturing business, cost is partially attributable to the direct touch-labor cost. This is modified by the direct support tasks performed by manufacturing engineering, production control, supervision, quality assurance, and liaison engineering. On some programs this may equal or exceed the touch-labor cost. Added to this is the more general overhead cost, which may double or triple the base cost. Also, since this in-house labor content may amount to a very small percentage of the overall product cost (which includes the cost paid to suppliers for material, parts, subassemblies, services, etc.), it may not be the only driving force in determining the price to the customer. Peter Drucker, in *The Practice of Management*, states that "hourly employees' wages as a factor in product cost are down from 23% to 18%, and that productivity is on the rise." General Motors' hourly employees' costs are still in the 30% range, partially due to restrictive work rules in their labor contracts. Some Japanese car manufacturers who produce in the United States pay similar wages but operate at hourly employee costs of less than 20%. The trend is toward 15%. In selecting a design approach, it is perhaps more valuable to look at total cost in calculating earnings than to look at the amount of direct labor involved. One measurement of cost and earnings is return on investment (ROI).

ROI is a relationship between bottom-line earnings, or profit, and the amount of money (assets) that must be spent (invested) to make these earnings. An example from the books of a medium-size manufacturer shows the following:

1. Accounts receivable — $6M
2. Inventory and work in process — $14M
3. Land — $2M
4. Buildings and equipment (net) — $15M
5. Total assets — $37M

By looking closer at each item, we can improve the bottom line.

1. Submit billings sooner to reduce the amount that customers owe us.
2. Reduce the amount of raw stock and hardware in inventory. Don't get the material in-house until it is needed. Reduce work in process by reducing the cycle time in the shop, which reduces the number of units in flow. Also, complete assemblies on schedule for delivery to permit billing the customer for final payment.
3. Try not to buy any more land, since it cannot be depreciated: it stays on the books forever at its acquisition cost.
4. Evaluate the payoff of any additional buildings and capital equipment. Of course, some new buildings and equipment may be needed in order to perform contract requirements. Additional investment may also be wise if it contributes to a lower product cost by reducing direct labor, support labor, scrap, units in process, cost of parts and assemblies, etc.
5. The bottom line in investment may not be to reduce investment, but to achieve a greater return (earnings). In this simple example, we could add $10 million in equipment and reduce accounts receivable, inventory, and work in process by $10 million and have the same net assets. If this additional equipment investment could save $4 million in costs, we would increase our earnings and double our ROI.

In today's real world, we need to consider one more factor. Items 1, 2, and 3 above require company money to be spent. There is a limit to borrowing, however, and a point where cash flow becomes the main driver. It is therefore essential to reduce overall costs by utilizing existing buildings and capital equipment and by doing the job with fewer people and less inventory. The sharing of facilities and equipment between products becomes very attractive to all programs. This would reduce our need for additional capital and reduce the depreciation-expense portion of our overhead. The design engineer and manufacturing engineer are certainly key members of the product development and design team, but the input from all other factory functions becomes more important as we look toward the future.

1.0.3   Project Design Teams

In an increasing number of companies across the country, both the designer and the manufacturing engineer are climbing over the wall that used to separate the two functions. In addition, a full team consisting of all the development-related functions should participate in the design of a new product or the improvement of an existing product. The team may include the following:

> Administration—senior management to empower the team
> Design engineering
> Process engineering
> Manufacturing engineering
> Manufacturing
> Quality assurance
> Marketing
> Sales
> Purchasing
> Accounting
> Production control

Profitability, and even survival, depends on working together to come up with a design that can be made easily and inexpensively into a quality product. Other appropriate team members may be representatives from:

> Distribution
> Accounting
> Human resources
> Suppliers
> Customers
> Suppliers chosen as partners in the product development process
> Shop technicians
> Assembly technicians
> Joint ventures
> A customer's accounting department who represent their company in an "open book" cost-sharing method

An excellent example of quality improvement and cost reduction is Ford Motor Company. Ford has adopted design for assembly (and design for manufacture and assembly) as one of their concurrent engineering approaches. The company acknowledges the importance of product improvement and process development before going into production. Assembly is a small part of overall product cost, about 10–15%. However, reducing this cost by a small amount results in big money saved in materials, work in process, warehousing, and floor space. Figure 1.1 shows how a small investment in good design has the greatest leverage on final product cost.

World-class manufacturers know that they cannot dedicate a single factory or production line to one model or product. The generic assembly line must have the flexibility to produce different models of similar products rapidly and entirely new products with a minimum changeover time. There are exceptions where the production quantities of a single product are sufficiently large, and the projected product life is great enough, that a production factory dedicated to a single product is the best choice.

1.0.4   Bibliography

Drucker, P., *The Practice of Management*, Harpercollins, New York, 1986.
Pahl, B. and Beitz, W., *Engineering Design: A Systematic Approach*, Pomerans, A. and Wallace, K., Trans., Springer-Thomas, 1993.

1.1   CONCURRENT ENGINEERING

1.1.0   Introduction to Concurrent Engineering

What is concurrent engineering? Why do we care about concurrent engineering? To remain competitive in industry, we must produce high-quality products and services

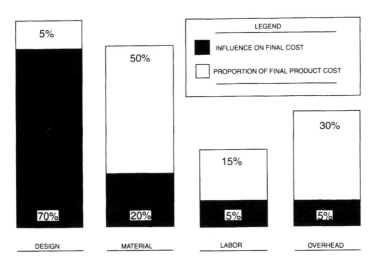

FIGURE 1.1   The cost of design in manufacturing at Ford. (Courtesy of Ford Motor Company. With permission.)

the first time. We can accomplish this by implementing concurrent engineering (CE) principles and tools in our programs. (The abbreviation "CE" in this text is not to be confused with "common European," used to denote compliance to British safety standards and engineering norms. For simplicity, it is used in this text only to mean "concurrent engineering.")

It does not matter whether we call the concept concurrent engineering, integrated product definition, or simultaneous engineering, as long as we consistently produce high-quality products on time for the best price. Agreeing on a common definition helps communication and understanding. We use the term concurrent engineering and the Department of Defense/Institute of Defense Analysis (DoD/IDA) definition because it has wide acceptance. IDA Report R-338 gives the following definition:

> Concurrent Engineering is a systematic approach to the integrated, concurrent design of products and their related processes, including manufacture and support. This approach is intended to cause the developers, from the outset, to consider all elements of the product life cycle from conception through disposal, including quality, cost, schedule and user requirements.

Four main elements have emerged as most important in implementing CE: the voice of the customer, multidisciplinary teams, automated tools, and process management. The underlying concept is not a new one: teamwork. The secret is to involve all the right people at the right time. We must increase our understanding of product requirements by being more effective in capturing the voice of the customer. We must increase our emphasis on product producibility and supportability. This requires the involvement of all related disciplines on our CE teams. We must acquire and use the best tools to permit the efficient development of "build-to" technical data packages. Finally, we must increase our emphasis on developing production processes in parallel with development of the product.

Industry is continuing to refine and improve the elements of concurrent engineering. All of us must contribute to this process. More than that, all of us must be willing to change, and concurrent engineering requires change. Concurrent engineering must be understood to be rooted in communication, communication through the complexity of corporate and geographical cultural differences. The technical and financial issues facing the manufacturing professional today are overwhelming; when combined with the difficulty of communication, the task of competing in the world market is made much harder. Someone has said, "All of us are smarter than one of us." We must learn to think like this: everyone has something to contribute when developing a new product.

1.1.1 Why Concurrent Engineering?

Concurrent engineering is a commonsense approach to product design, development, production, and support. By collecting and understanding all requirements that the product must satisfy throughout its life cycle at the start of concept definition, we can reduce costs, avoid costly redesign and rework, and shorten the development process. We do this by capturing all customer requirements and expectations and involving all related disciplines from the start. Working as teams on all product-related processes, we can provide for a smooth transition from development to production. Experience shows that concurrent engineering results in:

Well-understood user requirements
Reduced cycle times
First-time quality producible designs
Lower costs
Shorter development spans
A smoother transition to production
A new respect for other teammates
Highly satisfied customers

It is not surprising that some CE practices, namely the voice of the customer and process management, are rarely practiced. We need to leave our comfort zones in order to implement these practices effectively. Concurrent engineering pays off in:

Product development cycle time reduced 40–60%
Manufacturing costs reduced 30–40%
Engineering change orders reduced more than 50%
Scrap and rework reduced by as much as 75%

The primary elements of CE are:

Voice of the customer
Multidisciplinary teams
Automation tools and techniques
Process management

The voice of the customer includes the needs and expectations of the customer community, including end users. Concurrent engineering can best be characterized by the conviction that listening and responding to the voice of the customer can achieve product quality. The ideal way is to capture, at the outset, all requirements and expectations in the product specifications.

The most effective way we have found to accomplish this is to conduct focus group sessions with different elements of the customer's organization. Properly staffed multidisciplinary teams provide the means to enable all requirements, including producibility and supportability, to be an integral part of product design from the outset. CE teams are broadly based, including representatives from production, customer support, subcontract management and procurement, quality assurance, business systems, new business and marketing, and suppliers. Broadly based CE teams succeed because they can foresee downstream needs and build them into our products and processes.

Not all team members are necessarily full-time. In many cases, part-time participation is all that is needed and all that can be afforded.

Automation tools and techniques provide effective and efficient means of developing products and services. Computer-based tools such as Unigraphics can be used as an electronic development fixture (EDF) during prototyping, in lieu of mockups, to verify clearances and mechanism operations before hardware is fabricated.

There are a variety of home-grown and purchased CE tools in use on programs. Their importance to CE is an increased ability to communicate and transfer data readily among team members, customers, and suppliers. With reliable information sharing, we are able to review and comment on (or import and use) product and process data more rapidly, while eliminating sources of error.

Process management is the final key to controlling and improving the organization as well as the processes used to develop, build, and support a product. This is probably the newest and least practiced element of CE. Big gains can be made by defining the program work flows and processes and then improving them. Processes define the relationships of tasks and link the organization's mission with the detailed steps needed to accomplish it. Process management is an effective way of managing in a team environment. Program-wide processes provide a means of identifying the players who need to be involved and of indicating team interrelationships. Product processes are also a part of process management, requiring the definition and development of production processes in parallel with the definition and development of the product design.

1.1.2 Concurrent Engineering throughout the Acquisition Process

Concurrent engineering practices are applicable to all programs, old or new, regardless of program type or acquisition phase. Implementing CE at the very beginning of a program makes the biggest payoff. Approximately 80% of a product's cost is determined during the concept phase, so it is very important to have manufacturing, producibility, supportability, and suppliers involved then. There are benefits of CE to be realized during the later program phases, including reducing production costs for dual-source competitions or defining product improvements.

1.1.3 The Voice of the Customer

The voice of the customer (VOC) represents the needs and expectations by which the customer or user will perceive the quality of products and services. It is the program manager's responsibility to make sure that all CE participants understand the voice of the customer and that all program requirements can be traced back to those needs and expectations.

Quality is achieved by satisfying all customer needs and expectations; technical and legal documentation will not overcome bad impressions. Meeting the minimum contractual requirements will often not be enough, especially in a highly competitive environment. Customer quality expectations invariably increase based on exposure to "best-in-class" products and services. Nevertheless, the products and services provided must be consistent with contractual requirements and within the allotted budgets. This is not an easy task; it requires continuous attention and good judgment to satisfy the customer while staying within program constraints.

Requirements definition begins at the program's inception and continues throughout the product life cycle. It is essential that the right disciplines, including suppliers, are involved at all times in order to avoid an incomplete or biased outcome. Methodologies such as QFD can be used to enable the team to analyze customer requirements and to specify the appropriate product or service.

1.1.4 Capturing the Voice of the Customer

The concurrent engineering team must maintain a careful balance between satisfying customer needs and expectations and maintaining a reasonably scoped program. Early in the product life cycle, written requirements will be sparse and general. The team will use customer needs and expectations to interpret and expand the written requirements. It is at this time that the CE team has the greatest opportunity to influence the product life-cycle cost. Studies indicate that 80% of cumulative costs are already set by the end of the concept phase. Consequently, it is vital that all available information is used to choose the best product concept based on costs and user requirements.

Later in the program, the written requirements will become more detailed. The voice of the customer will then be used primarily to help clarify ambiguous requirements. In either case, the team will have to listen unceasingly to the voice of the customer in order to capture the needed requirements. The program manager will need to provide continuous support and encouragement, especially in light of the many obstacles the team might encounter. There are many real and perceived barriers to gathering the voice of the customer:

Restricted access to customers during competitions
User perception of VOC as a sales tactic
Confusion between higher quality and higher cost
Failure to identify all users
Lack of skills in analyzing the voice of the customer
Rush to design
Tradition and arrogance

1.1.5 Customer Focus Group Sessions

User focus group interview sessions can be very effective in gathering inputs from the users of a system. While there are many sources of customer needs and expectations, focus group sessions can uncover and amplify a broad range of requirements that might otherwise be overlooked. These requirements can range from the positive "We want this!" to the negative "Don't do that again!" These inputs can provide a better understanding of user needs and even a competitive edge for your program. See Figure 1.2.

FIGURE 1.2 Translating customer requirements into requirements for the designer. (Courtesy of Technicomp, Inc., Cleveland. Ohio. With permission.)

The use of focus group sessions is especially recommended at the beginning of product conceptualization and during the support phase of a deployed system. The latter sessions identify product improvement opportunities.

A well-conducted session uses group dynamics to encourage individuals to provide inputs in a synergistic manner. A moderator leads the discussion in order to maintain focus on the critical issues and to enable each panel member to participate as much as possible. Post-It Notes (© 3M) should be used to capture, sort, and prioritize the panel members' comments using a disciplined form of brainstorming. A limited number of concurrent engineering team members attend the session as nonparticipating observers to help capture and understand the voice of the customer.

The moderator should prepare open-ended questions to stimulate and focus the session:

Why is a new product or system needed?
What do you like about the existing system?
What do you dislike about the existing system?
What makes your current job difficult?

The results of the focus group sessions can be analyzed using one of the formal process management tools or design for manufacturability tools such as DFA or QFD.

1.1.6 Communicating Cross-Culturally

When communicating to groups of people, the way the information is communicated is of extreme importance. The delivery of information sometimes makes all the difference. Good news can often be easily communicated, but be sure that bad news has more than one way of being delivered. It is sometimes necessary to think about how we obtained the information that we ourselves possess. Someone has said that how we know what we know is very important to the persons to whom we deliver the information we have. The manufacturing engineer becomes an *information broker* when leading a product development team. Behind the assertions and explanations of the product development process, the leader of the team must be prepared to explain why the decisions were made and what benefits will be reaped from those decisions. This is a natural process that must not be feared or disregarded by the leader. Each and every team member must reach a point of agreement with the leader as to the direction the team is going. Time may not permit micromanaging by the team members of the leader, but the leader must understand that if others are to follow, they must have some degree of faith that the leader knows where he himself or she herself is going.

Motivation and decisions are made by people of all cultures based upon their ways of perceiving the world, the way they think, the way they express ideas, the way they act, the way they interact, how they channel their ideas, and how they make decisions. The successful team leader will seek to understand methods of successful communication with his team, his company, and his customer base.

1.1.7   Bibliography

Hesselgrave, D.J., *Communicating Christ Cross-Culturally,* 2nd ed., Zondervan Publishing Co., Grand Rapids, Mich., 1991.
Winner, R.I., Pennell, J.P., Bertrend, H.E., and Slusarczuk, M.M.G., "The Role of Concurrent Enginerring in Weapons System Acquisition," IDA Report R-338, Institute for Defense Analysis, Alexandra, Virginia, December 1988.

1.2   QUALITY FUNCTION DEPLOYMENT

1.2.0   Introduction to Quality Function Deployment

Quality function deployment (QFD) is a methodology used by teams to develop products and supporting processes based on identifying customer needs and wants, and comparing how your product and the competition meet these customer requirements.

Quality: What the customer/user needs or expects
Function: How those needs or expectations are to be satisfied
Deployment: Making it happen throughout the organization

QFD starts with the voice of the customer. Using all practical means, the team gathers customer needs and expectations. The team has to analyze these inputs carefully to avoid getting caught up in providing what they *expect* or *believe* that the customer needs. Conceptual models, such as the Kano model shown in Figure 1.3, can be used to help each team member understand better how the customer will evaluate the quality of the product. The team can then use QFD as a planning tool to derive product and process specifications that will satisfy the voice of the customer while remaining within business and technical guidelines and constraints. The program utilizes a series of matrices, starting with the customer wants on one axis

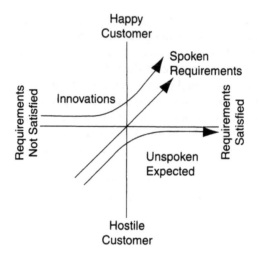

FIGURE 1.3   Kano model showing how the customer evaluates quality. (Courtesy of Technicomp, Inc., Cleveland, Ohio. With permission.)

and the company's technical interpretation of how to accomplish these customer expectations on the other axis. These elements are weighted and quantified, to permit focus on the most important issues or on the areas where the company product is most deficient.

1.2.1   Quality Function Deployment

QFD is defined as a *discipline* for product planning and development in which key customer wants and needs are deployed throughout an organization. It provides a structure for ensuring that customers' wants and needs are carefully heard, then translated directly into a company's internal technical requirements from component design through final assembly.

Strength of QFD

Helps minimize the effects of:

Communication problems
Differences in interpretation about product features, process requirements, or other aspects of development
Long development cycles and frequent design changes
Personnel changes

Provides a systematic means of evaluating how well you and your competitors meet customer needs, thus helping to identify opportunities for gaining a competitive edge (sort of mini-benchmarking)

Offers a great deal of flexibility, so it may be easily tailored to individual situations

Brings together a multifunctional group very early in the development process, when a product or service is only an idea

Helps a company focus time and effort on several key areas, which can provide a competitive edge

Applications of QFD

QFD is best applied to specific needs: areas in which significant improvements or breakthrough achievements are needed or desired. It can be used for virtually any type of product or service, including those from such areas as all types of discrete manufacturing, continuous and batch processes, software development, construction projects, and customer service activities in the airline, hotel, or other industries.

The Voice of the Customer

The first step in the QFD process is to determine what customers want and need from a company's product or service. This is best heard directly from customers themselves, and stated in their words as much as possible. This forms the basis for all design and development activities, to ensure that products or services are not developed only from the voice of the engineer, or are technology-driven.

Background of QFD

Technicomp, Inc., of Cleveland, Ohio, developed QFD. The first application as a structured discipline is generally credited to the Kobe Shipyard of Mitsubishi Heavy Industries Ltd. in 1972. It was introduced to the United States in a 1983 article in *Quality Progress,* a publication of the American Society of Quality Control (ASQC). Leaders in developing and applying QFD in Japan include Akao, Macabe, and Fukahara. U.S. companies that have utilized the Technicomp QFD program include Alcoa, Allen-Bradley, Bethlehem Steel, Boeing, Caterpillar, Chrysler, Dow Chemical, General Motors, Hexel, Lockheed, Magnavox, and others. The program consists of a series of videotapes, team member application guides, instructor guides, and other course materials.

Phases of QFD

QFD involves a series of phases (see Figure 1.4) in which customer requirements are translated into several levels of technical requirements. Phases are often documented by a series of linked matrices.

Phase 1: Product Planning. Customer requirements are translated into technical requirements or design specifications in the company's internal technical language.

Phase 2: Product Design. Technical requirements are translated into part characteristics.

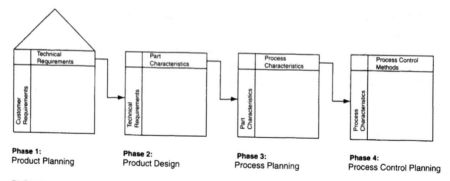

FIGURE 1.4   The four phases of quality function deployment (QFD). (Courtesy of Technicomp, Inc., Cleveland, Ohio. With permission.)

Phase 3: Process Planning. Part characteristics are translated into process characteristics.

Phase 4: Process Control Planning. Process characteristics are assigned specific control methods.

Potential Benefits

Some of the results achieved by companies that have implemented QFD include:

30–50% reduction in engineering changes
30–50% shorter design cycles
20–60% lower start-up costs
20–50% fewer warranty claims

Other results include:

Better, more systematic documentation of engineering knowledge, which can be more easily applied to future designs
Easier identification of specific competitive advantages
More competitive pricing of products or services, due to lower development and start-up costs
More satisfied customers

Requirements for Success

Management commitment to QFD is the minimum requirement; support by the entire organization is ideal. Participation on a project team is required by individuals who support QFD and represent all development-related functions, such as:

Design engineering
Process engineering

Manufacturing engineering
Manufacturing
Quality assurance
Marketing
Sales

Other appropriate members may be representatives from:

Purchasing
Distribution
Accounting
Human resources
Suppliers
Customers

1.2.2   The House of Quality

The QFD program introduces a chart in Phase 1 that is commonly called "the house of quality." The illustration shown in Figure 1.5 is a simplified example for large rolls of paper stock used in commercial printing. The following is a brief summary of the completed chart shown in Figure 1.6.

(A)   *Customer requirements:* Customers' wants and needs, expressed in their own words.
(B)   *Technical requirements:* Design specifications through which customers' needs may be met, expressed in the company's internal language.
(C)   *Relationship matrix:* Indicates with symbols where relationships exist between customer and technical requirements, and the strength of those relationships.
(D)   *Target values:* Show the quantifiable goals for each technical requirement.
(E)   *Importance to customer:* Indicates which requirements are most important to customers.
(F)   *Importance weighting:* Identifies which technical requirements are most important to achieve. In this chart, each weighting is calculated by multiplying the "importance to customer" rating times the value assigned to a relationship, then totaling the column.

The following are shown in Figure 1.6.

(G)   *Correlation matrix:* Indicates with symbols where relationships exist between pairs of technical requirements, and the strength of those relationships.

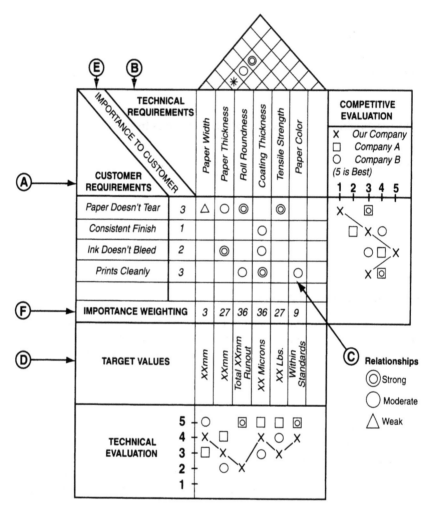

FIGURE 1.5 QFD house of quality chart for rolls of paper stock. (Courtesy of Technicomp, Inc., Cleveland, Ohio. With permission.)

(H) *Competitive evaluation:* Shows how well a company and its competitors meet customer requirements, according to customers.

(I) *Technical evaluation:* Shows how well a company and its competitors meet technical requirements.

Product Planning (Phase 1 of QFD)

Most companies begin their QFD studies with a product-planning phase. Briefly, this involves several broad activities, including collecting and organizing customer wants and needs, evaluating how well your product and competitive products meet

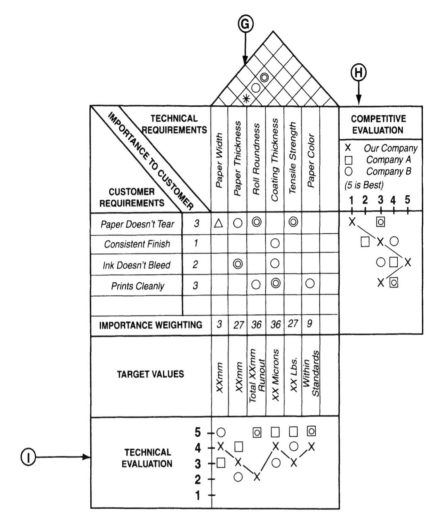

FIGURE 1.6   Completion of the house of quality chart for rolls of paper stock. (Courtesy of Technicomp, Inc., Cleveland, Ohio. With permission.)

those needs, and translating those needs into your company's specific technical language.

The House of Quality

In very simple terms, the house of quality can be thought of as a matrix of *what* and *how:*

   What do customers want and need from your product or service? (customer requirements)

How will your company achieve the *what*?
(technical requirements)

The matrix shows where relationships exist between *what* and *how,* and the strength of those relationships. Before starting a chart, the scope and goals of the study must be clearly identified. A typical chart enables you to:

Learn which requirements are most important to customers
Spot customer requirements that are not being met by technical requirements,
    and vice versa
Compare your product or service to the competition
Analyze potential sales points
Develop an initial product or service plan

Constructing a House of Quality

Suggested elements and a recommended sequence of construction for a house of quality are given below. Also keep in mind that the entire project team constructs the chart, unless noted otherwise. (See Figures 1.7, 1.8, and 1.9.)

(A) Customer requirements

Every chart must begin with the voice of the customer.
Identify all customer groups. Various groups will probably have some
    common needs, as well as some conflicting needs.
Collect accurate information from customers about their wants and needs.
Brainstorm to identify any additional customer requirements.
Use an affinity diagram to help group raw customer data into logical
    categories.
Transfer the list of customer requirements to the house of quality.

Do not assume that your company already knows everything about its
    customers. QFD teams have been astounded at the results of focused
    efforts to listen to the voice of the customer. Primary benefits of QFD
    often include clearing up misconceptions and gaining an accurate under-
    standing of customers' demands.

(B) Degree of importance

Identify the relative priority of each customer requirement.
Use customer input as the basis for determining values, whenever possible.
Use a scale of 1 to 10, with 10 indicating very important items.

A more rigorous statistical technique also can be effective for determining
    degrees of importance.

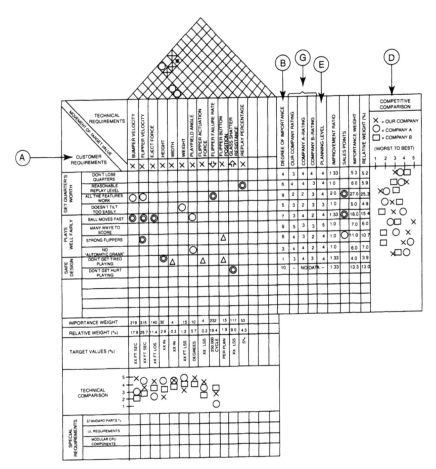

FIGURE 1.7 Construction of a house of quality: (A) customer requirements; (B) degree of importance; (C) competitive comparison; (D) competitors' ratings; (E) planned level to be achieved. (Courtesy of Technicomp, Inc., Cleveland, Ohio. With permission.)

(C) Competitive comparison

Identify how well your company and your competitors fulfill each of the customer requirements.
Use customer input as the basis for determining numeric ratings.
Use a scale of 1 to 5, with 5 being the best.

(D) Competitors' ratings

Use symbols to depict each company's rating, so that you can easily see how well, in your customers' view, your company compares to the competition.

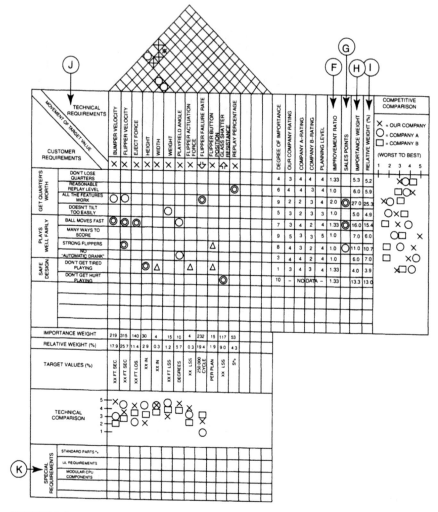

FIGURE 1.8 The house of quality (continued): (F) improvement ratio; (G) sales points; (H) importance weight of customer requirements; (I) relative weight of customer requirements; (J) technical requirements; (K) special requirements. (Courtesy of Technicomp, Inc., Cleveland, Ohio. With permission.)

(E)   Planned level

Determine what level you plan to achieve for each customer requirement.

Base the rating on the competitors comparison, and use the same rating scale.

Focus on matching or surpassing the competition on items that will give your product or service a competitive edge, or that are very important to customers. (It is not necessary to outdo the competition on every item.)

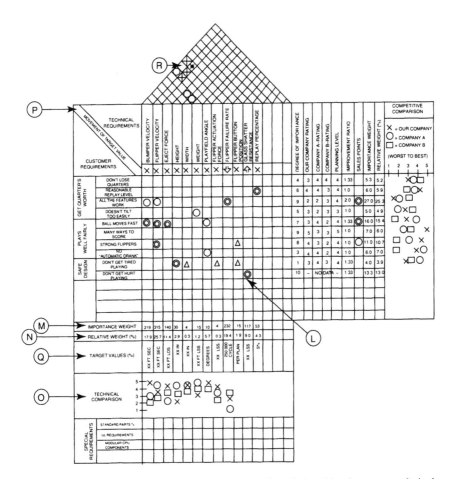

FIGURE 1.9 The house of quality (continued): (L) relationships between technical requirements and customer requirements; (M) importance weight of technical requirements; (N) relative weight of technical requirements; (O) technical comparison; (P) movement of target values; (Q) target values. (Courtesy of Technicomp, Inc., Cleveland, Ohio. With permission.)

(F) Improvement ratio

Quantify the amount of improvement planned for each customer requirement.

To calculate, divide the value of the planned level by the value of the current company rating:

improvement ratio = planned level ÷ current company rating

(G) Sales points

Identify major and minor sales points.
Limit the team to only a few points.

To indicate the importance of sales points, use symbols to which values are assigned:

$$\odot = \text{major point} = 1.5$$
$$O = \text{minor point} = 1.2$$

Major and minor sales points are often assigned values of 1.5 and 1.2, respectively.

Note that items with high degrees of importance are often logical sales points. Also, remember that any items that will be new and exciting to customers are likely sales points, although they probably will not have high degrees of importance.

(H) Importance weight (customer requirements)

Quantify the importance of each customer requirement to your company.

To calculate, multiply the degree of importance by the improvement ratio and by the sales point value (if applicable):

importance weight = degree of importance × improvement ratio
× sales point value

(I) Relative weight (%) (customer requirements)

Quantify the relative importance of each customer requirement by expressing it as a percentage.

To calculate:
   Total the importance weights.
   Divide the importance weight of an item by the total.
   Multiply by 100.
Use the relative weights as a guide for selecting the key customer requirements on which to concentrate time and resources.

(J) Technical requirements

Develop this list internally, using existing data and the combined experience of team members.

Begin by collecting available data.

Brainstorm to identify any additional requirements.

Follow general guidelines while developing the technical requirements:
   Address global requirements of the product or service, not lower-level performance specifications.
   Identify performance parameters, and avoid simply restating the components of the existing product or service.
   Try not to include parts of mechanisms.

Establish definitions that are understood and agreed upon by all team members.

Use terminology that is understood internally.

Transfer the technical requirements to the house of quality chart.

(K) Special requirements

List any unusual needs or demands, such as government standards, certification requirements, or special company objectives.

Note that these items must be considered during design and development, but normally do not appear in a list of customer demands.

(L) Relationship matrix

Identify any technical requirements that bear on satisfying a customer requirement.

Evaluate each pair of requirements by asking if the technical requirement in any way affects the customer requirement.

For large charts, consider dividing the matrix into "strips," assigning them to groups of only a few team members for evaluation, and then have the full team review the combined results.

To indicate the strengths of relationships, use symbols to which values are assigned:

$$\odot = \text{strong relationship} = 9$$
$$\bigcirc = \text{moderate relationship} = 3$$
$$\triangle = \text{weak relationship} = 1$$

(M) Importance weight (technical requirements)

Quantify the importance of each technical requirement.

To calculate this value, only those customer requirements that are related to a technical requirement are factored into the calculation:

Multiply the value of any relationships shown in the column of the technical requirement by the relative weight of the customer requirement.

Test the results.

(N) Relative weight (%) (technical requirements)

Quantify the relative importance of each technical requirement by expressing it as a percentage.

To calculate:

Total the importance weights.

> Divide the importance weight of an item by the total.
> Multiply by 100.

At this point, the house of quality contains enough information for you to choose several key technical requirements on which to focus additional planning and definition. As a general rule, especially for pilot QFD projects, limit the number of key requirements to three to five. Those with the highest importance or relative weights are good candidates. They should have a strong bearing on satisfying the highest-priority customer requirements, or many of the customer requirements. However, don't look at just the numbers when choosing the key technical requirements. Also consider:

> The potential difficulty of meeting a requirement
> Any technical breakthroughs that may be needed
> Requirements that are unfamiliar to your company

These factors can be particularly important if you must choose key requirements from several that have similar weightings.

(O)   Technical comparison

> Identify how well your company and your competitors fulfill each of the technical requirements.
> Use internal expertise from sources such as engineers, technicians, and field personnel to develop the comparisons.
> Consider evaluation techniques such as bench testing, laboratory analysis, product teardowns, field observations, and reviews of data from outside testing labs or agencies.
> Do not expect to be able to evaluate every technical requirement.
> Convert the test data into values that are appropriate for a rating scale of 1 to 5, with 5 being the best.
> Plot each company's performance using symbols, so that comparisons can be seen easily.

(P)   Movement of target values

> Use symbols to indicate the desired direction for each target value:

> > $\uparrow$ = increase the value
> > $\downarrow$ = decrease the value
> > x = meet a specified nominal value

(Q)   Target values

> Assign specific target values for as many technical requirements as possible, in order to:
> > Establish concrete goals for designers and engineers.

Help define further actions to ensure that customers' demands are carried through design and development.

Compare your current actual values against those of the competition:

For each technical requirement, look at competitive ratings for the related customer requirements.

If you rate lower than the competition in key areas, assign target values that equal or better their values.

Check historical test results or operating data to help assign target values, which reflect desired improvements to key items.

Consider using the results of designed experiments, Taguchi experiments, or other developmental work to achieve optimal values.

Try working backward from broader, less specific requirements to develop specific target values.

Be sure that each target value is measurable; if not, the team should develop an alternative that is measurable.

Make sure that the team agrees on exactly how target values will be measured.

Keep in mind that some of the values may not be feasible to achieve later in development; for now, they represent important starting points.

The vast majority of U.S. companies have completed only the first phase of QFD in their studies. A QFD study should proceed beyond phase 1 if any of the following are true:

Product or service requires more or better definition

Additional detail is needed to make significant improvements

Technical requirements shown on the house of quality cannot now be executed consistently

1.2.3   Product Design (Phase 2)

The overall goal of Phase 2 is to translate key technical requirements from the house of quality into specific parts characteristics. *Parts* are tangible items that eventually compose the product or service: its elements, raw materials, or subsystems. Inputs to this phase are:

Highest priority technical requirements from Phase 1

Target values and importance weights for those requirements

Functions of the product or service

Parts and mechanisms required

The outcome of the product design phase is the identification of key part characteristics, and the selection of the new or best design concepts. Figure 1.10 shows a typical product design matrix chart as utilized in Phase 2.

Product Development

			Mechanism									Mechanism									Mechanism								
			Part			Part			Part			Part			Part			Part			Part			Part			Part		
Key Technical Requirements	TR Target Values	TR Importance Weights	Part Characteristic	Part Characteristic	Part Characteristic	Part Characteristic	Part Characteristic	Part Characteristic	Part Characteristic	Part Characteristic	Part Characteristic	Part Characteristic	Part Characteristic	Part Characteristic	Part Characteristic	Part Characteristic	Part Characteristic	Part Characteristic	Part Characteristic	Part Characteristic	Part Characteristic	Part Characteristic	Part Characteristic	Part Characteristic	Part Characteristic	Part Characteristic	Part Characteristic	Part Characteristic	Part Characteristic
PC Importance Weights																													
PC Target Values																													

TR = Technical Requirement
PC = Part Characteristic

TECHNICOMP

FIGURE 1.10 Product design matrix used to identify key part characteristics and to select the best design concepts. (Courtesy of Technicomp, Inc., Cleveland, Ohio. With permission.)

1.2.4 Other Elements of the QFD Program

The Technicomp Corporation goes on to explain the other two phases shown in Figure 1.4.[1] In addition to QFD, the company has a large number of additional training programs.

1.3 DESIGN FOR ASSEMBLY

Geoffrey Boothroyd, Boothroyd Dewhurst, Inc., Wakefield, Rhode Island

and

Jack M. Walker, consultant, manufacturing engineering, Merritt Island, Florida

1. The editor wishes to thank Technicomp, Inc., for allowing us to utilize part of their QFD system in this handbook, since we consider it one of the important tools that should be in the manufacturing engineer's toolbox.

1.3.0   Introduction to Design for Assembly

One example of the effectiveness of the DFA concept is the Pro printer line of computer printers launched by IBM in 1985.[2] Before then, IBM got its personal printers from Japan's Seiko Epson Corp. By applying DFA analysis, IBM turned the tables on Japan, slashing assembly time from Epson's 30 minutes to only 3 minutes. Ford Motor Company trimmed more than $1.2 billion and in one year helped Ford edge out General Motors as Detroit's most profitable automaker. GM later started using the software, which has a database of assembly times established as the industry standard.

DFA is systematic in its approach and is a formalized step-by-step process. The techniques are concerned with simplifying the design in order to minimize the cost of assembly and the cost of the parts. The best way to achieve this minimization is first to reduce the number of parts to be assembled, and then to ensure that the remaining parts are easy to manufacture and assemble. It is important to have a measure of how efficient the design is in terms of assembly. The DFA process shows how to quantify this factor.

1.3.1   Choice of Manual or Machine Assembly System

It is important to decide at an early stage in design which type of assembly system is likely to be adopted, based on the system yielding the lowest costs. In product design requirements, manual assembly differs widely from automatic assembly due to the differences in ability between human operators and any mechanical method of assembly. An operation that is easy for an assembly worker to perform might be impossible for a robot or special-purpose workhead.

The cost of assembling a product is related to both the design of the product and the assembly system used for its production. The lowest assembly cost can be achieved by designing the product so that it can be economically assembled by the most appropriate assembly system. The three basic methods of assembly as shown in Figure 1.11 are

1. Manual assembly
2. Special-purpose machine assembly
3. Programmable machine assembly

2. The design for assembly (DFA) process was developed by Geoffrey Boothroyd at the University of Massachusetts in the mid-1970s. He envisioned a method that would stress the economic implications of design decisions. This is crucial, because while design is usually a minor factor in the total cost of a product, the design process fixes between 70% and 95% of all costs. The National Science Foundation got things rolling in 1977 with a $400,000 research grant, with further grants during the 1980s.

Private companies began supporting Boothroyd's research in 1978, led by Xerox and AMP, Inc. Then came Digital Equipment, General Electric, Westinghouse, and IBM. After Peter Dewhurst joined Boothroyd in 1981 at the University of Massachusetts, the two professors formed Boothroyd Dewhurst, Inc. (BDI), to commercialize the concept in the form of personal computer software. After they moved to the University of Rhode Island, Ford and IBM quickly became the biggest supporters, investing $660,000 into research that has since moved beyond just assembly to include process manufacturing. Ford embraced the concept with great fervor and has trained roughly 10,000 people in DFMA® (Design for Manufacture and Assembly; DFMA is the registered trademark of Boothroyd Dewhurst, Inc.).

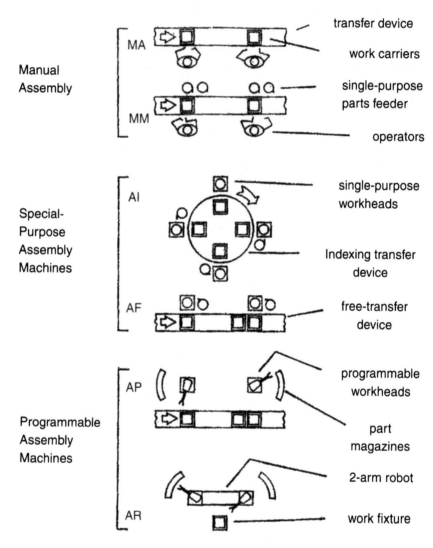

FIGURE 1.11 The three basic types of assembly systems. (From G. Boothroyd, *Assembly Automation and Product Design*, Marcel Dekker, New York, 1992. With permission.)

In manual assembly (MA), the tools required are generally simpler and less expensive than those employed on automatic assembly machines, and the downtime due to defective parts is usually negligible. Manual assembly costs remain relatively constant and somewhat independent of the production volume; also, manual assembly systems have considerable flexibility and adaptability. Sometimes it will be economical to provide the assembly operator with mechanical assistance (MM) in order to reduce assembly time.

Special-purpose assembly machines are those machines that have been built to assemble a specific product; they consist of a transfer device with single-purpose workheads and parts feeders at the various workstations. The transfer device can operate on an indexing (synchronous) principle (AI) or on a free-transfer (nonsynchronous) principle (AF). They often require considerable prove-in time, and defective parts can be a serious problem unless the parts are of relatively high quality.

Programmable assembly machines are similar to nonsynchronous special-purpose machines, except that some or all of the special-purpose workheads are replaced by programmable robots. This arrangement (AP) allows for more than one assembly operation to be performed at each workstation and provides for considerable flexibility in production volume and greater adaptability to design changes. For lower production volumes, a single robot workstation may be preferable (robotic assembly). Sometimes two robot arms will be working interactively at the same work fixture (AR). When considering the manufacture of a product, a company must take into account the many factors that affect the choice of assembly method. For a new product, the following considerations are generally important:

Suitability of the product design
Production rate required
Availability of labor
Market life of the product

Figure 1.12 shows for a particular set of conditions how the range of production volume and number of parts in the assembly affect the choice of system.

1.3.2 Design for Manual Assembly

Although there are many ways to increase manufacturing productivity (utilizing improved materials, tools, processes, plant layout, etc.), consideration of manufacturing and assembly *during product design* holds the greatest potential for significant reduction in production costs and increased productivity. Robert W. Militzer, onetime president of the Society of Manufacturing Engineers, has stated, "As manufacturing engineers, we could do far more to improve productivity if we had greater input to the design of the product itself—it is the product designer who establishes the production (manufacture and assembly) process. For the fact is that much of the production process is implicit in product design."

In other words, if the product is poorly designed for manufacture and assembly, techniques can be applied only to reduce to a minimum the impact of the poor design. Improving the design itself may not be worth considering at this late stage; usually too much time and money have already been expended in justifying the design to consider a completely new design or even major changes. Only when manufacture and assembly techniques are incorporated early in the design stage (i.e., product design for ease of manufacture and assembly) will productivity be significantly affected.

Product costs are largely determined at the design stage. The designer should be aware of the nature of assembly processes and should always have sound reasons for requiring separate parts, and hence longer assembly time, rather than combining

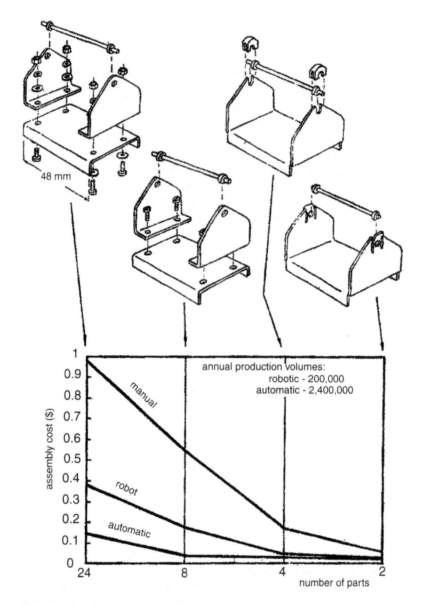

FIGURE 1.12 Example of the effect of design for assembly (DFA). (From G. Boothroyd, *Assembly Automation and Product Design*, Marcel Dekker, New York, 1992. With permission.)

several parts into one manufactured item. The designer should always keep in mind that each combination of two parts into one will eliminate an operation and a workstation in manual assembly, or usually an entire workstation on an automatic assembly machine. It will also tend to reduce the part costs. The following discussion on manual assembly will be used to introduce the basic DFA principles. This is always a necessary step, even if the decision is ultimately to employ some degree of automation, in order to compare the results against manual assembly. In addition, when automation is being seriously considered, some operations may have to be carried out manually, and it is necessary to include the cost of these operations in the analysis.

In the DFA technique embodied in the widely used DFA software, developed by Boothroyd Dewhurst, Inc., features of the design are examined in a systematic way, and a *design efficiency* or *DFA index* is calculated. This index can be used to compare different designs. There are two important steps for each part in the assembly:

1. A decision as to whether each part can be considered a candidate for elimination or combination with other parts in the assembly in order to obtain a theoretical minimum number of parts for the product
2. An estimation of the time taken to grasp, manipulate, and insert each part

Having obtained this information, it is then possible to obtain the total assembly time and to compare this figure with the assembly time for an ideal design. First you obtain the best information about the product or assembly. Useful items are:

Engineering drawings and sketches
Exploded three dimensional views
An existing version of the product
A prototype

Next we list the parts in assembly sequence. If the product contains subassemblies, treat these at first as parts, and analyze them later. In the example of the controller assembly in Figure 1.13, the tube with nuts at each end is treated as one item and named "tube assembly." If a prototype of the product is available, it is usually best to name the items as they are removed from the product. However, the software always lists them in a viable assembly order.

Finally, during the assembly (or reassembly) of the product, the minimum number of parts theoretically required is determined by answering the following questions:

1. Is the part or subassembly used only for fastening or securing other items?
2. Is the part or subassembly used only for connecting other items?

If the answer is yes to either question, then the part or subassembly is not considered theoretically necessary. If the answer is no to both questions, the following criteria questions are considered:

1. During operation of the product, does the part move relative to all other parts already assembled? Only gross motion should be considered—small motions that can be accommodated by elastic hinges, for example, are not sufficient for a positive answer.
2. Must the part be of a different material than, or be isolated from, all other parts already assembled? Only fundamental reasons concerned with material properties are acceptable.

3. Must the part be separate from all other parts already assembled because otherwise necessary assembly or disassembly of other separate parts would be impossible?

If the answer to all three criteria questions is no, the part cannot be considered theoretically necessary.

When these questions have been answered for all parts, a theoretical minimum part count for the product is calculated. It should be emphasized, however, that this

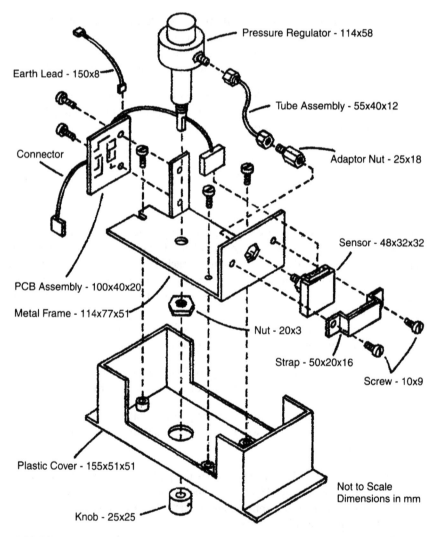

FIGURE 1.13 Exploded view of original controller assembly. (From G. Boothroyd, *Assembly Automation and Product Design*, Marcel Dekker, New York, 1992. With permission.)

theoretical minimum does not take into account practical considerations or cost considerations, but simply provides a basis for an independent measure of the quality of the design from an assembly viewpoint.

Also, during reassembly of the product, you answer questions that allow handling and insertion times to be determined for each part or subassembly. These times are obtained from time-standard databases developed specifically for the purpose. For estimation of the handling times (Figure 1.14) you must specify the dimensions of

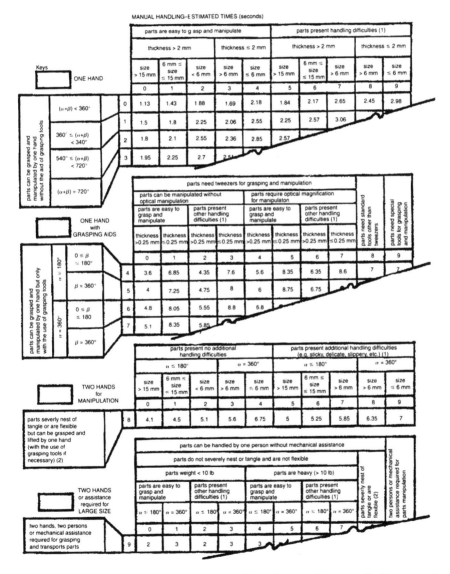

FIGURE 1.14 Classification, coding, and database for part features affecting manual assembly time. (From Boothroyd Dewhurst, Inc., Wakefield, R.I. With permission.)

the item; its thickness; whether it nests or tangles when in bulk; whether it is fragile, flexible, slippery, or sticky; and whether it needs two hands, grasping tools, optical magnification, or mechanical assistance.

For estimation of insertion and fastening as shown in Figure 1.15, it is important to know whether the assembly worker's vision or access is restricted, whether the item is difficult to align and position, whether there is resistance to insertion, and

MANUAL INSERTION–ESTIMATED TIMES (seconds)

PART ADDED but NOT SECURED

			after assembly no holding down required to maintain orientation and\ location (3)				holding down required during subsequent processes to maintain orientation or location (3)			
			easy to align and position during assembly (4)		not easy to align or position during assembly		easy to align and position during assembly (4)		not easy to align or position during assembly	
			no resistance to insertion	resistance to insertion (5)	no resistance to insertion	resistance to insertion (5)	no resistance to insertion	resistance to insertion (5)	no resistance to insertion	resistance to insertion (5)
			0	1	2	3	6	7	8	9
part snd associated tool including handel can easily reach the desired location		0	1.5	2.5	2.5	3.5	5.5	6.5		
		1	4	5	5	6				
due to obstructed access or restricted vision (2)		2	5.5	6.5						

PART SECURED IMMEDIATELY

no screwing operation or plastic deformation immediately after insertion (snap/press fits, circlips, spire nuts, etc.)

plastic deformation immediately after insertion

			plastic bending or torsion				riverting or similar operation				screw tightening immediately after insertion	
			easy to align and position with no resistance to insertion (4)	not easy to align or position during assembly and/or resistance to insertion (5)	easy to align and position during assembly (4)		not easy to align or position during assembly		easy to align and position during assembly (4)		not easy to align or position during assembly		easy to align and position with no torsional resistance (4)	not easy to align or position and/or torsional resistance(5)
					no resistance to insertion	resistance to insertion (5)			no resistance to insertion	resistance to insertion (5)			
parts and associated tool (including hands) can easily reach the desired location and the tool can be operated easily			0	1	2	3	4	5	6	7	8	9
due to obstructed access or restricted vision (2)		3	2	5	4	5	6	7	8	9		
due to obstructed access or restricted vision (2)		4	4.5	7.5	6.5	7.5	8.5	9.5				
due to obstructed access or restricted vision (2)		5	6	9	8	9						

SEPARATE OPERATION

			mechanical fastening processes (part(s) already in place but not secured immediately after insertion)				non-mechanical fastening processes (part(s) already in place but not secured immediately after insertion)					non-fastening processes	
			none or localized plastic deformation			bulk plastic deformation (large proportion of part is plastically deformed during fastening)	metallurgical processes						
			bending or similar processes	riveting or similar processes	screw tightening or other processes		no additional material required (e.g. resistance, friction welding, etc.)	additional material required (e.g. resistance, etc.)	soldering processes	weld/braze processes	chemical processes (e.g. adhesive bonding etc.)	manipulation of parts or sub-assembly (e.g. orienting, fitting or adjustment of part(s), etc.)	other processes (e.g. liquid insertion, etc.)
assembly processes where all solid parts are to place			0	1	2	3	4	5	6	7	8	9
		9	4	7	5	12	7				8	

FIGURE 1.15 Classification, coding, and database for part features affecting insertion and fastening. (From Boothroyd Dewhurst, Inc., Wakefield, R.I. With permission.)

whether it requires holding down in order to maintain its position for subsequent assembly operations.

For fastening operations, further questions may be required. For example, a portion of the database for threaded fastening is shown in Figure 1.16. Here, the type of tool used and the number of revolutions required are important in determination of the assembly time.

A further consideration relates to the location of the items that must be acquired. The database for handling shown in Figure 1.14 is valid only when the item is within easy reach of the assembly worker. If turning, bending, or walking is needed to acquire the item, a different database (Figure 1.17) is used.

It can be seen in Figures 1.14 through 1.17 that for each classification, an average acquisition and handling or insertion and fastening time is given. Thus, we have a set of time standards that can be used to estimate manual assembly times. These time standards were obtained from numerous experiments, some of which disagree with some of the commonly used predetermined time standards such as methods time measurement (MTM) and work factor (WF). For example, it was found that the effect of part symmetry on handling time depends on both alpha and beta rotational symmetries, illustrated in Figure 1.18. In the MTM system, the "maximum possible orientation" is employed, which is one half the beta rotational symmetry of a part. The effect of alpha symmetry is not considered in MTM. In the WF system, the symmetry of a part is classified by the ratio of the number of ways the part can be inserted to the number of ways the part can be grasped preparatory to insertion. In this system, account is taken of alpha symmetry, and some account is taken of beta symmetry. Unfortunately, these effects

Insertion and fastening times (seconds)
for threaded items using power tools

			easy to align and position		not easy to align and position	
			one part	several parts	one part	several parts
			0	1	2	3
autofeed	no vision or obstruction restrictions	0	1.9	2.3	3.6	4
screw or nut inserted into tool or assembly	no vision or obstruction restrictions	1	3.6	4	5.3	
	restircted vision only	2	6.3	6.7		
	obstructed access	3				

– Times do not include fastening tool acquisition and replacement.
– Times in column 0 – 5 assume screw or nut fastening requiring 5 revolutions. For operations requiring more than 5 revolutions, times for additional revolutions in column 6 should be added.

FIGURE 1.16 Classification, coding, and database for threaded fastening. (From Boothroyd Dewhurst, Inc., Wakefield, R.I. With permission.)

Acquisition and handling times (2) (seconds)
for one large part not within easy reach

average to distance to location of parts (ft.)		easy to grasp	difficult to grasp (1)	requires two persons (3)	fixed swing crane
		0	1	2	3
14" to 4	0	2.54	4.54	8.82	12
4 to 7	1	4.25	6.25		
7 to 10	2	5.54			

1) For large items, no features to allow easy grasping (e.g. no finger hold).
2) Times are for part acquisition only. Multiply by 2 if replacement time for a fixture
   is to be included. When cranes are used, the time to return the crane is included.
3) Times are equivalent times for one person.

FIGURE 1.17 Classification. coding. and database for turning, bending, and walking. (From Boothroyd Dewhurst, Inc., Wakefield, R.I. With permission.)

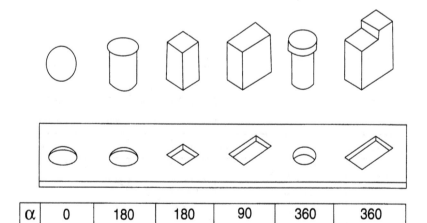

α	0	180	180	90	360	360
β	0	0	90	180	0	360

FIGURE 1.18 Alpha and beta rotational symmetries for various parts. (From G. Boothroyd, *Assembly Automation and Product Design*, Marcel Dekker, New York, 1992. With permission.)

are combined in such a way that the classification can be applied to only a limited range of part shapes.

Similar experimentation for DFA was carried out for part thickness, part size, weight, and many other factors which give a more precise basis for comparing one type of part design to another in the detailed analysis required by the DFA method.

1.3.3   Results of the Analysis

Once the analysis is complete, the totals can be obtained as shown in Figure 1.19. For the example of the controller, the total number of parts and subassemblies is 19, and there are 6 additional operations. The total assembly time is 227 seconds, the corresponding assembly cost is $1.90, and the theoretical minimum number of items is five.

If the product could be designed with only five items and each could be designed so that no assembly difficulties would be encountered, the average assembly time per item would be about 3 seconds, giving a theoretical minimum assembly time of 15 seconds. This theoretical minimum is divided by the estimated time of 227 seconds to give the design efficiency or DFA index of 6.6%. This index can be independently obtained and provides a basis for measuring design improvements.

The major problem areas can now be identified, especially those associated with the installation of parts that do not meet any of the criteria for separate parts. From the tabulation of results, it can be seen that attention should clearly be paid to combining the plastic cover with the metal frame. This would eliminate the assembly operation for the cover, the three screws, and the reorientation operation, representing a total time saving of 54.3 seconds, which constitutes 24% of the total assembly time. Of course, the designer must check that the cost of the combined plastic cover and frame is less than the total cost of the individual piece parts. A summary of the items that can be identified for elimination or combination and the appropriate assembly time savings are presented in Figure 1.20a.

Design changes now identified could result in saving of at least 149.4 seconds of assembly time, which constitutes 66% of the total. In addition, several items of hardware would be eliminated, resulting in reduced part costs. Figure 1.20b shows a conceptual redesign of the controller in which all the proposed design changes have been made, and Figure 1.21 presents the revised tabulation of results. The total assembly time is now 84 seconds, and the assembly efficiency is increased 18%, a fairly respectable figure for this type of assembly. Of course the design team must now consider the technical and economic consequences of the proposed designs. Some of the rapid prototyping techniques discussed in Subchapter 1.4 for rapidly producing a prototype part in a very short lead time may be able to furnish a part to the new design for evaluation or testing.

First, there is the effect on the cost of the parts. However, experience shows, and this example should be no exception, that the saving from parts cost reduction would be greater than the savings in assembly costs, which in this case is $1.20.

It should be realized that the documented savings in materials, parts fabrication, and assembly represent direct costs. To obtain a true picture, overheads must be

MANUAL - BENCH ASSEMBLY Name of Assembly - $MAIN SUB Item Name: Part, Sub of Pch assembly No. or Operation	Manual handling code No. of items		Handling time per item (s)	Manual Insertion code Insertion time per item (s)		Total oper'n time RP*(TH+TI) Total opertion cost-cents TA*OP		Figures for min. parts	Operator rate OP: 30.00 $/hr 0.83 c/s Description
	RP	HC	TH	IC	TI	TA	CA	HM	
1 $pressure regulator	1	30	1.95	00	1.5	3.5	2.9	1	place in fixture
2 metal frame	1	30	1.95	06	5.5	7.4	6.2	1	add
3 nut	1	00	1.13	39	8.0	9.1	7.6	0	add & screw fasten
4 Reorientation	1	-	-	98	9.0	9.0	7.5	-	reorient & adjust
5 $sensor	1	30	1.95	08	6.5	8.4	7.0	1	add
6 strap	1	20	1.80	08	6.5	8.3	6.9	0	add & hold down
7 Screw	2	11	1.80	39	8.0	19.6	16.3	0	add & screw fasten
8 Apply tape	1	-	-	99	12.0	12.0	10.0	-	special operation
9 adoptor nut	1	10	1.50	49	10.5	12.0	10.0	0	add & screw fasten
10 tube assembly	1	91	3.00	10	4.0	7.0	5.8	0	add & screw fasten
11 Screw fastening	1	-	-	92	5.0	5.0	4.2	-	standard operation
12 &PCB ASSEMBLY	1	83	5.60	08	6.5	12.1	10.1	1	add & hold down
13 Screw	2	11	1.80	39	8.0	19.6	16.3	0	add & screw fasten
14 connector	1	30	1.95	31	5.0	6.9	5.8	0	add & snap fit
15 earth lead	1	83	5.60	31	5.0	10.6	8.8	0	add & snap fit
16 Reorientation	1	-	-	98	9.0	9.0	7.5	-	reorient & adjust
17 $knob assembly	1	30	1.95	08	6.5	8.4	7.0	1	add & screw fasten
18 Screw fastening	1	-	-	92	5.0	5.0	4.2	-	standard operation
19 plastic cover	1	30	1.95	08	6.5	8.4	7.0	0	add & hold down
20 reorientation	1	-	-	98	9.0	9.0	7.5	-	reorient & adjust
21 screw	3	11	1.80	49	10.5	36.9	30.8	0	add & screw fasten

FIGURE 1.19 Completed design for assembly (DFA) worksheet for original controller design. (From G. Boothroyd, *Assembly Automation and Product Design*, Marcel Dekker, New York, 1992. With permission.)

added, and these can often amount to 200% or more. Total cost per shop man hour could be assumed to be $30 to $50 per hour in the mid 1990s. In addition, there are other savings more difficult to quantify. (See Subchapter 1.0.) For example, when a part such as the metal frame is eliminated, all associated documentation, including part drawings, is also eliminated. Also, the part cannot be misassembled or fail in service, factors that lead to improved reliability, maintainability, and quality of the

Design change	Items	Time saving, sec
1. Combine plastic cover with frame and eliminate 3 screws and reorientation	19,20,21	54.3
2. Eliminate strap and 2 screws (snaps in plastic frame to hold sensor, if necessary)	6,7	27.9
3. Eliminate screws holding PCB assembly (provide snaps in plastic frame)	13	19.6
4. Eliminate 2 reorientations	4,16	18.0
5. Eliminate tube assembly and 2 screwing operations (screw adapter nut and sensor direct to the pressure regulator)	10,11	12.0
6. Eliminate earth lead (not necessary with plastic frame)	15	10.6
7. Eliminate connector (Plug sensor into PCB)	14	7.0

(a)

FIGURE 1.20 (a) Summary of items to be considered for elimination or combination on the controller assembly. (Part b on next page.)

product. It is not surprising, therefore, that many U.S. companies have been able to report annual savings measured in millions of dollars as a result of the application of the DFA analysis method described here.

The DFA system is much more comprehensive than outlined here. There could be wide variations in the results, depending on the degree of automation eventually selected, and the results of detail analysis of the parts fabrication of the final design selected. However, the example of the basic system approach outlined in this handbook should serve as an introduction to the DFA system.

1.3.4 Bibliography

Boothroyd, G., *Assembly Automation and Product Design,* Marcel Dekker, New York, 1992.
Boothroyd, G., *Product Design for Manufacture and Assembly,* Marcel Dekker, New York, 1994.

1.4 RAPID PROTOTYPING

1.4.0 Introduction to Rapid Prototyping

Rapid prototyping is one of the names given to a new group of technologies for converting designs from computer representations directly into solid objects without human intervention. The technologies are sometimes known collectively as solid

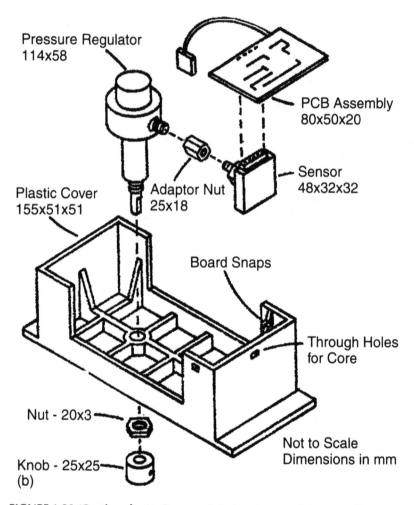

Pressure Regulator
114x58

PCB Assembly
80x50x20

Plastic Cover
155x51x51

Adaptor Nut
25x18

Sensor
48x32x32

Board Snaps

Through Holes
for Core

Nut - 20x3

Not to Scale
Dimensions in mm

Knob - 25x25
(b)

FIGURE 1.20 (Continued) (b) Conceptual design changes of the controller assembly. (From G. Boothroyd. *Assembly Automation and Product Design*, Marcel Dekker, New York. 1992. With permission.)

free-form fabrication or computer automated fabrication. No single one of the many technologies has yet proven that it can meet all market requirements, so those intending to be involved in this industry must know the fundamental processes, limits, and potentials of the competing machines.

One of the goals of all successful companies is to reduce the time between initial concept of a product and production deliveries. The other vital concept is the early involvement of manufacturing expertise with the product design team. This includes outside suppliers along with in-house experts. The outside suppliers that must be involved are those trades and those with skills involving parts that your company may buy rather than build in-house. This often includes castings, forgings, plastic

MANUAL - BENCH ASSEMBLY		Manual handling code		Manual Insertion code		Total oper'n time RP*(TH+TI)	Figures for min. parts		Operator rate OR: 30.00 $/hr 0.83 c/s	SUB ASSEMBLY OR PART COSTS	
Name of Assembly - $MAIN SUB		No. of items	Handling time per item (s)	Insertion time per item (s)		Total oper'n cost-cents TA*OP				Total item cost $	Total cooling cost kS
No.	Item Name: Part. Sub or Pcb assembly or Operation	RP	HC	TH	IC	TI	TA	CA	HM	Description	CT	CC
1	$pressure regulator	1	30	1.95	00	1.5	3.5	2.9	1	place in fixture	10.46	0.0
2	plastic cover	1	30	1.95	06	5.5	7.4	6.2	1	add & hold down	0.00	0.0
3	nut	1	00	1.13	39	8.0	9.1	7.6	0	add & screw fasten	0.20	0.0
4	$knob assembly	1	30	1.95	08	6.5	8.4	7.0	1	add & screw fasten	-	-
5	Screw fastening	1	-	-	92	5.0	5.0	4.2	-	standard operation	-	-
6	Reorientation	1	-	-	98	9.0	9.0	7.5	-	reorient & adjust	-	-
7	Apply tape	1	-	-	99	12.0	12.0	10.0	-	special operation	-	-
8	adaptor nut	1	10	1.50	49	10.5	12.0	10.0	0	add & screw fasten	0.30	0.0
9	$SENSOR	1	30	1.95	39	8.0	9.9	8.3	1	add & screw fasten	1.50	0.0
10	$PCB ASSEMBLY	1	83	5.60	30	2.0	7.6	6.3	1	add & snap fit	0.00	0.0

FIGURE 1.21 Completed DFA analysis for the redesigned controller assembly. (From G. Boothroyd, *Assembly Automation and Product Design*, Marcel Dekker, New York, 1992. With permission.)

moldings, specialty sheet metal, and machining. Their knowledge can assist the original equipment manufacturer (OEM) in producing a better product, in a shorter time, at a lower cost.

Other firms that may be of great assistance are firms that concentrate on producing prototype parts on an accelerated schedule to aid in the design and process selection. They may also produce the early production parts while you are waiting for your "hard tooling" to be completed. They may even be your eventual production supplier, or a backup source if your production supplier has problems.

As we shall see in Chapter 2, on product design, often the small details of a part or process can greatly influence the design. In the product design and development phase of a project, it is often valuable to have a prototype part made of a proposed design or of alternative designs to permit members of your team to get a better feel for the design concept. In the automobile industry, for example, the appearance and feel of door handles, controls, etc. are important to the ultimate customer. We can look at drawings or computer screens of a concept design but few of us can assess what a Shore Durometer hardness number means on paper as well as we can by handling several copies of a part with different flexibility or color, or made from different materials. Subchapter 6.1, on ergonomics, discusses the importance of considering the human capabilities of the user in the design of a product. The design of Logitec's new ergonomic mouse for computers is a recent success story. Most good designs are an iterative process, and most companies do not have the in-house capability to quickly make accurate prototypes in their production shops. A prototype part is more than a model. It is as much like the eventual production part as possible in a short period of time!

There are a few specialty companies in the United States that have the ability to produce models and patterns using the latest rapid prototyping techniques, including stereolithography, laminated object manufacturing (LOM) and CAM hog-outs from solid stock, as well as traditional pattern making and craft skills for producing prototype plaster-mold metal castings and plastic parts. Some specialize in photo etching metal blanks, others in quickly producing sheet metal parts with temporary, low-cost tooling. Complex investment castings can be made without tooling. With these skills and methods, they can create premium-quality, economical parts in the shortest time possible. While 2 or 3 weeks is fairly common lead time, they sometimes can deliver prototype parts in less than a week.

In this subchapter we will introduce some of the techniques available, and the capabilities of some typical firms specializing in this field. Some of these companies build only prototype quantities, but most can produce the initial start-up quantities of parts while your hard tooling is being built. Many of these companies can produce parts directly from CAD files, as well as from drawings or sketches. This allows us to send data via modem, rather than relying on other communication systems. For example, Armstrong Mold Company, in East Syracuse, New York, uses IGES, DFX, CADL, and STL file formats directly, or can work from floppy disk or DC 2000 mini data tape cartridges to produce prototype parts with short lead times.

Advantages of Rapid Prototyping

Paperless manufacturing (working directly from CAD files or sketches)
Functional parts in as little as 1 week
Produce models, which can be used as masters for making metal castings and
    plastic parts in small or large quantities
Fit and function models used to detect design flaws early in development
Allows prototype part to be used during qualification testing of a product

1.4.1   Prototype Models and Patterns

Stereolithography

Stereolithography is used to create three-dimensional objects of any complexity from CAD data. Solid or surfaced CAD data is sliced into cross sections. A laser generating an ultraviolet beam is moved across the top of a vat of photosensitive liquid polymer by a computer-controlled scanning system. The laser draws each cross-section, changing the liquid polymer to a solid. An elevator lowers the newly formed layer to recoat, and establishes the next layer thickness. Successive cross sections are built layer by layer one on top of another, to form the a three-dimensional plastic model. Armstrong, along with a few other firms, takes advantage of this process.

Laminated Object Manufacturing

Laminated object manufacturing (LOM) is also used to create three dimensional objects of any complexity from CAD data. Solid or surfaced CAD data is sliced into cross sections. A single laser beam cuts the section outline of each specific layer out of paper from 0.002 to 0.020 in. thick coated with heat-seal adhesives. A second layer is then bonded to the first, and the trimming process is repeated. This process continues until all layers are cut and laminated, creating a three-dimensional, multilayered wood-like model. One LOM machine observed had a 32 in. × 22 in. × 20 in. working envelope.

Machined Prototypes

Three-dimensional objects can be cut from a variety of solid stocks (metal, resin, plastic, wood) directly from two- or three-dimensional CAD data using multiaxis CNC machining centers. Applicable machines can be utilized for making models from drawings or sketches.

1.4.2   Casting Metal Parts

Plaster Molds

Aluminum or zinc castings can be produced by pouring liquid metal into plaster (gypsum) molds. Typical applications at Armstrong include:

Castings for business machines, medical equipment, computers, automotive, aerospace, electronics, engineering

Molds for the plastics industry: rotational, vacuum form, expanded polystyrene, kirksite, and injection molds

The first step in the process is to make a model or master pattern. This can be done from a drawing or CAD file. The material can be wood, plastic, or machined from brass or other metal. The model is either hand-crafted or machined, using multiaxis CNC or tool room machines, depending on the complexity and availability of CAD data. The model will have shrinkage factors built in, as well as draft, and machining allowance if desired. A customer-furnished model or sample part can be adapted to serve as the master pattern (see Figure 1.22).

The second step is to make a negative mold, and core plugs if required, from the master model. A positive resin cope and drag pattern is now made from the negative molds. Core boxes are built up, and the gating and runner system, with flasks as necessary, are added.

Next, a liquid plaster slurry is poured around the cope and drag pattern and into the core boxes. After this sets, the plaster mold is removed from the cope and drag patterns. The plaster mold and cores are then baked to remove moisture.

Molten metal is prepared by degassing, and spectrographic analysis of a sample checks the chemical composition of the material. The molten metal is then poured into the assembled plaster mold. The plaster is removed by mechanical knockout and high-pressure water jet. When the casting has cooled, the gates and risers are removed.

The raw castings are then inspected and serialized. Any flash and excess metal is removed (snagged). Castings may then require heat treatment, x-ray, or penetrant

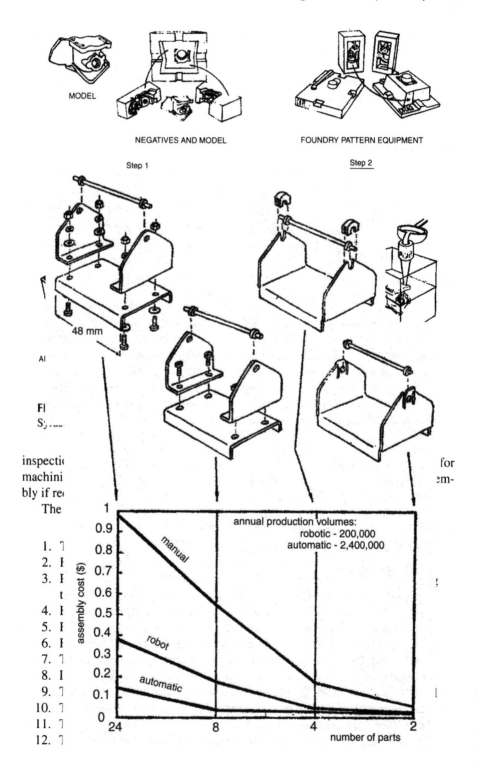

inspecti⟨   for
machini   ⟨m-
bly if re⟨

The

1. ⊺
2. I
3. I
   t
4. I
5. I
6. I
7. ⊺
8. I
9. ⊺
10. ⊺
11. ⊺
12. ⊺

Design and Technical Information

Size: No limitation, but best range is within 2 in. to 36 in. cube
Finish: Can hold 63 μin. but normally 90 μin.
Shape: Considerable design freedom for unusual and complex shapes
Wall thickness:
    Thin: 0.030–0.060 in.
    Average: 0.080–0.120 in.
    Thick: 0.180–0.500 in.
General tolerances:
    0–2 ± 0.010 in.   2–3 ± 0.012 in.
    3–6 ± 0.015 in.   6–12 ± 0.020 in.
    12–18 ± 0.030 in.   1–30 ± 0.040 in.
    Note: Tighter tolerances can be negotiated.

The process is limited to nonferrous metals with pouring temperatures below 2000°F. This includes all aluminum and zinc casting alloys and some copper-based alloys. All aluminum and zinc casting alloys are per commercial and military specifications (see Figure 1.23). This includes the appropriate mechanical properties. Small holes, 1/4 in. or less, are not economical to cast unless they are an odd shape or inaccessible for machining. Zero draft is possible in specified areas, but is typically 1/2 to 2°. Corner radii and fillets are as required, but are typically 1/16 in.

A variety of models and patterns can be made, depending on the customer's
    request:
    Wood: to expedite for up to 20 pieces
    Epoxy resin: usually up to 500 pieces
    Metal (aluminum or brass): used to obtain best tolerances and quality
    Rubber: for quantities up to 1000 pieces; tooling can be duplicated easily
        from master tooling to expedite delivery or for higher volumes
A rule of thumb for costs of complex shapes in 15 in. cubes is:
    Tooling: 10% of die-cast tools
    Piece price: 10 times die-casting price

Normal delivery time is 1 to 2 weeks for simple parts, and 6 to 8 weeks for complex parts.

Plaster- and Sand-Molding Combination

A combination of plaster and sand molding is used for castings that require high metallurgical integrity as verified by radiographic or fluorescent penetrant inspection. When used in combination with no-bake sand molds, the properties of plaster mold castings can be enhanced considerably by taking advantage of the faster cooling rates inherent in sand molds in combination with the insulating aspect of plaster molds. (Material still pours and fills, maintaining its liquidity, but then cools quickly.)

Aluminum Alloys

		Mechanical Properties					
Alloy	Heat Treating	Ultimate Tensile Strength KSI	Min. Yield Strength Set @ 2% KSI	Elongation % in 2"	Hardness Rockwell E Scale 1/16" Ball	Pressure Tightness	Machin-ability
319.0	T6	36	24	2.0	85.5	B	A
355.0	T51	28	23	1.5	76.0	A	C
355.0	T6	35	5	3.0	85.5	A	B
C355.0	T6	39	29	5.0	89	A	B
356.0	T51	25	20	2.0	71	A	C
356.0	T6	33	24	3.5	79	A	A
A356.0	T51	26	18	3.0	71	A	C
A356.0	T6	40	30	6.0	82.5	A	A
357.0	T51	26	17	3.0	80	B	C
357.0	T6	45	35	3.0	91	B	B
A357.0	T6	45	35	3.0	89	B	B
380.0	NONE	46	23	3.5	85.5	B	C
712.0	*	35*	25*	5.0*	82.5	C	A

712.0 also known as D712, D612, and 40E

Zinc Alloys

ZA3	NONE	41	-	10.0	87	C	C
ZA8	NONE	34	30	1.5	95.5	C	C
ZA12	NONE	43	30	2.0	93	C	B
ZA27	NONE	61	46	9.5	100	C	B

These values are for separately cast test bars, and are typical values.

* Test 30 days after casting

A=Excellent   B=Good   C=Fair

Applicable Military Specifications

- Mechanical/Chemical Inspection to MIL-A-21180 QQ-A-601, AMS-4217
- System Control to MIL-1-45208
- Gage Control to MIL-STD-45662
- Penetrant Inspection to MIL-STD-6866
- Radiographic Inspection to MIL-STD-453, MIL-STD-2175
- Heat Treat to MIL-H-6088
- N.D.T. to MIL-STD-410

FIGURE 1.23 Aluminum alloy casting mechanical properties. (Courtesy of Armstrong Mold Corp., East Syracuse, N.Y. With permission.)

Investment Casting

Investment casting prototypes can be produced from plastic patterns without permanent tooling. In brief terms, the supplier replaces the wax pattern that would have been produced from an injection die with a low-melting-point plastic pattern assembled from multiple pieces. A radius made from wax is then formed on the assembly by hand, and from this point on, the process continues in the same manner as the typical investment casting process. This gives the OEM the opportunity to qualify a casting as part of the initial testing program, eliminating the need for requalification had a machined part or a weldment been used as a substitute in qualification test hardware. Depending on the complexity of the design, the supplier can deliver a prototype investment casting in 2 weeks or so, at less cost than a machining. Uni-Cast, Inc., in Nashua, New Hampshire, will provide engineering assistance to build a large, thin-walled initial plastic pattern in your model shop under the supervision of your design team. This pattern can then be taken back to their shop for the casting process (see Figure 1.24).

1.4.3 Manufacture of Plastic Parts

Castable Urethane

Castable urethane is a liquid, two-component, unfilled urethane resin system that is poured into a closed mold. The result after solidification is a tough, high-impact-resistant, dimensionally stable part. This process is also used by Armstrong.

With this process:

1. Prototype and low-volume plastic parts can be produced.
2. Injection-molded parts to check fit and function can be simulated, and designs can be debugged.
3. As in injection molding, there is considerable design freedom.
4. Molds can be made from room temperature cure silicone (RTV) resin, sprayed metal, or, for precision parts, machined aluminum.
5. Tooling is low in cost and allows ease of modification.
6. A variety of inserts can be molded or post installed.
7. Stereolithographic molds can be duplicated.
8. Delivery depends on size and complexity, and can be 1 to 7 weeks after receipt of order.
9. There is UL traceability.

Design and Technical Information

Size: Up to 24 in.
Finish: Molded color white (no-bake primer and paint available if required)
Wall thickness: Can vary but should be kept less than 1/4 in.
General tolerances:
    0–3 ± 0.010 in.
    3–6 ± 0.015 in.
    6–12 ± 0.020 in.
    12–24 ± 0.030 in.
    Note: Tighter tolerances can be negotiated.
Holes: All holes molded to size; if precision required can be machined
Draft: Recommended 1–2°, but no draft is possible if required
Corner and fillet radii: Can be sharp, but some radius (1/16 in.) preferred

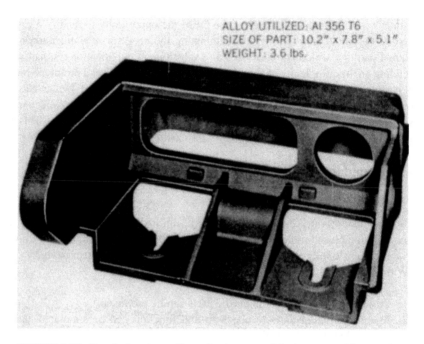

ALLOY UTILIZED: Al 356 T6
SIZE OF PART: 10.2" x 7.8" x 5.1"
WEIGHT: 3.6 lbs.

FIGURE 1.24 Part designed as a die casting because of the large quantities required. Smaller repeat orders at a later date are more economical as an investment casting. (Courtesy of Uni-Cast, Inc., Nashua, N.H. With permission.)

See Figure 1.25 for more information on properties of castable urethane. A variety of materials are available for this process, which offer different properties.

Reaction Injection Molding

In the reaction injection molding (RIM) process, two liquid components, component A (a formulated polymeric isocyanate) and component B (a formulated polyol blend), are mixed in a high-pressure head and then pumped into a mold cavity. A reaction then occurs in the mold, resulting in a high-density polyurethane structural foam part.

Paul Armstrong of Armstrong Mold says that this process has the following features:

1. Can be used to produce covers, bases, keyboards, bezels, and housings for computers, business machines, and medical applications.
2. For prototype through medium-volume applications (1 to 2000 pieces).
3. Considerable design freedom to produce complex shapes.
4. Molds can be resin, sprayed metal, cast aluminum, or machined aluminum.
5. Tooling is low in cost and allows for ease of modification.

Typical Properties

		Test Method
Color	Dark Amber	Visual
Hardness, Shore D	83 ± 5	ASTM D22/40
Tensile Strength, psi	7966	ASTM D412
		CONAP
Linear Shrinkage (in/in)	0.0001	CONAP
Heat Distortion Temperature	127.4°F	ASTM D648 (*Surface Stress 66 psi)
Elongation	13%	ASTM D412
Flexural Modulus (psi)	174,593	ASTM D790
Flexural Strength (psi)	8,095	ASTM D790
Specific Gravity	1.23	
Flammability 1/16" Sample	94V-0	UL-94

The above properties are intended as a guide only and may vary depending on thickness and shape.

FIGURE 1.25 Typical properties of castable urethane. (Courtesy of Armstrong Mold Corp., East Syracuse, N.Y. With permission.)

6. A variety of castings, sheet metal, and threaded inserts can be molded in place.
7. Alternative foams are available which offer a range of insulating properties.
8. UL traceability.

Design and Technical Information

Size: Up to 48 in.
Weight: Up to 18 lb.
Finish (if required):
  No-bake prime and paint
  RF shielding
  Silkscreen
Wall thickness: Can vary, but should not be less than 1/4 in., although local thicknesses of 1/8 in. can be produced
General tolerances:
  0–3 ± 0.010 in.
  3–6 ± 0.015 in.
  6–12 ± 0.020 in.
  12–18 ± 0.030 in.
  18–48 ± 0.040 in.
  Note: Tighter tolerances can be negotiated.
Holes: All holes molded to size
Draft: Recommended 1–2°, but no draft is possible when specified
Radii and fillets: Should be as liberal as possible—1/8 to1/4 in.
Density: Can be varied depending on foam used to offer a variety of insulation properties

1.4.4 Photochemical Machining

Blanking

Photochemical machining (PCM) allows burr-free, stress-free blanking on virtually all ferrous and nonferrous metals with no chemical or physical alteration to the materials. With precise multiple imaging, multiple parts blanks can be produced on sheets up to 30 in. wide. The process can work with metals of any temper with thickness ranging from 0.001 in. to heavy gauge. The tolerance range is $\forall$ 0.001 to $\forall$ 0.010 in., depending on material type and thickness. See Figure 1.26 for some examples of parts made by this process.

The process starts with CAD drawings, either furnished by your company or made by the supplier. This is then copied multiple times to make a master negative containing the total parts required in the initial order, and checked for accurate registration. A metal sheet is sheared to size, cleaned, and coated with a photoresist. A contact print is made from the master sheet and developed. The entire sheet is then immersed in an acid bath and the unwanted metal is eaten away, leaving the finished part. The photoresist is then removed. Machining, forming, or whatever subsequent process is required can then follow this.

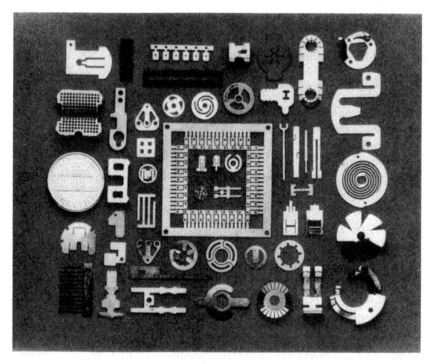

FIGURE 1.26 Examples of some parts made by photochemical machining. (Courtesy of Microphoto, Inc., Roseville, Mich. With permission.)

Forming and Fabrication

Many companies offer complete photochemical engineering services, using state-of-the-art computer-aided design. Precision plotting and photographic equipment allow for the design of complex and intricate parts with economical photo-tooling costs. In general, these companies are very willing to work with your design team to bring your concept to production quickly and accurately. Because tooling is produced photographically, lead time can be reduced from weeks to hours, and design changes can be made quickly, without scrapping expensive dies. Typical lead time is 1 to 2 weeks using readily available materials, but this time can be shortened if the schedule demands it. An example of a firm with a complete in-house tooling, forming, and fabrication department, Microphoto, Inc. (Detroit, Michigan), can economically produce quality-finished parts. Using both mechanical and hydraulic presses, they produce precision parts to most print tolerances after PCM blanking.

1.4.5 Short-Run Stampings

Tooling costs for short-run stampings are kept to a minimum. You share the tools and holders with all of the suppliers' customers. Standard round and square punches meet many demands. Standard forming tools make a wide variety of shapes. Other standard tools permit many companies to build parts quickly at a minimum cost. Design changes are most easily incorporated with short-run tooling. This minimizes costly scrap and retooling. Tooling techniques for the short-run process at Dayton Rogers (Minneapolis, Minnesota) are unique. Their large tool rooms use the Andrew Linemaster, Electrical Discharge Machine, and other special tools to speed the production of dies. Short runs do not create unnecessary penalties in either quality or cost.

The quality of stamped parts is usually more reliable than from most other methods. The part-to-part repeatability is unusually good. A good firm's quality control department will be well equipped to assure that your parts meet specifications.

Delivery of sheet metal prototype parts is usually assured. Dayton Rogers, for example, has over 150 presses up to 300 tons. Their large metal inventory makes 2 million pounds of stock immediately available. Production control methods include a daily check of order status against schedule. Their "scoreboard performance" indicates that over 99% of all orders are shipped on time or ahead of schedule.

How to Use Short-Run Stampings

Short-run stampings offer an economical way to produce parts in quantities from prototype to 100,000 pieces with a short lead time. It is an ideal method for checking the design, assembly process, and market acceptance of a new product, all with minimum investment. The investment in short-run tooling is always a secure one. Even if the product exceeds sales forecasts, the tooling can be a major asset as a second source in the event of high-rate production die breakdown. Subcontractors review thousands of prints each year, searching for ways to save customers money. Their suggestions involve material specifications, dimensions, and tolerances. Use their expertise!

Standard warehouse materials save both time and money. Price and delivery are both much more favorable when the supplier has a large metals inventory. For small lots, you can save on part price by using metal bought in large lots. If thickness tolerance is critical, double disk grinding is an option. If a part is extremely thick for its shape, sometimes two pieces laminated together is a solution.

Tolerances

Consider the necessity of specific tempers or closer material tolerances. If a part requires a specific temper or closer material tolerances, strip steel is desirable. However, if temper or closer material tolerances are not a factor, sheet steel is available at a lesser cost. Tolerances should be no tighter than necessary to make a part functional. Close or tight tolerances result in high cost to build tools, high tool maintenance, higher run cost, and costly in-process inspection to assure specifications.

1.4.6   Bibliography

Armstrong Mold Company, 6910 Manlius Road, East Syracuse, NY 13057.
Uni-Cast, Inc., 45 East Hollis Street, Nashua, NH 03060.
Microphoto, Inc., 6130 Casmere, Detroit, MI 48212.
Dayton Rogers Mfg. Company, 2824 13th Avenue South, Minneapolis, MN 55407.

2 Product Design

Jack M. Walker

and

Richard D. Crowson

2.0 INTRODUCTION TO PRODUCT DESIGN

There is no magic formula for guiding the design team of a new or redesigned product, or for support by the manufacturing engineering member of the team representing the factory. The considerations and techniques in Chapter 1 are valuable guides in (1) working together as an integrated team for product development and design, (2) examining the customer's wants and needs, (3) identifying the problems with assembly of the final product, and (4) developing the product cost.

This chapter is devoted to discussion of some of the major manufacturing processes used in making detail parts. A discussion of each of the processes is presented in order to better understand the tolerances, lead times, costs, and do's and don'ts of designing parts using each process. Subchapter 2.1 covers design for machined parts, Subchapter 2.2 goes into designing for castings and forgings, Subchapter 2.3 discusses designing parts from sheet metal, and Subchapter 2.4 is a summary to aid in selecting the best process for a particular design application.

Experience tells us that it is necessary to involve production process expertise early in the design and development phases of all new programs. The best product design for manufacturing will be one that:

Considers all elements of the fabrication, subassembly, and assembly operations leading to a completed product

Is able to accept the manufacturing process tolerances

Uses straightforward, understandable manufacturing processes that lead to first-time quality during production start-up

Can be built with automated process equipment using programmable process controllers (when appropriate)

Does not require separate inspection operations, but empowers the operator with the responsibility for his or her work

Allows statistical process control (SPC) to monitor the production operations and permits operator adjustment of the process to ensure 100% quality every time

Contributes to a shorter manufacturing cycle with minimum work in process
Can be built for the lowest cost

2.1 DESIGN FOR MACHINING

2.1.0 Introduction to Design for Machining

An early comment on designing parts for machining should be to question the necessity
of machining the part. With the options available for other processes to consider, this
is no simple task. Leo Alting, in *Manufacturing Engineering Processes* (1982), discusses "mass conserving" processes such as rolling, extrusion, forming, swaging, cold
heading, and the like. The idea is that a piece of metal may be modified to produce a
finished part, without cutting chips. Many times, a part using "mass reducing" techniques, such as cutting chips on some type of machining center, may retain only 20%
of the original metal, while 80% winds up as chips or scrap. In many competitive
job shops that use production screw machines or other high-rate machines, the profit
comes from the sale of the scrap chips! During the concurrent engineering process,
whether you utilize design for assembly (DFA) or some other formal technique to
assist in design analysis, all the various production options must be explored. Quantity
of parts to be made, the required quality, and the overall design requirements will
influence the decision. Figure 2.1 shows two examples where conventional machining
is not the correct choice. In Figure 2.1a, the part was changed to an extrusion, with a
simple cutoff operation; and in Figure 2.1b, the large quantities permitted use of cold
heading plus standard parts. New casting alloys and techniques, increased pressures
available for forging and pressing, and the progress made in powder metal and engineering plastics make "near net shape" worth strong consideration.

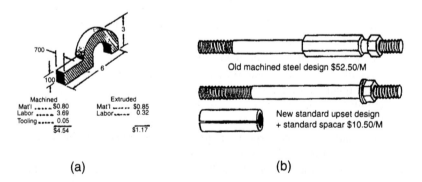

	(a)		(b)

FIGURE 2.1   (a) Change from machined plate to extruded material produced at 74%
savings on quantities of 2000 pieces. (b) Stud used by General Electric requiring 200,000
pieces per year. An automatic upset and rolled, threaded stud with a standard roll-formed
spacer offered worthwhile savings. (From R. A. Bolz, *Production Processes*, Industrial
Press, New York, 1963. With permission.)

Machine Fundamentals

In order to acquire a sound working knowledge of the design of parts to be made by machining, it is essential to begin by becoming familiar with a few of the fundamentals. There are several advantages to studying the elements of a subject as complex as today's machine tools. In the first place, it provides us a chance to get acquainted with the main points before having to cope with the details. Once learned, these main points will serve as a filing system for mentally sorting and cataloging the details as they are brought up and discussed. It is then easier to study the subject in a logical fashion, progressing step by step from yesterday's equipment and processes to tomorrow's.

To begin, *machine tool* in the context of this chapter is defined as follows:

A machine is a device to transmit and modify force and motion to some desired kind of work.

Therefore, we could say:

A machine tool is a mechanism for applying force and motion to a cutting instrument called a tool.

There are several qualifications to this broad definition. First, a machine tool does not always apply force and motion to the tool, but sometimes applies them to the material to be cut, while it is in contact with the fixed tool.

A further limitation of the term "machine tool" is usually observed in practice by not including such power tools as cutoff saws, shears, punch presses, routers, and the like. Although such machines do apply mechanical force to a cutting tool, they do not machine metals by "cutting chips" in the same way, as does a lathe or milling machine.

Besides employing power and a cutting tool, true machine tools also provide means for accurately controlling the tool to take as deep or as shallow a cut as desired, as well as a means for applying this action consecutively in the direction needed to obtain the desired finished shape. Further, a true machine tool must provide means for interrupting this action whenever necessary, to allow the operator to check the progress of the work, or for some other function to occur.

Our chapter has defined all machine tool operations to be confined to the surface of the workpiece; therefore, the shape of any workpiece is determined by the nature of its surfaces.

Shapes can thus be divided into three broad classifications:

1. Those composed primarily of flat surfaces
2. Those composed primarily of surfaces curving in one direction
3. Those composed primarily of surfaces curving in more than one direction

The cube is a typical example of the first class, the cylinder of the second class, and the sphere of the third class. Of course, we can machine the interior surface of a part as well as the exterior surface.

All machine tools are basically alike in that their purpose is to change the shape, finish, or size of a piece of material. All machines accomplish this objective by bringing a cutting edge or edges into contact with the workpiece when one or both of them are in motion. The variations in machine design, tool design, operation, and procedure will be better understood through the knowledge that the major function of a machine tool is to move a cutting edge or a piece of material while they are being held in contact with each other in such a way that one cut or a series of cuts can be made for the purpose of changing its size, shape, or finish.

Machine Motion

Fundamentally, the motion of either the material or the tool can occur in two forms (reciprocating or rotating), but this motion can take place in two directions in the vertical plane (up or down), and four directions in the horizontal plane (left, right, forward, or backward). See Figure 2.2. (We will see later that there are additional axes for moving the material or the cutter, but we should start with the above definition.)

2.1.1   The Lathe

In a lathe, the material is rotated and the tool is reciprocated back and forth along the length of the surface of the cylindrical part, with each pass of the cutting tool moved toward the centerline of the part to increase the depth of cut (see Figure 2.3a). Therefore a lathe must have some means of holding the material and rotating it and some means of holding the cutting tool and controlling its movement, as shown in the schematic of a simple lathe in Figure 2.3b. Most lathes are provided with tool holders of various types and styles for holding turning tools, drills, reamers, boring bars, etc.

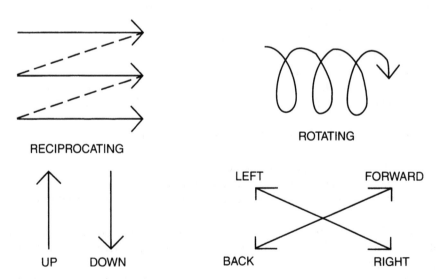

FIGURE 2.2   Fundamentals of motion.

They are also provided with collets, chucks, centers, and face plates for holding and driving the material from the headstock spindle. When a casting or forging requires machining, specially designed holding fixtures are needed to grasp the material and hold it in position to be presented to the cutting tools.

2.1.2   The Milling Machine

Much of the previous discussion on lathes applies equally to the milling machine process. In a sense, the milling machine is the opposite of the lathe, in that it provides cutting action by rotating the tool while the sequence of cuts is achieved by reciprocating the workpiece. When the sequence of consecutive cuts is produced by moving the workpiece in a straight line, the surface produced by a milling machine will normally be straight, in one direction at least. A milling machine, however, uses a multiple-edged tool, and the surface produced by a multiple-edged tool conforms to the contour of the cutting edges. Figure 2.4a shows the horizontal

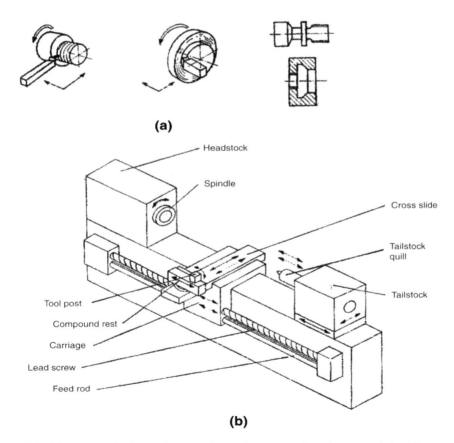

(a)

(b)

FIGURE 2.3   (a) Cutting action on a lathe. (b) Pattern of motions on a lathe. (From L. Alting, *Production Processes*, Marcel Dekker, New York, 1982. With permission.)

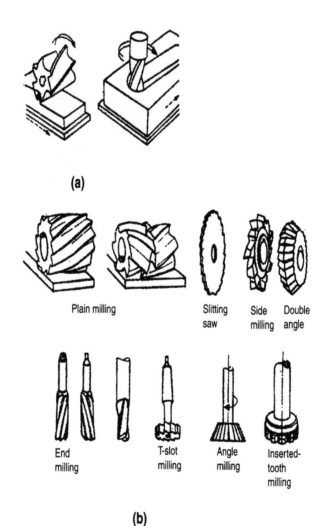

(a)

Plain milling   Slitting   Side   Double
                                  saw   milling   angle

End   T-slot   Angle   Inserted-
milling   milling   milling   tooth
                                                   milling

(b)

FIGURE 2.4 (a) Horizontal (left) and vertical (right) milling patterns. (b) Typical milling cutters. (From L. Alting, *Production Processes*, Marcel Dekker, New York, 1982. With permission.)

milling cutter on the left and a vertical milling cutter on the right, with additional types of milling cutters shown in Figure 2.4b. If the milling cutter has a straight cutting edge, a flat surface can be produced in both directions. The workpiece is usually held securely on the table of the machine, or in a fixture clamped to the table. It is fed to the cutter or cutters by the motion of the table. Multiple cutters can be arranged on the spindle, separated by precision spacers, permitting several parallel cuts to be made simultaneously. Figure 2.5 shows a schematic of a horizontal milling machine.

2.1.3 Drilling

In the construction of practically all products, a great many holes must be made, owing to the extensive use of bolts, machine screws, and studs for holding the various parts together.

In drilling, the material is fixed and the tool is rotated and advanced to complete a given sequence of cuts. (An exception is drilling on a lathe, with the drill held fixed in the tailstock while the work is being rotated.)

Drilling Machines

Drilling machines, or "drill presses," as they are often called, which are used for drilling these holes, are made in many different types designed for handling different

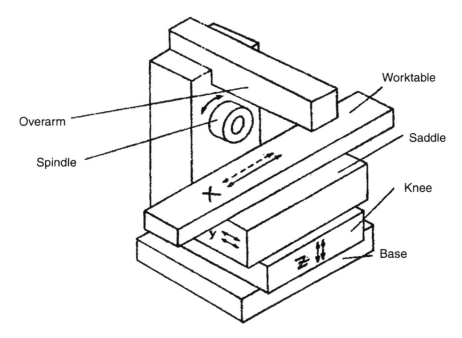

FIGURE 2.5 Plain column-and-knee milling machine, showing motion patterns. (From L. Alting, *Production Processes*, Marcel Dekker, New York, 1982. With permission.)

classes of work to the best advantage. The various types are also built in a great variety of sizes, as the most efficient results can be obtained with a machine that is neither too small nor too large and unwieldy for the work it performs. The upright drill press is one type that is used extensively. As the name indicates, the general design of the machine is vertical, and the drill spindle is in a vertical position. Figure 2.6 shows a schematic design of a single-spindle drill press. The heavy-duty machine is a powerful tool for heavy drilling. It has an adjustable knee in front of the column, and is supported by an adjusting screw, somewhat like a vertical milling machine. (See Figure 2.7.) The more common drill presses, for medium or lighter work, where the spindle can be moved up and down and the table adjusted to any height, can be arranged with multiple drill heads. Special-purpose drill presses include a sensitive drilling machine in which the

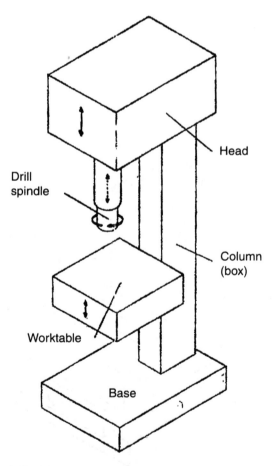

FIGURE 2.6   Vertical drilling machine. (From L. Alting, *Production Processes*, Marcel Dekker, New York, 1982. With permission.)

spindle rotates very quickly, making small-diameter holes, as in printed circuit boards. There are multiple-head machines that may have from 4 to 48 spindles, driven off the same spindle drive gear in the same head.

Drills

The twist drill (originally made from twisted bar stock) is a tool generally formed by milling, or forging followed by milling, two equal and diametrically opposite spiral grooves in a cylindrical piece of steel. The spiral grooves, or flutes, have the correct shape to form suitable cutting edges on the cone-shaped end, to provide channels for the free egress of chips and to permit the lubricant to get down to the cutting edges when the drill is working in a hole. It is probably the most efficient tool used, for in

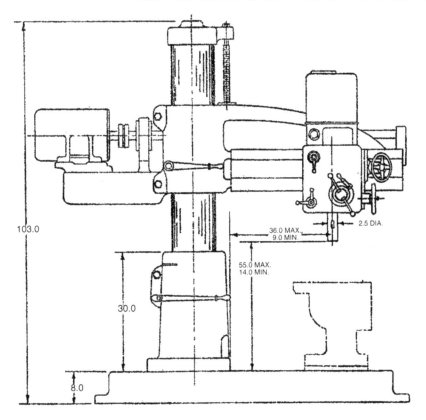

FIGURE 2.7 Heavy-duty radial-arm drill. (Courtesy of McDonnell Douglas, St. Louis, Mo. With permission.)

no other tool is the cutting surface so large in proportion to the cross-sectional area of the body that is its real support. (See Figure 2.8.)

Drill sizes are designated under four headings:

1. Numerical: no. 80 to no. 1 (0.0135 to 0.228 in.)
2. Alphabetical: A to Z (0.234 to 0.413 in.)

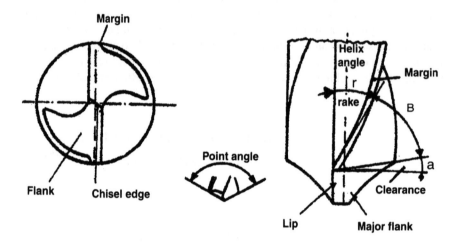

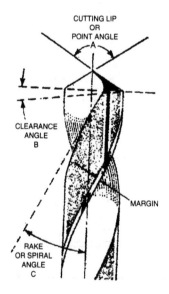

FIGURE 2.8 Cutting angles and margins on a twist drill. (From L. Alting, *Production Processes*, Marcel Dekker, New York, 1982. With permission.)

3. Fractional: 1/64 in. to 4 in. and over, by 64ths
4. Millimeter: 0.5 to 10.0 mm by 0.1 mm increments, and larger than 10.0 mm by 0.5 mm increments

Odd sizes may be ground to any specified diameter, but are only available on special order, at a higher cost.

Speeds and Feeds

When we refer to the speed of a drill, we mean the speed at the circumference, called "peripheral speed." Speed refers to the distance the drill would travel if, for example, it rolled 30 feet a minute. Speed usually does not refer to revolutions per minute unless specifically stated.

Except under certain rather rare conditions, a drill does not pull itself into the work like a corkscrew; it requires constant pressure to advance it. This advance (measured in fractions or decimals of an inch per revolution) is called "feed." Feed pressure, however, is the pressure required to maintain the feed. Some typical feeds and speeds are shown in Figure 2.9.

TYPICAL SPEEDS FOR HIGH-SPEED STEEL DRILLS

Drill Diameter Inches	Peripheral Speed Feet Per Minute
Less than 1	600
1 to 1½	550
Over 1½	450

TYPICAL DRILL FEEDS FOR FIRST DRILL ENTRY [*]

Drill Diameter Inch	Tolerance Inch±	Feed for 2011-73 Inch/Rev	Feed for Other Alloys Inch/Rev
.0625	.0015	.004	.004
.125	.002	.012	.010
.187	.002	.014	.012
.250	.002	.017	.014
.375	.0025	.020	.017
.500	.0025	.020	.017
.750	.003	.020	.017

[*] For multiple drill entries these feeds should be reduced about 15 percent for each succeeding entry. Lower feeds should also be used for thin-wall parts.

FIGURE 2.9 Typical feeds and speeds for drilling aluminum. (Courtesy of Reynolds Metals Company, Richmond, Va. With permission.)

Tolerances

Under average conditions drilled holes, preferably not over about four times the diameter in depth, can be held within the following tolerances:

#80 to #71	+0.002–0.001
#70 to #52	+0.003–0.001
#51 to #31	+0.004–0.001
1/8" to #3	+0.005–0.001
1/32" to R	+0.006–0.001
11/32" to 1/2"	+0.007–0.001
33/64" to 23/32"	+0.008–0.001
47/64" to 63/64"	+0.009–0.001
1" to 2"	+0.010–0.001

These tolerances are sufficiently liberal for general production, and although a definite set of limits suitable for all cases is out of the question, will provide a good conservative working base. Where more exacting specifications are desirable, redrilling or reaming will usually be necessary. Tolerances on drilled holes are normally plus, owing to the fact that drills invariably cut oversize from one to five thousandths, depending on the accuracy of sharpening. For this reason, the above tolerances are often specified as all minus 0.000 in. Straightness is dependent on the hole depth and the homogeneity of the material.

A tolerance of 0.001 in. can be held on ordinary reaming operations. With additional boring operations proceeding, reaming tolerances can be held to 0.0005 in. Depending on the type material being cut, the surface quality left in turret lathe operations will be about 63 μin., rms (root mean square), or less. Surface patterns depend on the type of tooling and cuts employed.

2.1.4   Design for Machining

In manufacturing, the product produced is always a compromise between acceptable quality and acceptable cost. The seldom-reached goal is always one of maximum attainable quality for minimum cost. Close tolerances and fine finishes may have very little influence on the functional acceptance by the ultimate user, who is the only one who can really judge quality. Many times it is very useful to prepare a chart with the help of the accounting staff and the shop technicians that defines the cost of close tolerance dimensions and surface finishes. For example, dimensions that have two-decimal accuracy may cost twice as much to produce as single-digit finish dimensions cost. The reason for higher cost with closer tolerance is easy to define when one follows the logic the craftsman uses in meeting the requirements specified by the engineer. As the workpiece approaches the finished dimension, the craftsman will slow the process to insure that the critical dimension is not missed. Overrunning or underrunning a dimension is very costly in scrap. Higher-precision dimensions require repeated checking by the craftsman and the quality control technician. Once a cost is defined and known by the engineering staff, then the decision to add or subtract critical dimensions will be viewed with a more discerning eye. With these few general guidelines, we should examine some of the dos and don'ts of design features for parts that will be machined by cutting chips.

Error Budgets

The manufacturing engineer must invest time in reviewing the design in order to determine where the precision that is required from the finished product will be most affected by tolerance errors; errors due to kinematic, dynamic, structural, thermal, vibration, wear, and age; and a number of other issues too numerous to mention in this book. Dr. Alex Slocum, professor of mechanical engineering at MIT, has published a book entitled *Precision Machine Design* that develops concepts for an error budget. Such a spreadsheet approach to tabulating the possible error sources will help the product development engineer understand what issues are critical to the performance of this product and will allow a certain amount of forgiveness in areas that do not contribute greatly to the functionality of the machine. For example, if a base plate is chosen to mount a machine that is leveled on three adjusting screws which prevent the machine from contacting the base plate, that plate will likely not require a high degree of finished machining and three-decimal accuracy to define its size. The material selection may be general and cold-rolled steel may replace the choice of a high alloy hardened plate that may mistakenly be chosen without the consideration of the end application. More care should be placed on the choice of thread pitch and the thread class to insure a good fit, and the type of thread used will make a difference with the adjustment of this machine. Such an investment in the big picture before cutting chips will sometimes save a lot of money in the machining of product and will eventually pay large dividends when the machine is required to deliver its precise function.

Cost Targets

The previous target is related to this subject. No product development project should be launched without correct understanding of the cost targets well in sight of the manufacturing engineer. The product development engineer or manufacturing engineer should always bring the cost of every process and every component up at the appropriate times to judge the best use of resources in the development of the product. Cost-effectiveness is equally important to precise function.

Machinability

When an economical machining operation is to be established, the interaction among geometry, the material, and the process must be appreciated. Figure 2.10 shows the natural finish of several materials. The lowest cost would obviously be not to machine at all. It is not sufficient to choose a material for a part that fulfills the required functional properties; its technological properties describing the suitability of the material for a particular process must also be considered. In mass-conserving processes, the material must possess a certain ductility (formability); and in mass-reducing processes, such as machining, it must have properties permitting machining to take place in a reasonable

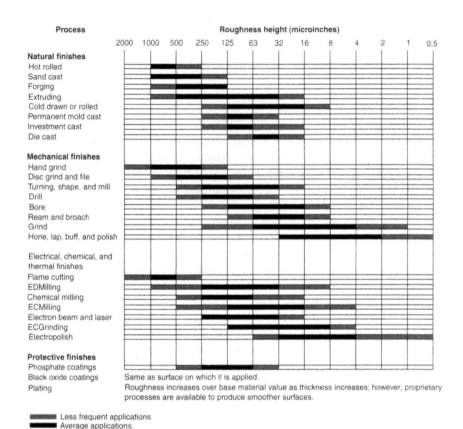

FIGURE 2.10 Surface finishes versus manufacturing process. (From J.P. Tanner. *Manufacturing Engineering*, Marcel Dekker. New York, 1991. With permission.)

way. The technological properties describing the suitability of a material for machining processes are collectively called its "machinability."

Machinability cannot be completely described by a single number, as it depends on a complex combination of properties which can be found only by studying the machining process in detail. The term "machinability" describes, in general, how the material performs when cutting is taking place. This performance can be measured by the wear on the tool, the surface quality of the product, the cutting forces, and the types of chip produced. In many cases, tool wear is considered the most important factor, which means that a machinability index can be defined as the cutting speed giving a specified tool life. When a component is to be machined by several processes, the machinability index corresponding to the process most used is chosen.

The machinability of a material greatly influences the production costs for a given component. Poor machinability results in high costs, and vice versa. In Figure 2.11 the machinability for the different materials groups is expressed as the removal rate per millimeter depth of cut when turning with carbide cutters.

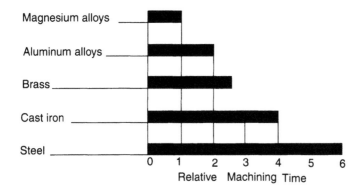

FIGURE 2.11 General comparative machining rates of various materials. (From R. A. Bolz, *Production Processes*, Industrial Press, New York, 1963. With permission.)

The chart can be used only as a general comparative guideline; in actual situations, accurate values must be obtained for the particular material chosen for the product design. The machinability of a particular material is affected primarily by its hardness, composition, and heat treatment. For most steels, the hardness has a major influence on machinability. A hardness range of HB 170 to 200 is generally optimal. Low hardness tends to lead to built-up edge formation on the cutters at low speeds. High hardness, i.e., above HB 200, leads to increased tool wear. The heat treatment of the work material can have a significant influence on its machinability. A coarse-grained structure generally has a better machinability than does a fine-grained structure. Hardened, plain carbon steels (0.35% C) with a martensitic structure are very difficult to machine. Inclusions, hard constituents, scale, oxides, and so on have a deteriorating effect on the machinability, as the abrasive wear on the cutting tool is increased.

2.1.5   Cost versus Accuracy

Surface roughness is also a factor in design of machined parts. Typical roughness values (arithmetical mean value, R_a) for different processes are shown in Figure 2.10. It can be said that the roughness decreases when feed is decreased, the nose radius is increased, and the major cutting-edge angle and the minor cutting-edge angles are reduced on the cutter. Furthermore, increasing cutting speeds and effective cutting lubrications can improve surface quality.

It is difficult to equate minor design features with their proper influence on ultimate product cost. Figure 2.12 shows charts of general cost relationship on various degrees of accuracy. Also shown are some examples of tolerance versus cost. These data are plotted from a variety of sources, and the trend is only indicative of the cost of refinement and does not really take into account the possibilities offered by judicious designing for production. The profound effect of tolerances on cost can be seen by the examples in Figures 2.13, 2.14, and 2.15, which show a variety of turned and milled parts.

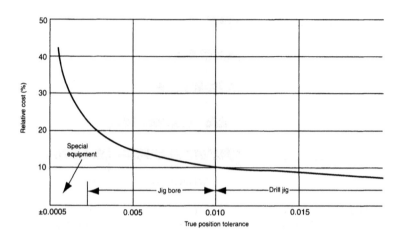

Relative machining cost vs tolerance and surface finish

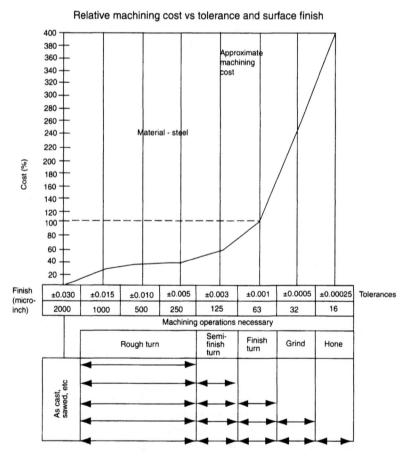

FIGURE 2.12 General cost relationship of various degrees of accuracy and surface finish. (From J. P. Tanner, *Manufacturing Engineering*, Marcel Dekker, New York, 1991. With permission.)

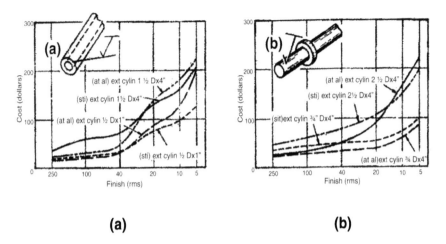

(a)   (b)

FIGURE 2.13 (a) Comparative costs for internal holes with varying degrees of surface requirement. (b) Plot of surface refinement versus cost for some turned and ground surfaces. (From R. A. Bolz, *Production Processes*, Industrial Press, New York, 1963. With permission.)

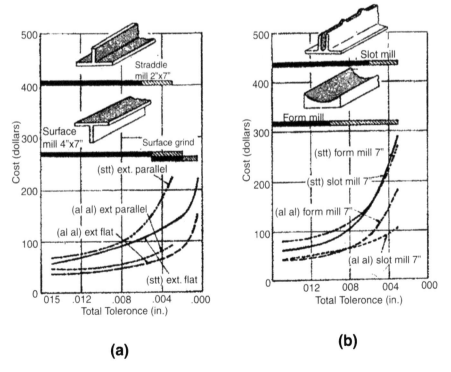

(a)   (b)

FIGURE 2.14 (a) Straddle and face milling costs plotted against tolerance requirements. (b) Tolerance effect of slotting and form milling costs. (From R. A. Bolz, *Production Processes*, Industrial Press, New York, 1963. With permission.)

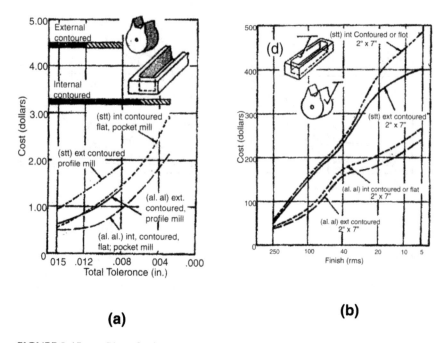

FIGURE 2.15 (a) Plot of tolerances versus costs for profiled and pocket milled parts. (b) Surface finish versus costs for profiled and milled parts. (From R. A. Bolz, *Production Processes*, Industrial Press, New York, 1963. With permission.)

One trap that is easy to fall into today is the new machines and their ability to achieve very tight tolerances. Do not permit a tight tolerance to appear on the product drawings just because it is possible to achieve it. As machines and tools wear, the new-machine tolerances may not be held so easily and the cost will be higher than necessary. Also, the looser tolerances will permit greater choice of machine utilization in the shop, making the new, tight machines available for work that must have the closer tolerances.

2.1.6 General Rules for Design

John P. Tanner, president of Tanner and Associates, Orlando, Florida, suggests that when addressing surface tolerances and tolerances for machined parts, the following cautionary steps should be followed:

Avoid surface finishes that cannot be achieved in the first operation of fabrication (i.e., a forging requiring a 32 μin. finish will need a subsequent machining operation).

Generally, surface finishes of 63 μin. or less will require additional machine operations to meet specifications and should be avoided.

Avoid tolerances on machined parts that require several operations to meet specifications. Use the most economical design tolerances indicated for the type of process selected.

Avoid hole depths that are three times greater than the hole diameter.

Generally, avoid tolerances tighter than 0.004 in. in drilling operations.

Generally, avoid tolerances tighter than 0.0025 in. in milling operations.

Generally, avoid tolerances tighter than 0.0015 in. in turning operations.

An additional checklist for machined parts design follows:

1. Ensure maximum simplicity in overall design. Develop product for utmost simplicity in physical and functional characteristics. Production problems and costs have a direct relation to complexity of design, and any efforts expended in reducing a product to its lowest common denominator will result in real savings. Always ask, "Is it absolutely required?" The most obvious way to cut costs is to eliminate or combine parts to obviate the expense of attaching or assembling. Refinements of design and analysis are the key to sharp reductions in costs, not revolutionary improvements in production techniques.

2. Analyze carefully all materials to be used. Select materials for suitability as well as lowest cost and availability. For instance, though aluminum alloys and aluminum-alloy forgings cost somewhat more than steel for parts of similar design, time required for machining is often sufficiently lower (about one third that required for similar steel parts) to offset the original additional material cost.

3. Use the widest possible tolerances and finishes on parts. The manufacturing engineer should have a set of process capability charts available on every machine in the shop, including any contemplated new equipment purchases. Our goal should be 6-sigma quality, which says that SPC will eliminate any inspection operations so that we won't make any bad parts. However, the machine capability and tolerances produced must be tighter than any of the part tolerances. Be sure that surface roughness and accuracies specified are reasonable and in keeping with the product and its function. The roughest acceptable finish should be contemplated and so specified. There is a direct relationship between surface finish and dimensional tolerancing. To expect a tolerance of 0.0001 in. to be held on a part turned or milled to an average of 500 μin. is rather foolish. Likewise, to specify a finish of 10 to 15 μin. for a surface which is merely intended to provide proper size for locating subsequent operations is also needless. A 40- to 60-μin. finish would be satisfactory and the cost will be at least 50 to 60% less.

4. Standardize to the greatest extent possible. Use standard available parts and standardize basic parts for one product as well as across a maximum number of products. Evolving the basic idea for a product constitutes about 5% of the overall effort in bringing it onto the market; some 20% of the effort goes

toward general design of the product, and about 50% of the effort is necessary to bring about economic production of the product, while the remainder goes toward removal of bugs from design and production, including servicing. The big job is in making the product economically, through design for more economic production.

5. Determine the best production or fabricating process. Tailor the product to specifically suit the production method or methods selected. It is possible to tailor the details of a product to assure economic production, and it is possible to have a basically similar part, which is excessively expensive to produce. Small but extremely important details can create this difference. Figure 2.16 shows the difference in a design with a tangent radius. A concave cutter usually creates a nick on one or both sides of the piece shown in (a). The nontangent design shown in (b) offers considerable production economy. Every production method has a well-established level of precision, which can be maintained in continuous production without exceeding normal costs. The practice of specifying tolerances according to empirical rules should be abandoned. Practical design for production demands careful observances of the natural tolerances available with specific processes. (See Figure 2.17.)

6. Minimize production steps. Plan for the smallest possible number or lowest-cost combination of separate operations in manufacturing the product. The primary aim here is to ensure the elimination of needless processing steps. Even though the processing may be automatic, useless steps require time and equipment. Reducing the number of separate processing operations required to complete a part will usually yield the greatest savings. Next in importance is reduction in the number of parts used, and following that, savings through the use of stock parts or interchangeable parts.

7. Eliminate handling problems. Ensure ease in locating, setting up, orienting, feeding, chuting, holding, and transferring parts. Where parts must be held or located in a jig for a processing operation, consideration must be given to the means by which the operation will be accomplished. Lugs may be added, contours changed, or sections varied slightly to make handling a simple rather than a costly operation. The difficulty of handling certain parts in production is often overlooked. Pads or locating positions are almost invariably required in processing, regardless of quantity. Use of false lugs for locating and clamping often may be a necessity. While ingenious handling methods and mechanical production aids are continuously being developed, the designer who can eliminate the need for such equipment simplifies production problems and effectively reduces overall cost.

Drawing Notes

Each part machined must have a starting point in its process.

The old story in process engineering was that the first operation on any machined part started with "Slab mill the base." This is still true of most parts.

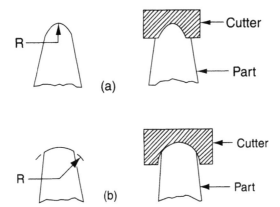

FIGURE 2.16 Tolerances have a profound effect on cost: (a) requires turning and finish grinding; (b) requires turning only. (From R. A. Bolz, *Production Processes*, Industrial Press. New York, 1963. With permission.)

Range of sizes		Tolerances								
From	Through									
0.000	0.599	0.00015	0.0002	0.0003	0.0005	0.0008	0.0012	0.002	0.003	0.005
0.600	0.999	0.00015	0.00025	0.0004	0.0006	0.001	0.0015	0.0025	0.004	0.006
1.000	1.499	0.0002	0.0003	0.0005	0.0008	0.0012	0.002	0.003	0.005	0.008
1.500	2.799	0.00025	0.0004	0.0006	0.001	0.0015	0.0025	0.004	0.006	0.010
2.800	4.499	0.0003	0.0005	0.0008	0.0012	0.002	0.003	0.005	0.008	0.012
4.500	7.799	0.0004	0.0006	0.001	0.0015	0.0025	0.004	0.006	0.010	0.015
7.800	13.599	0.0005	0.0008	0.0012	0.002	0.003	0.005	0.008	0.012	0.020
13.600	20.999	0.0006	0.001	0.0015	0.0025	0.004	0.006	0.010	0.015	0.025
Lapping & honing										
Grinding, diamond, turning boring										
Broaching										
Reaming										
Turning, boring, slotting & planing										
Milling										
Drilling										

FIGURE 2.17 "Natural" tolerances available with various processes. (From J. P. Tanner, *Manufacturing Engineering*, Marcel Dekker, New York, 1991. With permission.)

The part is then turned over, and all subsequent operations are taken from the base as the common reference.

Three points define a plane; therefore there must be some reference on the blueprints as to the location of these points. The pattern maker, the die maker, and the casting or forging fabricator uses these tooling points. Any subsequent holding fixtures required for machining use these same reference points. Final inspection (if required) will also start with these common reference points.

2.1.7 Bibliography

Alting, L., *Manufacturing Engineering Processes,* Marcel Dekker, New York, 1982.
Bolz, R.A., *Production Processes,* Industrial Press, New York, 1963.
Slocum, A., *Precision Machine Design,* Prentice Hall, 1992.
Tanner, J.P., *Manufacturing Engineering,* Marcel Dekker, New York, 1991.

2.2   DESIGN FOR CASTING AND FORGING

2.2.0   Introduction to Design for Casting and Forging

In casting, the liquid material is poured into a cavity (die or mold) corresponding to the desired geometry. The shape obtained in the liquid material is now stabilized by solidification (cooling) and removed from the cavity as a solid component. Casting is the oldest known process for producing metallic components. The main stages are producing a suitable mold cavity, melting the material, pouring the liquid material into the cavity, stabilizing the shape by solidification, removing or extracting the solid component from the mold, cutting off sprues and risers, and cleaning the component.

Forging, or its close cousin, pressing, consists of preheating a solid piece of wrought metal and placing it in a lower die. The upper die is then closed, squeezing the material into the closed die cavity and forming the desired shape of the part. The pressure may be applied slowly, as in pressing, or rapidly, with one or more hammer actions of the upper die, as in forging. Any excess material is squeezed out between the die halves as flash. The die is opened and the part ejected. The main stages are cutting the blank to the correct volume or shape, preheating the blank, forging the part, opening the die and ejecting the part, trimming off the flash (usually in a punch press die), and cleaning the part.

In both processes, depending on the properties of the metal and the desired requirements of the product, the finishing from this point on is similar. Both may require heat treatment, straightening, machining, plating or painting, etc. Aluminum and steel, as well as most other metals, may be cast or forged. Different alloying ingredients are added to improve the processing characteristics, making the precise alloy selection important in both processes. Some alloys can only be cast and others can only be forged. In general, forging materials require a high ductility at the elevated temperatures required for forging. Forgings also have better mechanical properties in the finished part, since the grain structure of the material becomes somewhat flattened or stretched, as in other wrought alloys. Castings, on the other hand, have a more nonoriented grain structure, with the size of the grains being a function of the rate of cooling. The casting alloying ingredients usually have more effect on the liquidity at the pouring temperature. After heat treatment and the accompanying changes in the internal structure and properties of the metal, castings are generally of lower strength and ductility than forgings.

The design problems of casting and forging involve comparing the cost of one process against another, coupled with the limitations of web thicknesses, radii, strength, and the like. The quantity of parts to be produced is also an important factor. Lead times for various processes differ greatly. For example, the tooling lead time could vary by several months for a sand casting versus a die casting. Subchapter 1.4 discusses some short-time options that are available for use until the final production tooling is available. In nearly all cases, a part that will ultimately be cast or forged can be hogged-out of solid stock, although the cost may be very high. The designer has more leeway in the physical shape of a part by utilizing one of the casting processes rather than a forging. Castings are usually lower in cost than forgings. Forging are superior in strength, ductility, and fatigue resistance, while castings may be better for absorbing vibration and are generally more rigid.

The introduction to casting and forging processes in this section is not meant to make the manufacturing engineer or product designer an expert in the field. The processes are described in enough detail to provide a reference as to the limitations of the various processes available, in order for the design team to consider the strength, tolerances, surface finish, machinability, lead time for tooling, and cost.

2.2.1   Sand Casting

Sand casting is the oldest method of casting. The process is basically a simple one, with a few universal requirements. A mold cavity is formed around the pattern in a two-part box of sand; the halves are called the *cope* (upper) and the *drag*. (See Figure 2.18.) Fitted together, these form the complete mold. To withdraw the

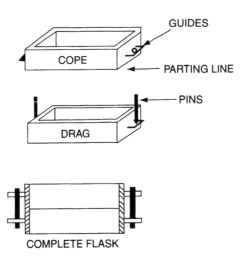

FIGURE 2.18 Basic flask for sand casting. (From C. W. Ammen, *Complete Handbook of Sand Casting*, Tab Books, Blue Ridge Summit, Pa., 1981. With permission.)

pattern from the mold, the pattern must be tapered and have no undercuts. This taper is called *draft* and usually has an angle of 3°. The most common molding material used in sand casting is green sand, which is a mixture, usually of ordinary silica sand, clay, water, and other binding materials. Sometimes this binder is resin. The molding sand is placed in the lower half of a two-part form and packed around half the pattern (up to the parting line). Then a dry molding powder is sprinkled over the pattern and the sand in the cope as a release agent, the upper half of the form is put in place on top of the drag, and the remaining sand is packed into place. When the sand is compacted sufficiently, the cope and drag are separated and the pattern is removed (see Figure 2.19). The two halves are then rejoined, and the metal is poured in through an opening (sprue) into the cavity. The size of the pattern must be adjusted to allow for the shrinkage of the metal as it cools. Aluminum and steel shrinks about 1/4 in./ft., depending on the thickness of the part. Thicker parts shrink less. Molten metal may flow from the sprue to various entry ports (gates) in the cavity. Risers at each gate act as reservoirs of molten metal that feed the casting during solidification. A sprue and riser can be seen in Figure 2.20. Aluminum castings need larger gates and risers than other metal castings; thus the average weight of metal poured is about two or three times the weight of the finished casting. Figure 2.21 shows the weight comparison of equal volumes of various metals. The importance of proper gating has been demonstrated by the fact that redesign and relocation on the gates feeding a previously improperly gated casting have increased the strength of the casting as much as 50 to 100%.

Cores

Inserting a core of sand into the cavity before pouring makes a hollow casting. Figure 2.22 shows the making and use of cores in a sand-casting mold. Any leakage between the halves of the mold cavity forms thin ridges of flash. Protuberances left

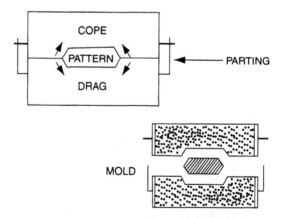

FIGURE 2.19   Development of split pattern mold. (From C. W. Ammen, *Complete Handbook of Sand Casting*, Tab Books, Blue Ridge Summit, Pa., 1981. With permission.)

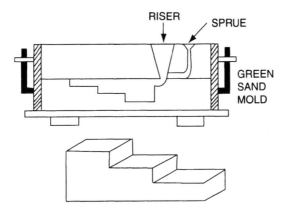

FIGURE 2.20 Casting a step pattern. (From C. W. Ammen, *Complete Handbook of Sand Casting*, Tab Books, Blue Ridge Summit, Pa., 1981. With permission.)

Metal	Relative Weight Factor
Magnesium	.64
Aluminum	1.00
Titanium	1.68
Cast Iron	2.63
Zinc	2.64
Tin	2.70
Cast Steel	2.90
Cast Brass (35% Zinc)	3.14
Monel Metal	3.25
Cast Bronze (5% Tin)	3.28
Copper	3.32
Lead	4.20
Uranium	6.93

FIGURE 2.21 Comparison of weight of aluminum with equal volumes of other metals. (From Reynolds Metals Company, Richmond, Va., 1965. With permission.)

from risers and sprues must be removed from the part, along with the flash, after it has solidified and the sand has been cleaned off.

Patterns

When more than one part is to be made, the pattern is split along the parting line and each half is mounted on a pattern board or match plate, as shown in Figure 2.23. One half of the impression is made in the cope, the other half in the drag. You can see the locating holes for the core in both the cope and drag. The mold is then opened up and the pattern board removed, the core installed, and the mold reclosed for pouring.

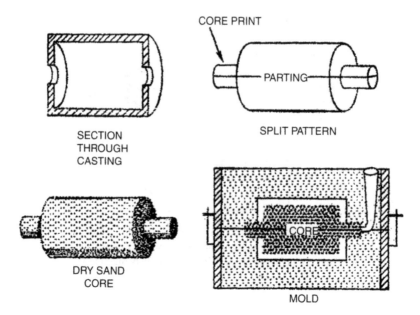

FIGURE 2.22 Core prints. (From C. W. Ammen, *Complete Handbook of Sand Casting*, Tab Books, Blue Ridge Summit, Pa., 1981. With permission.)

General

Sand casting is the most common casting process used with aluminum alloys. Casting has the lowest initial cost and permits flexibility in design, choice of alloy, and design changes during production. For these reasons, sand casting is used to make a small number of cast pieces, or to make a moderate number that require fast delivery with the likelihood of repeat production. A sand-cast part will not be as uniform in dimensions as one produced by other casting methods, so greater machining allowances must be made. Surface finish can be controlled somewhat by varying the quality of sand in contact with the metal and by applying special sealers to the sand surfaces. Hot shortness of some aluminum alloys is also important. Certain alloys have low strength at temperatures just below solidification. The casting may crack if these alloys are cast in a mold that offers resistance to contraction as the metal solidifies. This is called "hot cracking." Hot shortness varies with the alloy used. Aluminum-silicon alloys show considerably less hot shortness than aluminum-copper alloys. The wide range of aluminum alloys available enables the designer to choose an aluminum alloy and avoid hot cracking when this factor is important.

Basic Design Considerations

Often someone who does not have the experience in how the casting is to be produced designs a casting, and the foundry will then inherit problems caused by

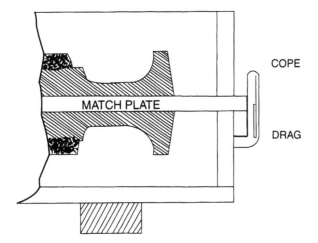

FIGURE 2.23 Split pattern mounted on pattern board. (From C.W. Ammen, *Complete Handbook of Sand Casting*, Tab Books, Blue Ridge Summit, Pa., 1981. With permission.)

poor design. Someone who has no knowledge of the foundry and its requirements should talk with a local casting expert or the foundry before starting design.

Some basic considerations are the following:

1. On castings that must present a cosmetic appearance along with function, the exterior should be designed to follow simple, flowing lines with a minimum of projections.
2. Avoid irregular or complicated parting lines whenever possible. Design for partings to be in one plane.
3. Use ample but not excessive draft. Avoid any no-draft vertical surfaces unless there is no other way out.
4. Avoid long, slender cores through heavy metal sections or long spans. When unavoidable, they should be straight and well anchored to the mold.
5. Avoid the use of pattern letters on any surface other than one parallel to the parting.
6. Avoid sudden changes in section thicknesses that will create hot spots.
7. Use ribs to stiffen or strengthen castings, thus reducing weight.
8. Avoid all sharp corners and fillet all junctions.
9. Stagger crossing ribs so that the junction will not create a hot spot, which could shrink.

Figure 2.24 shows some examples of good and bad sand-casting designs.

Volume of Parts

Since pattern equipment for sand casting is much less expensive than dies for permanent-mold or pressure die casting, sand casting is indicated where small quantities are wanted. As larger quantities are called for, the point where it becomes economical to go to permanent-mold or die casting depends on the size and complexity of the casting and other factors.

2.2.2 Investment Casting

Investment castings have some of the advantages of the sand casting and some of the advantages of the closed-mold processes. The process begins by making a metal die (mold), often of aluminum, that is the shape of the finished part plus normal shrinkage factors for the material. The mold is injected with wax or low-melting-point plastic. Several of these wax parts are attached to a central shaft, or sprue, forming a cluster of parts attached to the sprue. Each wax pattern has one or more gates, which are used to attach it to the sprue. The entire cluster of parts is then coated with ceramic slurry or placed in a metal flask and the flask filled with mold slurry. After the mold material has set and cured, the mold is heated, allowing the wax pattern to run out, leaving a hollow core. This is the origin of the term "lost wax" method. Molten aluminum (or other metal) is then poured into the cluster through the sprue and all cavities are filled. Upon solidification, the mold material is broken

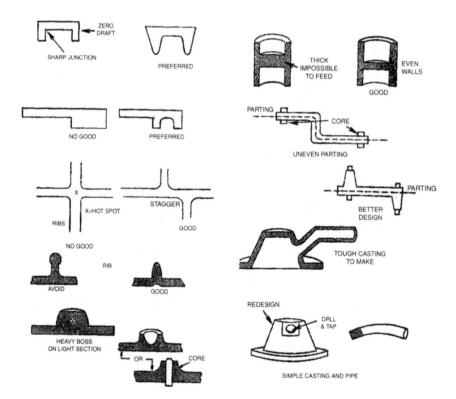

FIGURE 2.24 Examples of good and bad castings. (From C.W. Ammen, *Complete Handbook of Sand Casting*, Tab Books, Blue Ridge Summit, Pa., 1981. With permission.)

away and the parts broken off the sprue. Figure 2.25 shows both the investment-flask and investment-shell processes.

Design for Investment Casting

Wax or plastic temperature, pressure, die temperature, mold or shell composition, back-up sand, firing temperature, rate of cooling, position of the part on the tree, and heat treatment temperature all bear directly on tolerances required in the investment casting process. The amount of tolerance required to cover each process step is dependent, basically, on the size and shape of the casting and will vary from foundry to foundry. This is because one foundry may specialize in thin-walled, highly sophisticated castings, another in mass-production requirements, and yet another in high-integrity aerospace or aircraft applications. One factor, however, is constant. The cost of any casting increases in proportion to the preciseness of specifications, whether on chemistry, nondestructive testing, or tolerance bands.

Tolerances

As a general rule, normal linear tolerance on investment castings can be up to 0.010 in. For each additional inch thereafter, allow 0.003 in. Figure 2.26a shows a chart that indicates expected normal and premium tolerances. Normal tolerances are tolerances that can be expected for production repeatability of all casting dimensions.

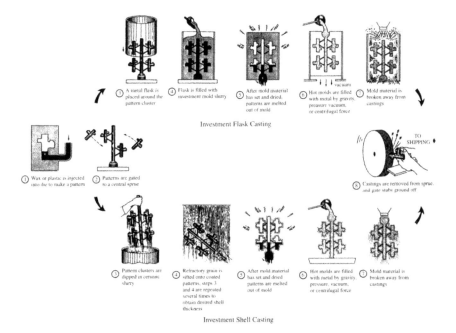

FIGURE 2.25 Basic production techniques for investment castings. (From *Investment Casting Handbook*, Investment Casting Institute, Chicago, 1968. With permission.)

LINEAR TOLERANCE		
DIMENSIONS	NORMAL	PREMIUM
up to ½"	± .007"	± .003"
up to 1"	± .010"	± .005"
up to 2"	± .013"	± .008"
up to 3"	± .016"	± .010"
up to 4"	± .019"	± .012"
up to 5"	± .022"	± .014"
up to 6"	± .025"	± .015"
up to 7"	± .028"	± .016"
up to 8"	± .031"	± .017"
up to 9"	± .034"	± .018"
up to 10"	± .037"	± .019"
maximum Variation	± .040"	

An exception to the Standard Linear Tolerance exists on thin wall thickness where the tolerance must be a minimum of ± .020".

(a)

SECTION THICKNESS	POSSIBLE DISH PER FACE OF CASTING
up to ¼"	not significant
¼" to ½"	0.002"
½" to 1	0.004"
over 1"	0.006"

(b)

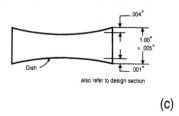

(c)

FIGURE 2.26 Investment casting tolerances. (From *Investment Casting Handbook*, Investment Casting Institute, Chicago, 1968. With permission.)

Premium tolerances are those that require added operations at extra cost, and which provide for closer tolerances on selected dimensions.

Flatness, or the effect of dishing, tolerances cannot be quoted, as they vary with configuration and alloy used. Figure 2.26b is a rough guide in areas under 6 in.[2] The amount of dishing allowed is in addition to the basic tolerance. Thus, on a block 0.005 in. thick, the tolerances as shown in Figure 2.26c would apply.

Roundness, or "out of round," is defined as the radial difference between a true circle and a given circumference. Tolerances are shown in Figure 2.27a. Figure 2.27b shows the relationship between roundness and concentricity. Two cylindrical surfaces sharing a common point or axis as their center are concentric. Any dimensional difference in the location of one center with respect to the other is the extent of eccentricity. When the length of a bar or tube does not exceed its component diameters by a factor of more than two, the component diameters will be concentric within 0.005 in. per inch of separation. The roundness of a cast hole as shown in Figure 2.27c is affected by the mass of surrounding metal. If an uneven mass is adjacent, the hole will be pulled out of round. If the surrounding metal is symmetrical, holes up to 1/2 in. in diameter can be held to 0.003 in. when checked with a plug gauge. Larger holes may be affected by interior shrinkage or pulling, and the foundry should be consulted. The longer the hole or the more mass of the section around it, the more pronounced will be the effect. Although the above notes are mostly taken from the *Investment Casting Handbook*, which is published by the Investment Casting Institute, it is strongly recommended that the design team

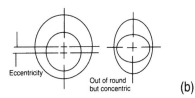

OUT OF ROUNDNESS	
Diameter	TIR or ½ difference between diameters
½"	.010"
1"	.015"
1½"	.020"
2"	.025"
On larger diameters, linear tolerances apply.	(a)

Eccentricity

Out of round but concentric   (b)

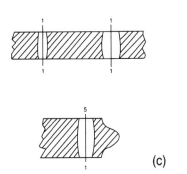

(c)

FIGURE 2.27 Roundness and concentricity tolerances of investment castings. (From *Investment Casting Handbook*, Investment Casting Institute, Chicago, 1968. With permission.)

consult with the supplier for a particular part if there might be a question regarding a particular feature not described above. The general castability rating of investment casting alloys is shown in Figure 2.28.

Drawing Notes

Linear tolerances, unless otherwise specified, 0.010 in. for first inch plus
 0.003 in. per inch for every inch thereafter
Corner radii 0.030 in. max., fillet radii 0.060 in. max, unless otherwise
 specified
Wall thickness 0.080, +0.015−0.000 in.
Surface finish 125 rms
Add 0.040 to 0.060 in. stock to surfaces identified by machining mark
Material: aluminum alloy, MIL-C-11866-T6 Comp. 2 (356-T6)

2.2.3   Die Casting

Permanent-Mold Casting

Permanent-mold casting employs split molds made from cast iron or steel. These molds are usually constructed with thick walls, which provide sufficient thermal capacity to assure rapid chilling of cast metal. In semi-permanent-mold casting, sand

ALLOY	CASTABILITY	RATING	FLUIDITY SHRINKAGE	RESISTANCE TO HOT TEARING
Silicon Irons (Electrical alloys of pure iron and silicon)				
0.5% Si.	75	3	3	2
1.2% Si.	80	3	3	2
1.5% Si.	80	3	3	2
1.8% Si.	75	3	3	2
2.5% Si.	70	3	2	2
Carbon & Sulphur Steels (A.I.S.I. designations)				
1015	80	3	3	3
1018	80	3	3	3
1020	80	3	3	3
1025	85	3	3	2
1030	85	3	2	2
1035	85	3	2	2
1040	85	2	2	2
1045	85	2	2	2
1050	85	2	2	2
1060	85	2	1	2
1117	75	2	3	2
1140	80	2	2	3
Low Alloy Steel (A.I.S.I. designations)				
2345	90	2	2	2
3120	85	2	3	2
4130	90	2	2	2
4140	90	2	2	2
4150	90	2	2	2
4340	90	2	2	2
4615	85	2	2	2
4620	85	2	3	2
4640	90	2	2	2
5130	85	2	2	2
6150	90	2	3	2
8620	85	2	2	2
8630	85	2	3	2
8640	90	2	2	2
8645	90	2	2	2
8730	85	2	2	2
8740	90	2	3	2
52100	80	1	2	2
Nitralloy	75	2	3	2
400 Series Stainless (A.IS.I. designations)				
405	90	2	3	2
410	95	1	3	2
415	85	1	3	2
420	90	1	3	2
430	90	1	3	2
430 F	90	1	3	2
431	90	1	3	2
440 A	85	1	3	2
440 C	85	1	3	2
440 F	85	1	3	3
AMS 5355 (Armco 17-4 PH)	85	1	3	2
AMS 5354	85	1	3	2
300 Series Stainless (A.I.S.I. designations)				
302	100	1	1	1
303	95	1	1	2
304	100	1	1	1
310	90	1	1	1
312	90	1	1	1
316	100	1	1	1

ALLOY	CASTABILITY	RATING	FLUIDITY SHRINKAGE	RESISTANCE TO HOT TEARING
347	95	1	1	1
CF-8M (ACI)	100	1	1	1
CN-7M (ACI)	95	1	1	1
High Nickel Alloys				
Monel (QQ-N-288-A)	85	1	2	2
Monel, R.H,	75	1	2	3
Monel, S (QQ-N-288-C)	75	1	2	3
Inconel (AMS-5665)	85	1	2	2
47-50 (47% NT-50% Fe)	80	3	2	2
Invar	75	1	2	2
Cobalt Alloys				
Cobalt J	80	1	1	3
Cobalt 3	80	1	1	3
Cobalt 6 (AMS-5387)	80	1	1	3
Cobalt 19	85	1	1	3
Cobalt 21 (AMS-5385C)	90	1	1	2
Cobalt 31 (AMS-5382B)	90	1	1	2
Cobalt 93	70	1	1	3
N-155 (AMS-5531)	80	1	1	3
Tool Steels (A.I.S.I. designations)				
A-2	85	2	2	2
A-6	80	2	2	2
D-2	85	3	2	2
D-3	85	3	2	2
D-6	80	3	2	2
D-7 (BR-4)	80	3	2	3
BR-4 FM	80	3	2	2
F-2	75	3	2	2
H-13	85	2	2	2
L-6	80	2	2	3
M-2	80	2	2	3
M-4	75	2	1	3
O-1	80	2	1	2
O-2	80	2	1	2
O-7	80	2	1	2
S-1	90	2	2	2
S-2	90	2	2	2
S-4	90	2	2	2
S-5	90	2	2	2
T-1	80	2	1	3
Aluminum				
13	85	2	2	2
40E	75	3	3	3
43	90	2	2	2
356-A356	100	1	1	1
355-C355	95	1	1	1
B-195	85	1	2	2
Copper Base				
Al. Bronze Gr. C	80	1	3	1
Al. Bronze Gr. D	80	1	3	1
88-10-2(G BR. & Gun Metal)	85	1	3	1
Mn Bronze	80	2	3	1
Hi Tensile Mn Bronze	80	2	3	1
Naval Brass (Yellow Bronze)	85	2	2	1
Navy "M"	85	2	2	1
Navy "G"	85	2	2	1
Phosphor Bronze (SAE 65)	85	2	2	1
85-5-5-5 (Red Bronze)	90	1	1	1
Silicon Brass	100	1	1	1
Be Cu 10C	90	1	1	1
Be Cu 20C	100	1	1	1
Be Cu 275C	90	1	1	1

Castability ratings are based on a casting of relatively simple configuration and upon comparisons with three alloys having excellent foundry characteristics and assigned castability rating of 100. These are 302 stainless steel (ferrous), 20 C beryllium copper (non-ferrous) and aluminum alloy 356. The fluidity, shrinkage and hot tear ratings of each alloy are based on 1—best, 2—good, and 3—poor.

FIGURE 2.28 Castability rating of investment casting alloys. (From *Casting Aluminum*, Reynolds Aluminum Company, Richmond, Va., 1965. With permission.)

or plaster cores are used in a cast iron or steel mold to produce more intricate and complex shapes. The molten metal is poured by gravity into heated metal molds. Sand casting is essentially a batch process, while permanent-mold casting is suitable for quantity production of a continuous nature. This production-line approach requires a different arrangement of foundry equipment, metal-handling methods, and production procedures compared to those used with sand casting.

With permanent-mold casting, a carefully established and rigidly maintained sequence of operations is essential. Every step in the foundry, from charging the furnace to removal of the cast piece from the mold, must be systematized. If any of the factors (pouring temperature, mold temperature, pouring rate, solidification rate) are thrown out of balance, the resultant castings may end up as scrap.

General

The improved mechanical properties of permanent-mold castings are the result of the advantageous crystalline structure that develops in the cast alloy as it cools and solidifies rapidly in the metal mold. Permanent-mold castings have closer dimensional tolerances and smoother surfaces than can be obtained in sand castings, so less machining and finishing is needed on permanent-mold castings. Thinner sections can be used, due to the higher mechanical properties.

Pressure Die Casting

Pressure die casting was developed because of the need to provide large numbers of cheap parts for assembly-line production methods. An automobile may have as many as 250 parts in it which are die cast, from door handles to aluminum-alloy engine blocks. Some simple die cast items can be produced on automatic machinery at the rate of hundreds per minute. Pressure die-casting machinery is expensive and complicated to design, because it has three functions: to inject the molten metal into the die, to clamp the two halves of the die together at forces which may be as high as several thousand tons, and to eject the finished casting when it has solidified. The costs of such machinery can only be recouped over long production runs. Since a small die of only moderate complexity may cost several thousand dollars, the process should be considered only for a large production run.

The molds used in automated die-casting methods are designed with water-circulation tubing in them to cool the castings rapidly. The machine opens the mold so that the half of the casting which cools quickest is uncovered first, then ejecting pins push the casting out of the other half of the mold. The two halves of the mold are coated with a lubricant to facilitate the ejection of the finished casting. A typical pressure die-casting die is shown in Figure 2.29. The fixed cover half of the die on the left shows the gate where the liquid shot of molten metal is injected. The ejector half of the die, on the right, shows the large alignment pins, the sprue pin, the cores for blind holes, and the ejector pins.

Die-casting processes are divided into two types: cold-chamber and hot-chamber methods. In the cold-chamber method, the molten metal is ladled into a cylinder and then shot into the mold by a hydraulic or pneumatic plunger. In

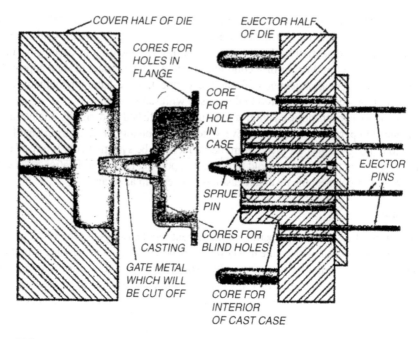

FIGURE 2.29 Typical pressure-die-casting die. All the fixed cores are aligned to permit ejection of the casting. (From *Designing for Alcoa Die Castings*, Aluminum Corporation of America, Pittsburgh, Pa., 1949. With permission.)

the hot-chamber method, shown in Figure 2.30, the injection cylinder itself is immersed in the molten alloy and successive shots are made either by a plunger or compressed air. There are also high-pressure and low-pressure techniques. The low-pressure system is newer and was developed to extend the versatility of die casting to alloys with higher melting points. Another development is the vacuum mold, in which a vacuum is produced in the mold, enabling the pressure shot to fill the mold faster.

Most die-casting alloys have a low melting point. Almost the entire production of zinc-based alloys is used in the die-casting industry. The higher the melting point, the shorter will be the useful life of the mold. A tool steel die may be able to produce up to a half million castings in zinc, but the total for brass may be only a few thousand. Typical mechanical properties for an aluminum die-cast part are shown in Figure 2.31.

Design and Tolerances

The cast surfaces that slide on the die cavity during die opening or ejection must have draft or taper to free the casting without binding. Insufficient draft causes sticking in the die, distortion of the castings, and galling of cast surfaces. Ample draft assures smoothness of cast surfaces. In general, outer walls or surfaces should have a minimum draft of 0.5° (0.009 in. per linear inch of draw). Inner walls which exert higher

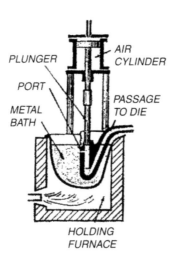

FIGURE 2.30 The submerged type of injection system is suitable for use with low-melting-temperature alloys. (From *Designing for Alcoa Die Castings*, Aluminum Corporation of America, Pittsburgh, Pa., 1949. With permission.)

	Alcoa Alloy	13	43	A360	360	85	A380	380	384	218
Room Temperature 68°F.[2]	Tensile Strength, psi[1]	39,000	30,000	41,000	44,000	40,000	46,000	45,000	46,000	45,000
	Tensile Yield Strength, psi[2]	21,000	16,000	23,000	27,000	24,000	25,000	26,000	27,000	27,000
	Elongation in 2 Inches, %	2.0	9.0	5.0	3.0	5.0	3.0	2.0	1.0	8.0
	Shear Strength, psi	25,000	19,000	26,000	28,000	26,000	29,000	29,000	29,000	27,000
	Endurance Limit, psi[4]	19,000	17,000	18,000	19,000	22,000	19,000	20,000	21,000	23,000
212°F.[2]	Tensile Strength, psi	34,000	26,000	36,000	39,000	38,000	44,000	43,000	44,000	40,000
	Tensile Yield Strength, psi	21,000	16,000	23,000	27,000	24,000	26,000	27,000	28,000	25,000
	Elongation in 2 Inches, %	5.0	13.0	6.0	3.0	5.0	5.0	3.0	2.0	12.0
300°F.	Tensile Strength, psi	30,000	23,000	32,000	34,000	33,000	38,000	37,000	38,000	34,000
	Tensile Yield Strength, psi	19,000	15,000	21,000	25,000	22,000	23,000	24,000	25,000	23,000
	Elongation in 2 Inches, %	7.0	16.0	8.0	4.0	7.0	7.0	4.0	5.0	15.0
400°F.	Tensile Strength, psi	25,000	19,000	27,000	29,000	24,000	28,000	28,000	28,000	25,000
	Tensile Yield Strength, psi	18,000	14,000	20,000	24,000	19,000	20,000	21,000	22,000	19,000
	Elongation in 2 Inches, %	12.0	19.0	11.0	5.0	10.0	12.0	6.0	8.0	18.0
500°F.	Tensile Strength, psi	19,000	14,000	20,000	21,000	17,000	20,000	20,000	20,000	16,000
	Tensile Yield Strength, psi	13,000	10,000	15,000	17,000	13,000	14,000	15,000	16,000	12,000
	Elongation in 2 Inches, %	20.0	23.0	15.0	9.0	16.0	15.0	10.0	12.0	22.0

FIGURE 2.31 Typical mechanical properties of Alcoa aluminum die-casting alloys. (From *Designing for Alcoa Die Castings*, Aluminum Corporation of America, Pittsburgh, Pa., 1949. With permission.)

shrinkage forces on the die as the casting cools require greater draft allowance, as much as 2° or 3°, depending on the design. Surfaces which are on the outer part of the casting but which tend to bind against the die during contraction of the casting are treated as inner walls in determining draft.

Where the greatest degree of smoothness is required on die-cast surfaces, draft should be increased to several times the allowable minimum. This will ensure rapid production and make it possible to maintain excellent surface smoothness on the castings. Generous fillets and rounded corners can be used to advantage as means to reduce the areas subject to sliding on the die.

The practical minimum wall thickness is usually governed by the necessity of filling the die properly. Under the most difficult conditions, a wall 9/64 to 5/32 in. thick may be needed for metal flow in the die. Where the casting is small (less than about 6 in. by 6 in. overall) and its form promotes good casting conditions, walls may be thin as 0.050 to 0.065 in.

Draft in a cored hole allows the core to be withdrawn from the casting, or the casting to be ejected from the core. The amount of draft needed is so slight that many cored holes perform the functions of straight cylindrical holes without the necessity of reaming. They often serve as pivot bearings, screw holes, tap holes, or seats for mating parts. In such cases, where a close approach to parallel sides is desired, it is well to specify the minimum allowable draft. The minimum draft for cored holes is shown in Figure 2.32a, and the recommended depth of blind-cored holes is shown in Figure 2.32b.

In addition to the commonly employed processes mentioned above, others include centrifugal, plaster-mold, and shell-mold casting. Figure 2.33 compares design and cost features of basic casting methods.

2.2.4 Forging and Pressing

This section provides a brief introduction to the forging and pressing industry. In earlier days, forging was the process of shaping metal by heating and hammering. Today, metal is not always heated for forging, and several types of heavy machines that apply impact or squeeze pressure with swift precision may perform the work. In today's forging industry machines of modern technology that produce metal parts of unparalleled strength and utility enhance the skill and seasoned judgment of the forgeman.

Forging permits the structure of metals to be refined and controlled to provide improved mechanical properties. Also, forging produces a continuous "grain flow" in the metal which can be oriented to follow the shape of the part and result in maximum strength efficiency of the material. Since virtually all metals, from aluminum to zirconium, can be forged, extensive combinations of mechanical and physical properties are available to meet demanding industrial applications. Figure 2.34 shows the dollar sales by commercial forging industry to major end-use markets.

Materials

In most cases the stock to be forged has been preworked by the material supplier to refine the dendritic structure of the ingot, remove defects inherent in the casting

MINIMUM DRAFT FOR CORED HOLES

Diameter or Narrowest Section of Hole (Inches)	Draft on Diameter or Cross-Sectional Dimension per Inch of Depth (Inches)	
	Aluminum	Magnesium
1/10 to 1/8	.020	.010
1/8 to 1/4	.016	.008
1/4 to 1	.012	.006
Over 1	†.012 + .002 per inch of diameter over 1 inch.	†.006 + .001 per inch of diameter over 1 inch.

†As an example, a 3-inch deep cored hold of 1½ inches diameter in aluminum would require .012 plus ½ times .002, or .013 inch per Inch of depth. The draft would be three times .013, or 0.039 inch.

(a)

DEPTHS OF CORED HOLES

Diameter or Narrowest Section of Hole (Inches)	Maximum Recommended Depth (Inches)	Diameter or Narrowest Section of Hole (Inches)	Maximum Recommended Depth (Inches)
1/10 (min.)	8/8	1/4	2
1/8	1/2	1/2	4
5/32	3/4	3/4	6
3/16	1 1/8	1	8

(b)

FIGURE 2.32 Recommended minimum draft and depth for cored holes. (From *Designing for Alcoa Die Castings*, Aluminum Corporation of America, Pittsburgh, Pa., 1949. With permission.)

process, and further improve the structural quality. This is usually accomplished in successive rolling operations which reduce the cross section of the material under pressure, eliminating possible porosity, refining the crystalline structure of the base metal, and orienting any nonmetallic and alloy segregation in the direction of working. This directional alignment is called "grain flow." The forging process employs this directionality to provide a unique and important advantage by orienting grain flow within the component so that it lies in the direction requiring maximum strength.

Casting Method	Cost		Produc-tion Quan-tities	Mechan-ical Proper-ties	Surface Finish	Limitations		Dimen-sional Accu-racy	Minimum Section Thick-ness inch
	Produc-tion Equip-ment	Unit Casting				Casting Size	Alloy		
Green Sand	Low	High	Medium	Medium	Fair	None	None	Fair	1/8
Baked Sand	Medium	High	Medium	Medium	Fair	None	None	Fair	1/8
Semi-Permanent Mold	High	Low	Medium Large	High	Fair to Good	To Medium	Medium to High Fluidity	Good	1/16
Permanent Mold	High	Low	Medium Large	High	Good	To Medium	Medium to High Fluidity	Very Good	1/16
Die Casting	Very High	Lowest	Very Large	High	Very Good	To Medium	High Fluidity	Excellent	1/32
Centrifugal	High	Low	Medium	Medium High	Good	Limited by Shape	None	Good	1/8
Investment	Medium	High	Small	Medium	Excellent	Small	High Fluidity	Excellent	1/8-1/32
Plaster	Medium	High	Small	Low	Excellent	Small	High Fluidity	Excellent	1/8-1/32

FIGURE 2.33 Design and cost features of basic casting methods. (From *Designing for Alcoa Die Castings*, Aluminum Corporation of America, Pittsburgh, Pa., 1949. With permission.)

Properly developed grain flow in forgings closely follows the outline of the component. In contrast, bar stock and plate have grain flow in only one direction, and changes in contour require that flow lines be cut, exposing grain ends and rendering the material more liable to fatigue and more sensitive to stress corrosion. Figure 2.35 compares the grain structure in casting, bar stock machining, and forging. The forging process (through proper orientation of grain flow) develops maximum impact strength and fatigue resistance in a material, with greater values in these properties than are obtainable by any other metalworking process. Thus forgings provide greater life expectancy than is possible with other metal products.

Consistency

The consistency of material from one forging to the next, and between separate quantities of forgings produced months or even years apart, is extremely high. Forge plants can usually rely on acceptance testing for the quality control procedures and certified reports of major metal producers, assuring close control of metal composition.

Dimensional Uniformity

The dimensional characteristics of impression die forgings are remarkably stable from piece to piece throughout an entire production run. This exceptional continuity of shape from the first to the last piece is due primarily to the fact that impressions in which forgings are shaped are contained in permanent dies of special steel. Successive forgings are produced from the same impression.

MAJOR END-USE INDUSTRIES

PERCENT OF FORGING INDUSTRY SALES

Aerospace

Aircraft Engines and Engine Parts	15.4%	
Airframes, Aircraft Parts & Aux. Equip.	13.2%	
Missiles and Missile Parts	3.0%	31.6%
Automotive and Truck		20.5%

Off-Highway Equipment

Construction, Mining & Materials Handling	12.2%
Ordnance (except Missiles)	8.4%
Agricultural	4.2%
Plumbing Fixtures, Valves & Fittings	3.1%
Railroad	2.8%
Petrochemical	1.7%
Mechanical Power Transmission Equip. incl. Bearings	1.4%
Internal Combustion Engines (Stationary)	1.1%
Metalworking & Special Industry Machinery	1.0%
Pumps & Compressors	0.9%
Steam Engines & Turbines (except Locomotives)	0.8%
Refrigeration & Air-Conditioning	0.6%
Motors and Generators	0.5%
Motorcycles, Bicycles & Misc. Equipment	0.4%
Other	8.8%
TOTAL	100.0%

FIGURE 2.34 Major end-use industries: percentage of forging industry sales. (From *Forging Industry Handbook*, Forging Industry Association, Cleveland, Ohio, 1970. With permission.)

GRAIN

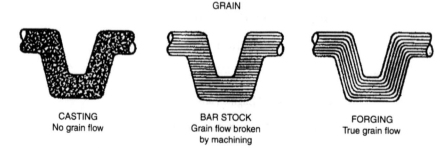

CASTING	BAR STOCK	FORGING
No grain flow	Grain flow broken by machining	True grain flow

FIGURE 2.35 Grain structure of casting, bar stock, and forging. (From *Forging Industry Handbook*, Forging Industry Association, Cleveland, Ohio, 1970. With permission.)

Materials

Since virtually all metals can be forged, the range of physical and mechanical properties available in forged products spans the entire spectrum of ferrous and nonferrous metallurgy.

Aluminum alloys are readily forged, primarily for structural and engine applications in the aircraft and transportation industries, where temperature environments do not exceed approximately 400°F. Forged aluminum combines low density with good strength-to-weight ratio.

Magnesium forgings are usually employed for applications at service temperatures lower than 500°F, although certain alloys (magnesium-thorium, for example) are employed for short-time service up to 700°F. Magnesium forgings are used efficiently in lightweight structures and have the lowest density of any commercial metal.

Low-carbon and low-alloy steels comprise the greatest volume of forgings produced for service applications up to 800 to 900°F. These materials provide advantages of relatively low material cost, ease of processing, and good mechanical properties. The design flexibility of steel is due in great part to its varied response to heat treatment, giving the designer a wide choice of properties in the finished forging.

Sizes and Shapes

Forgings are produced in an extremely broad range of sizes, from parts measuring less than an inch to those well over 23 ft. in length. Accordingly, the designer has considerable freedom in developing mechanical components in sizes required for most applications in machines and conveyances.

Metallurgy

The cast state of metal and conversion to the wrought state is shown for aluminum in Figure 2.36. This shows the striking instability of various microstructures at

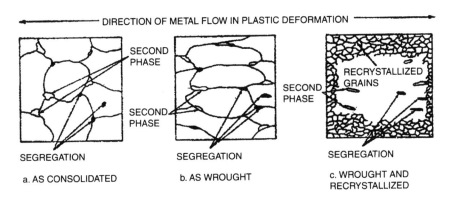

FIGURE 2.36 Typical development of grain flow in metal. (From *Forging Industry Handbook*, Forging Industry Association, Cleveland, Ohio, 1970. With permission.)

elevated temperatures and the eventual disappearance of all the modes of hardening at sufficiently high temperatures. The solidification process (transformation from the liquid state to the solid state) produces a metallurgical structure which can be characterized by (1) grain size and shape, (2) segregation, (3) porosity, and (4) nonmetallic inclusions. In the cast state, the grain size in shape castings and ingots is relatively coarse and may be more or less spherical in shape, or cigar-shaped. Mechanical properties are diminished, in general, by coarse grain size. Segregation, as also seen in Figure 2.36, leads to variability in properties and can result in embitterment by low-melting constituents in the grain boundaries (hot shortness). There is no clear-cut point at which the cast state is converted into the wrought state by the action of plastic deformation. It seems reasonable to mark this point at the degree of deformation that seals up the microporosity and eliminates the internal notch effect arising form this source. The typical transformation from the wrought condition to the final wrought and recrystallized grain development is shown in Figure 2.36c, and is the result of deformation of the wrought alloy.

Basic Forging Processes

For purposes of understanding forging design, it is helpful to consider several of the various forging operations in terms of the resulting characteristic metal flow. This allows breaking down complex operations into basic components and permits better understanding of the forging process.

Compression between flat dies is also called "upsetting" when an oblong workpiece is placed on end on a lower die, and its height is reduced by downward movement of the top die. Friction between the end faces of the workpiece and the dies is unavoidable. It prevents the free lateral spread of the ends of the workpiece and results in a barrel shape, as shown in Figure 2.37. This operation is the basic type of deformation employed in flat-die forging. The nature of local deformation can be followed in prepared test specimens. Bisected specimens show that, as deformation proceeds, material adjoining the dies remains almost stationary (area I in

Figure 2.37). Material near the outer surface of the cylinder is deformed as a result of the center material moving radially outward (area III), and the bulk of the deformation is concentrated in the remaining area (area II). Within this latter zone, heaviest deformation occurs at the center of the test piece, forcing material out of the center and developing circumferential tensile stresses. Material nearest the dies behaves somewhat like a rigid cone penetrating the rest of the specimen. Slip occurs at the faces of this cone.

It should be noted that while barreling is usually attributed to friction effects, the cooling of the specimen has much to do with it. Contact with the cool die surface chills the end faces of the specimen, increasing its resistance to deformation and enhancing the barreling. In this type of upsetting, the ratio of workpiece length to diameter (or square) is subject to limitations if one is to avoid buckling of the workpiece. For successful upsetting of unsupported material, ratios of 2.5:1 to 3:1 are generally considered appropriate, although some work requires a smaller ratio.

Compression between Narrow Dies

While upsetting between parallel flat dies is one of the most important of the *basic* forging operations, it is limited to deformation that is always symmetrical around a vertical axis. If preferential elongation is desired, the dies must be made rela-

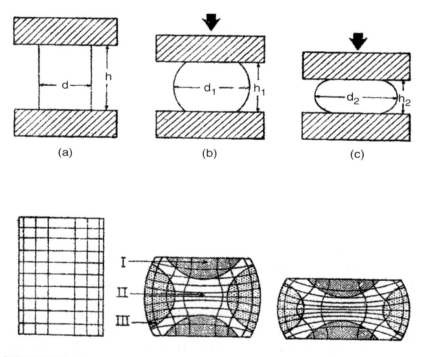

FIGURE 2.37 Compression of a cylindrical workpiece between flat dies. (From *Forging Industry Handbook*, Forging Industry Association, Cleveland, Ohio, 1970. With permission.)

tively narrow, as shown in Figure 2.38. When compressing a rectangular (or round) bar, frictional forces in the axial direction of the bar are smaller than in the direction perpendicular to it. Therefore, most of the material flow is axial, as shown in Figures 2.38b and 2.38c. Since the width of the tool still presents frictional restraint, spread is unavoidable. The proportion of elongation to spread depends on the width of the die. A narrow die will elongate better, but die width cannot be reduced indefinitely, because cutting instead of elongation will result (see Figure 2.38c).

Fullering and Edging

Using dies with specially shaped surfaces can more positively influence the direction of material flow. As shown in Figure 2.39, material can be made to flow away from the center of the dies or to gather there. In "fullering," the vertical force of deformation always possesses (except for the exact center) a component that tends to displace material away from the center, as seen in Figure 2.39a. An "edging" or "roller" die has concave surfaces and forces material to flow toward the center, where the thickness can be increased above that of the original bar dimensions, as seen in Figure 2.39b.

Impression-Die Forming

In the cases discussed thus far, the material was free to move in at least one direction. Thus, all dimensions of the workpiece could not be closely controlled. Three-dimensional control requires impressions in the dies, and thus the term "impression die" forging is derived. In the simplest example of impression-die forging, a cylindrical (or rectangular) workpiece is placed in the bottom die, as seen in Figure 2.40a. The dies contain no provision for controlling the flow of excess material. As the two dies are brought together, the workpiece undergoes plastic deformation until its enlarged sides touch the side walls of the die impressions, as in Figure 2.40b. At this point, a

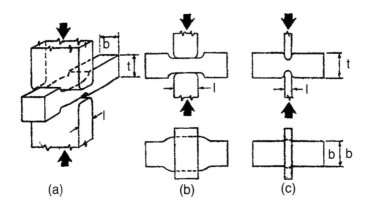

(a)   (b)   (c)

FIGURE 2.38 Compression between flat, narrow dies. (From *Forging Industry Handbook*, Forging Industry Association, Cleveland, Ohio, 1970. With permission.)

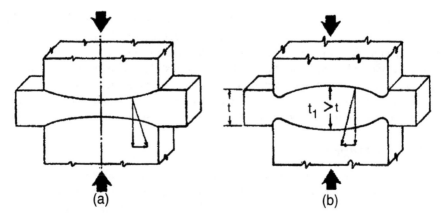

FIGURE 2.39 Compression between (a) convex and (b) concave die surfaces (fullering and edging or rolling, respectively). (From *Forging Industry Handbook*, Forging Industry Association, Cleveland, Ohio, 1970. With permission.)

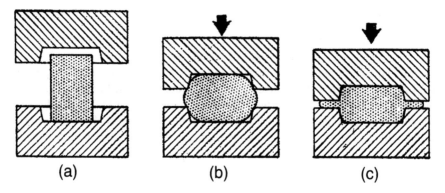

FIGURE 2.40 Compression in simple impression dies without special provision for flash formation. (From *Forging Industry Handbook*, Forging Industry Association, Cleveland, Ohio, 1970. With permission.)

small amount of material begins to flow outside the die impressions, forming what is known as "flash." In the further course of die approach, this flash is gradually thinned. As a consequence, it cools rapidly and presents increased resistance to deformation. In this sense, then, the flash becomes a part of the tool and helps to build up high pressure inside the bulk of the workpiece. This pressure can aid material flow into parts of the impressions hitherto unfilled so that, at the end of the stroke, the die impressions are nearly filled with the workpiece material, as in Figure 2.40c.

The process of flash formation is also illustrated in Figure 2.41. Flow conditions are complex even in the case of the very simple shape shown. During compression, the material is first forced into the impression that will later form the hub, then the material begins to spread and the excess begins to form the flash, as seen in

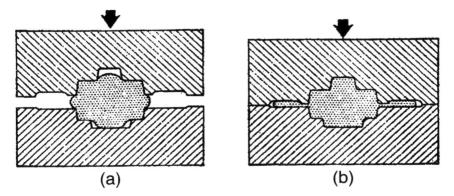

FIGURE 2.41 Formation of flash in a conventional flash gutter in impression dies. (From *Forging Industry Handbook*, Forging Industry Association, Cleveland, Ohio, 1970. With permission.)

Figure 2.41a. At this stage the cavity is not completely filled, and success depends on whether a high enough stress can be built up in the flash to force the material (by now gradually cooling) into the more intricate details of the impressions. Dimensional control in the vertical direction is generally achieved by bringing opposing die faces together, as shown in Figure 2.41b. The flash presents a formidable barrier to flow on several counts. First, the gradual thinning of the flash causes an increase in the pressure required for its further deformation. Second, the relatively thin flash cools rapidly in intimate contact with the die faces, and its yield stress rises correspondingly. Third, the further deformation of the flash entails a widening of the already formed ringlike flash, which in itself requires great forces.

Thus the flash can ensure complete filling even under unfavorable conditions, but only at the expense of extremely high die pressures in the flash area. Such dependence on excessive die pressures to achieve filling is usually undesirable, because it shortens die life and creates high power requirements. Forging design aims at easing metal flow by various means, making extremely high pressures in the flash area superfluous. In order to reduce the pressure exerted on the die faces, it is usual to form a flash gutter, as shown in Figure 2.41b, which accommodates the flash and also permits the dies to come together to control the vertical (or thickness) dimension. The flash will be removed later in the trimming operation and thus represents a loss of material. While flash is usually a by-product that is necessary for proper filling of the die cavity, too large a flash is not only wasteful, it imposes very high stresses on the dies. Impression die forging, as just described, accounts for the vast majority of commercial forging production.

Extrusion

In extrusion, the workpiece is placed in a container and compressed by the movement of the ram until pressure inside the workpiece reaches the flow stress. At this point the workpiece is upset and completely fills the container. As the pressure is further increased, material begins to leave through an orifice and forms the extruded product. In principle, the orifice can be anywhere in the container or the ram, as seen in Figure 2.42. Depending of the relative direction of motion between ram and extruded product, it is usual to speak of forward or direct (Figure 2.42a) and reverse or backward (Figure 2.42c) extrusion.

The extruded product may be solid or hollow. In the latter case, the outer diameter of the workpiece can be reduced, as in Figure 2.43a, or allowed to remain at its original dimension, as in Figure 2.43b. Tube extrusion is typical of forward extrusion of hollow shapes, as in (a), whereas reverse or back extrusion is used for the mass production of containers, as in (b).

A common feature of all processes discussed thus far is that the material is purely plastically deformed and no part of it is removed in any form. In punching and trimming, as seen in Figure 2.44, the mechanism is entirely different. The

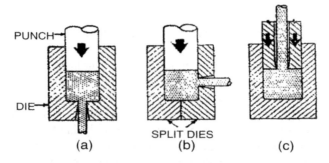

FIGURE 2.42 Extrusion of solid shapes: (a) forward (direct); (b) radial; (c) backward (reverse). (From *Forging Industry Handbook*, Forging Industry Association, Cleveland, Ohio, 1970. With permission.)

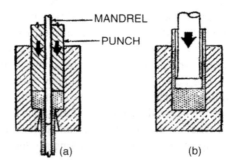

FIGURE 2.43 Extrusion of hollow shapes: (a) forward; (b) backward. (From *Forging Industry Handbook*, Forging Industry Association, Cleveland, Ohio, 1970. With permission.)

material is subjected to shear stresses between the punch and die until the stress exceeds the yield stress of the material in shear and cracks penetrate the material. Proper clearance between punch and trimming blade is important in obtaining best quality of the cut surface.

Figure 2.45 shows the four-step forging of a gear blank. It starts with a cut billet (Figure 2.45a), which is preheated. The first preform is shown in Figure 2.45b, the second preform in Figure 2.45c, and the finished forging in Figure 2.45d, ready for the punching operation. This part required three sets of dies for forming, with subsequent heating and forming, plus a punch die to remove the excess in the center of the part.

Forging Design Principles

Establishing the location and shape of the parting line (or flash line) is one of the most important tasks of forging design. Aside from the obvious need to remove the forging from the dies, the parting line influences many other factors, including: selection of the forging equipment; the design, construction, and cost of dies; grain flow; the trimming procedure; material utilization; and the position of locating surfaces for subsequent machining. Figure 2.46 illustrates a variety of simple shapes showing undesirable and desirable parting lines for hammer and press forgings.

Reasons for the preferred locations are as follows.

Case 1. The preferred choice avoids deep impressions that might otherwise promote die breakage.

Cases 2 and 3. The preferred choices avoid side thrust, which could cause the dies to shift sideways.

Case 4. Preference is based on grain-flow considerations. The satisfactory location provides the least expensive method of parting, since only one impression die is needed. The preferred location, however, produces the most desirable grain-flow pattern.

Case 5. The choice in this case is also based on grain-flow considerations. However, the desirable location usually introduces manufacturing problems and is used only when grain flow is an extremely critical factor in design. This, in turn, depends on the directional properties and cleanliness of the material being forged.

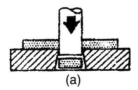   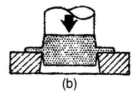
(a)   (b)

FIGURE 2.44 Shearing operations: (a) punching; (b) trimming. (From *Forging Industry Handbook*, Forging Industry Association, Cleveland, Ohio, 1970. With permission.)

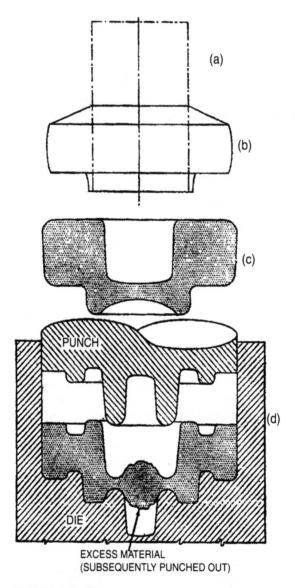

EXCESS MATERIAL
(SUBSEQUENTLY PUNCHED OUT)

FIGURE 2.45 Forging of a gear blank with internal flash:
(a) cut billet; (b) first preform; (c) second preform; (d) finished forging, ready for punching operation. (From *Forging
Industry Handbook*, Forging Industry Association, Cleveland,
Ohio, 1970. With permission.)

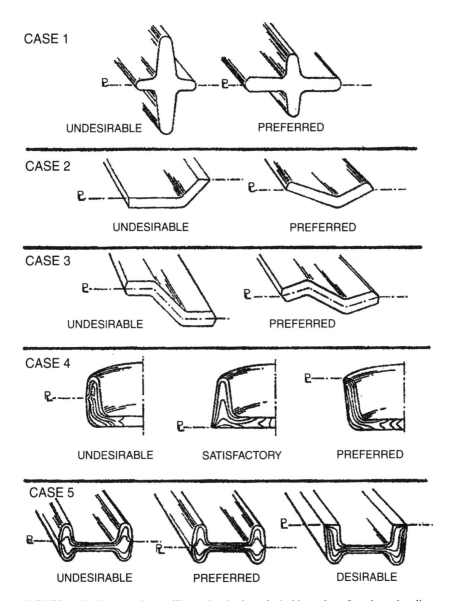

FIGURE 2.46 Forging shapes illustrating both undesirable and preferred parting line locations. (From *Forging Industry Handbook,* Forging Industry Association, Cleveland, Ohio, 1970. With permission.)

The most common draft angles are 3° for materials such as aluminum and magnesium and 5 to 7° for steels and titanium.

Draft angles less that these can be used, but the cost will increase.

Webs and Ribs

A *web* is a thin part of the forging lying in or parallel to the forging plane, while a *rib* is perpendicular to it. Since virtually all designs can be broken down into component sections formed either parallel or perpendicular to the fundamental forging plane, a discussion of webs and ribs has meaning for a great variety of forgings. Figure 2.47 shows examples of four basic types of rib designs.

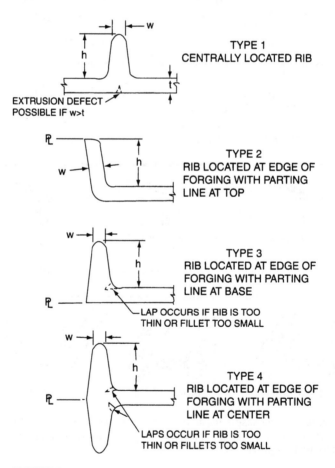

FIGURE 2.47 Four basic types of rib designs. (From *Forging Industry Handbook*, Forging Industry Association, Cleveland, Ohio, 1970. With permission.)

Type 1. Centrally located ribs are formed by extruding metal from the body of a forging. As a general rule, the web thickness should be equal to or greater than the rib thickness to avoid defects.

Type 2. Ribs at the edges of forgings with the parting line at the top are formed by metal flow similar to that in reverse extrusion. Ribs thinner than the other three types are possible with this design.

Type 3. Ribs at the edges of forgings with the parting line at the base are also formed by an extrusion type of flow. Section thickness limits are similar to type 1.

Type 4. Ribs and bosses at the edge of a forging with a central web are the most difficult to forge. Such rib designs almost always require preliminary tooling to gather the volume of metal necessary for filling the final die. Minimum thicknesses for type-4 ribs are generally larger than are those for the other three types.

There are no hard and fast rules that apply to the dimensions of ribs. In general, the rib height should not exceed eight times the rib width. Most forging companies prefer to work with rib height-to-width ratios between 4:1 and 6:1. Figure 2.48 summarizes, for several alloys, suggested minimum section-size limits for ribs and webs with conventional fillet and corner radii and draft angles. Fillet and corner radii are shown in Figure 2.49. In Figure 2.49a, the influence of impression depth on corner radii of ribs, bosses, and other edges for steel and aluminum forgings of average proportions is seen; in Figure 2.49b, the influence of rib height on minimum fillet radius for steel and aluminum is shown; in Figure 2.49c, representative fillet and corner radii for forgings of several materials with 1-in.-high ribs are given; and in Figure 2.49d, representative minimum fillet and corner radii of steel forgings on a weight basis are given. Figure 2.50a shows various die fillets for corner radii for steel and aluminum

Materials	Minimum Rib Thickness, in., for Forgings of Given Plan Area, sq. in.		Minimum Web Thickness, in., for Forgings of Given Plan Area, sq. in.			
	Up to 10	10-100	Up to 10	10-100	100-300	Over 300
2014 (aluminum)	0.12	0.19	0.10	0.19	0.31	0.50
AISI 4340 (alloy steel)	0.12	0.19	0.19	0.31	0.50	0.75
H-11 (hot-work die steel)	0.12	0.19	0.20	0.38	0.60	1.00
17-7PH (stainless steel)		0.25		0.38	0.60	1.00
A-286 (super alloy)		0.25		0.38	0.75	1.25
Ti-6Al-4V (titanium alloy)		0.25		0.31	0.55	0.75
Unalloyed Mo		0.38		0.38		

Note: This table is based on data provided by several forging companies. In many cases, the companies did not agree on minimum values for rib widths and web thicknesses. The values presented here indicate the most advanced state of commercial forging practice. There was general agreement on the need for more liberal dimensions with the alloys requiring the greater forging pressures.

FIGURE 2.48 Representative minimum section thickness for rib-and-web forgings of several materials. (From *Forging Industry Handbook*, Forging Industry Association, Cleveland, Ohio, 1970. With permission.)

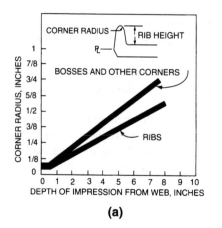

(a)

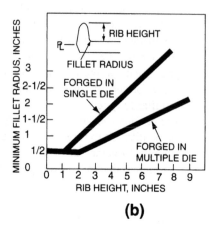

(b)

Material	Fillet Radius, inch		Corner Radius, inch	
	Preferred	Minimum	Preferred	Minimum
2014	1/4	3/16	1/8	1/16
AISI 4340	3/8-1/2	1/4	1/8	1/12
H-11	3/8-1/2	1/4	3/16	1/16
17-7PH	1/4-1/2	3/16	3/16	3/32
A-286	1/2-3/4	1/4-3/8	1/4	1/8
Ti-5Al-4V	1/2-5/8	3/8	1/4	1/8
Unalloyed Mo	1/2	—	Full radius up to 1/2 in.	3/8

(c)

Forging Weight, lb	Fillet Radius, inch	Corner Radius, inch
1	3/64-1/8	3/64-1/8
2	1/14-1/3	1/16-2/8
5	1/8-1/4	1/8
10	1/8-1/4	1/32-1/8
30	1/4-1/2	1/8-1/4
100	1/2	1/4

(d)

FIGURE 2.49 Representative minimum fillet and corner radii for forgings. (From *Forging Industry Handbook*, Forging Industry Association, Cleveland, Ohio, 1970. With permission.)

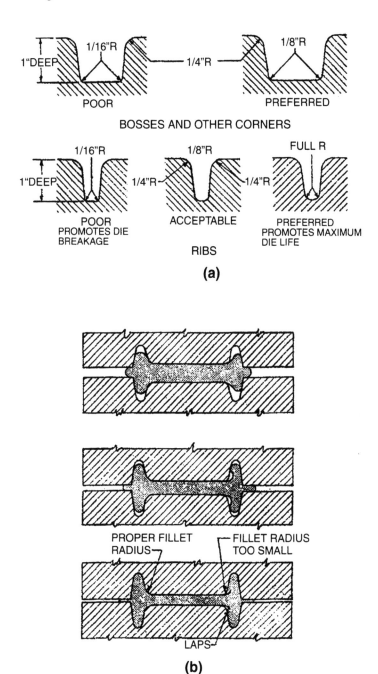

FIGURE 2.50 (a) Influence of fillet radius on metal flow during progressive stages of die closure. (b) Various die fillet designs for corner radii for steel and aluminum forgings, showing the order of preference for 1-in.-high bosses and ribs. (From *Forging Industry Handbook*, Forging Industry Association, Cleveland, Ohio, 1970. With permission.)

forgings, showing order of preference for 1-in.-high bosses and ribs. The schematic diagram (Figure 2.50b) shows the influence of fillet radius on metal flow during progressive stages of die closure.

Many forgings are used with little or no machining, where the normal forging variations and the surface condition typical for the material are acceptable. Finish allowance for machining varies with cost considerations, size of part, the way the machinist sets it up, and the oxidation and forgeability characteristics of the metal. Figure 2.51 shows representative finish allowances for impression die forgings of various materials. (Values are in addition to dimensional tolerances.) Finish allowances at corners are usually greater than the values shown in the figure.

Tolerances

Tolerances, on the whole, represent a compromise: the desired accuracy achievable within reason. Therefore, tolerances cannot be regarded as rigid standards. Their values change with the development of technology and design concepts. The tolerances

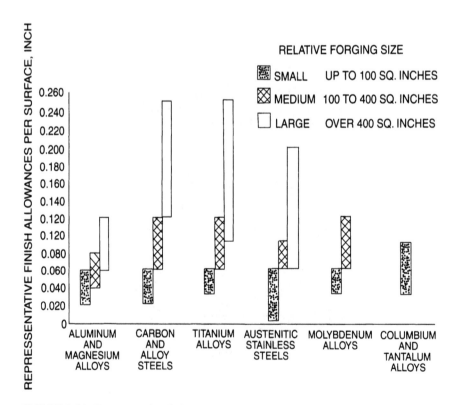

FIGURE 2.51 Representative finish allowances for impression die forgings of various materials. (Values are in addition to dimensional tolerances.) Finish allowances at corners are usually greater than values shown. (From *Forging Industry Handbook*, Forging Industry Association, Cleveland, Ohio, 1970. With permission.)

expressed in this handbook, the *Forging Industry Handbook,* and others represent reasonable average performance that can be expected from a forge plant. Tolerances closer than this obviously require greater care and skill, more expensive equipment, or a more expensive technology, and therefore carry a premium. When tolerances are excessively tight, cost of production may become disproportionately high. This makes the cost of any additional machining important to the design of the product. It may be worthwhile to pay additional costs for a forging that has a "near-net-shape" requiring little if any machining. On the other hand, a lower-cost forging with greater tolerances in some dimensions and requiring more machining may be a better design. These are the trade-offs that a manufacturing engineer must consider during the product design and development phases of a project. Figure 2.52a shows the standard aluminum forging of this aircraft part, 26 in. long and weighing over 21 lb. It required 11.8 lb. of metal machined off to obtain the machined fitting (Figure 2.52b), weighing 9.8 lb. The precision forging (Figure 2.52c), weighing 10.8 lb, is shown as it was received by the contractor, with flash and knockout lugs removed by the forging vendor with a saw. No weight-removal machining was required, though some

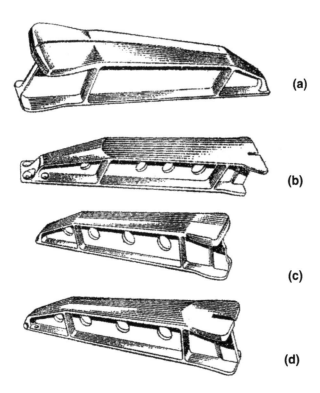

(a)

(b)

(c)

(d)

FIGURE 2.52 Standard aircraft forging (a), weighing 21 lb., and the same part as a precision forging (c), weighing 10.8 lb. (From *Forging Product Information,* Kaiser Aluminum and Chemical Sales, Cleveland, Ohio, 1959. With permission.)

hand finishing was necessary to smooth the rough surface finish at the parting line left by the saw cut. The finished forging, weighing 9.1 lb, after 1.7 lb. had been removed, is shown in Figure 2.52d.

Flash and Mismatch

Conventional forging dies are designed with a cavity for flash, as shown in Figure 2.53. This sketch of a completed forging in the closed die shows a typical gutter design. Flash extension tolerances are based on the weight of the forging after trimming and relate to the amounts of flash extension. Flash is measured from the body of the forging to the trimmed edge of the flash. Flash tolerances are expressed in fractions of an inch, as shown in the figure. Mismatch, often the result of die wear, and the expected tolerances are shown in Figure 2.54.

Radii Tolerances

Radii tolerances relate to variation from the purchaser's radii specifications on fillet radii, and on corner radii where draft is not subsequently removed by trimming, broaching, or punching. Radii tolerances are plus or minus one half the specified radii, except where corner radii are affected by subsequent removal of draft.

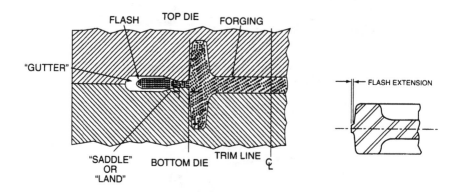

Tabulated figures are ranges of flash extension, expressed in inches.

Weights of forgings After Trimming, in Pounds	Carbon and Low Alloy	Stainless	Superalloys and Titanium	Aluminum and Magnesium	Refractory Alloys
10 and Under	0 to 1/32	0 to 1/14	0 to 1/16	0 to 1/32	0 to 1/8
Over 10 to 25 incl.	0 to 1/14	0 to 1/32	0 to 3/32	0 to 1/14	0 to 1/4
Over 25 to 50 incl.	0 to 1/12	0 to 1/8	0 to 1/8	0 to 3/32	0 to 1/4
Over 50 to 100 incl.	0 to 1/8	0 to 3/14	0 to 1/14	0 to 1/8	0 to 1/18
Over 100 to 200 incl.	0 to 3/16	0 to 1/14	0 to 1/4	0 to 3/14	0 to 3/8
Over 200 to 500 incl.	0 to 1/4	0 to 3/16	0 to 1/16	0 to 1/4	0 to 1/2
Over 500 to 1000 incl.	0 to 5/16	0 to 3/05	0 to 3/8	0 to 3/14	0 to 5/8
Over 1000	0 to 3/8	0 to 1/2	0 to 1/2	0 to 3/8	0 to 3/4

FIGURE 2.53 Conventional flash gutter design. (From *Forging Industry Handbook*, Forging Industry Association, Cleveland, Ohio, 1970. With permission.)

Materials	Weights of Forgings After Trimming, in pounds								
	Less than 2 lbs.	Over 2 to 5 incl.	Over 5 to 25 incl.	Over 25 to 50 incl.	Over 50 to 100 incl.	Over 100 to 200 incl.	Over 200 to 500 incl.	Over 500 to 1000 incl.	Over 1000
Carbon, Low Alloys	①	1/64	1/32	3/64	1/16 1/16	3/32	1/8	5/32	3/16
Stainless Steels	①	1/32	3/64	1/16	3/32	1/8	5/32	3/16	1/4
Super Alloys, Titanium	①	1/32	3/64	1/16	3/32	1/8	5/32	3/16	1/4
Aluminum, Magnesium	①	3/64	1/32	3/64	1/16	3/32	1/8	5/32	3/16
Refractory Alloys	①	1/16	3/32	1/8	5/32	3/14	1/4	5/16	3/8

① = Customarily negotiated with purchaser

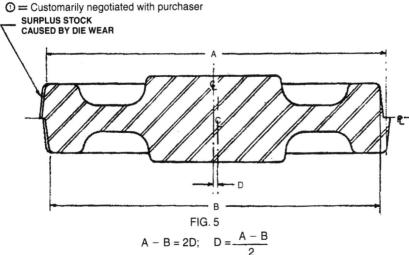

SURPLUS STOCK
CAUSED BY DIE WEAR

FIG. 5

$$A - B = 2D; \quad D = \frac{A - B}{2}$$

A = Projected maximum over-all dimensions measured parallel to the main parting line of the dies.
B = Projected minimum over-all dimensions measured parallel to the main parting line of the dies.
D = Displacement.

FIGURE 2.54 Forging die mismatch. (From *Forging Industry Handbook*, Forging Industry Association, Cleveland, Ohio, 1970. With permission.)

Straightness

Straightness tolerances allow for slight and gradual deviations of surfaces and centerlines from the specified contour, which may result from post-forging operations such as trimming, cooling from forging, and heat treating. Long, thin forgings are more susceptible to warpage than massive shapes; similarly, thin disks warp more than thick disks of comparable diameter. (See Figure 2.55.)

Materials	Area at the Trim Line, expressed in sq. in.—Flash not included						
	10 and under	Over 10 to 30 incl.	Over 30 to 50 incl.	Over 50 to 100 incl.	Over 100 to 500 incl.	Over 500 to 1000 incl.	1000 Over
Carbon, Low Alloy	1/32	1/16	3/32	1/8	5/32	3/16	1/4
400 Series Stainless	1/32	1/16	1/32	1/8	3/15	1/4	5/16
300 Series Stainless	1/16	3/32	1/8	5/32	3/14	1/4	5/16
Aluminum, Magnesium	1/32	1/32	1/16	3/32	1/8	3/16	1/4

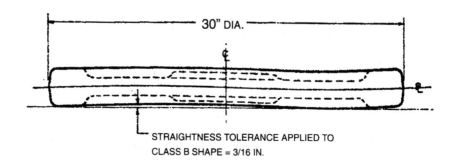

STRAIGHTNESS TOLERANCE APPLIED TO
CLASS B SHAPE = 3/16 IN.

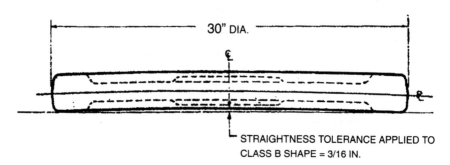

STRAIGHTNESS TOLERANCE APPLIED TO
CLASS B SHAPE = 3/16 IN.

FIGURE 2.55 Straightness tolerances, class B shapes. (From *Forging Industry Handbook*, Forging Industry Association, Cleveland, Ohio, 1970. With permission.)

Figure 2.56 shows the influence of forgeability and flow strength on die filling. Ease of filling decreases from upper left to lower right.

2.2.5   Bibliography

Forging Industry Handbook, Forging Industry Association, Cleveland, Ohio, 1970.
Investment Casting Handbook, Investment Casting Institute, Chicago, 1968.

INCREASING FLOW STRENGTH OR FORGING PRESSURE

	LOW GROUP	MODERATE GROUP	HIGH GROUP
GOOD	1030 (CARBON STEEL) 4340 (ALLOY STEEL) H-11 (TOOL STEEL) 6061 (Al ALLOY)	TYPE 304 (STAINLESS) Ti-6Al-4V	MOLYBDENUM 16-25-6 (STAINLESS)
MODERATE	AZ80 (Mg ALLOY) 7075 (Al ALLOY)	A-286 (STAINLESS) INCO 901 (NI ALLOY) Ti-5Al-2.5Sn	WASPALOY (NI ALLOY) Ti-13V-11Cr-3Al N 155 (Ni-Cr-Co ALLOY)
FAIR	RESULPHURIZED STEELS (e. g. 1130)	TYPE 321 (STAINLESS) 15-7 Mo (STAINLESS)	RENE 4l (Ni ALLOY) HASTELLOY C (Ni ALLOY) HASTELLOY B (Ni ALLOY)

(vertical axis label: DECREASING FORGEABILITY)

FIGURE 2.56 Influence of forgeability and flow strength on die filling. Ease of filling decreases from upper left to lower right. (From *Forging Industry Handbook*, Forging Industry Association, Cleveland, Ohio, 1970. With permission.)

2.3 DESIGN FOR SHEET METAL

2.3.0 Introduction to Design for Sheet Metal

More parts are made from sheet metal than from any other material in the world. Most of the processes are well established, and design limits are well known and accepted. Many finishes are available, and customer acceptance and product life are well proven. More steel is produced than any other metal, with aluminum second in volume and in use in fabricating products. The weight of aluminum is approximately one third the weight of steel. In this subchapter we will concentrate our discussions on parts fabricated from these two materials. Although precision cold-rolled sheet is more expensive per pound that billets or bars, this can seldom offset the total part cost since sheet metal stampings are generally the lowest-cost parts to produce. Both the machinery and labor are relatively low in cost, and the production rates can be quite high.

One of the problems the authors had in preparing this subchapter was separating design dos and don'ts from the nuts and bolts of the materials and fabrication processes. For the most part, the equipment and processes are described in Subchapter 15.1. Active participation by a manufacturing engineer (or other factory representative) in the design of a new product requires knowledge of materials, processes, and costs as their most important contribution to the design team. This subchapter therefore will primarily provide guidance to the design features of sheet metal parts as seen from the manufacturing point of view. However, we must introduce sufficient material and process information to understand why the design comments are applicable.

Stampings

Stamping parts from sheet metal is a relatively straightforward process in which the metal is shaped through deformation by shearing, punching, drawing, stretching, bending, or coining. Production rates are fairly high, and little if any secondary machining is required to produce finished parts within tolerances. In most cases such parts are fairly simple in shape and do not tolerate extreme loads. However, stamped parts can be produced in complex shapes and sizes, and with load-carrying capabilities that rival parts made by other production processes. In some cases, unusual techniques have been developed to deform the metal to its required shape without exceeding forming limits. In other instances, ingenuity combined with good design practice has produced sheet metal parts that would be impossible to fabricate by other methods. Comparing cost impact of the several design variables is difficult. Figure 2.57 is an example of minor differences in the design of a sheet metal part which can double the part cost.

Quantity Considerations

The simple example shown in Figure 2.58 shows the influence of quantity production on the design and the manufacturing processes for producing a large tubular shape of large diameter and thin wall. Lowest unit cost is practical where substantial quantities are required for each production run. Welding may not be required in either the low or the high production design if locking tabs can be utilized.

2.3.1   Design for Blanking

The first step in forming a product from sheet, plate, or foil is usually the cutting of a blank. A blank may be any size or pattern, if it represents the most suitable initial shape for the manufacture of the end product. Circular blanks are employed for roundware and most cylindrical-end products. Blanks with straight or curved edges and rounded corners are used for rectangular products. Irregularly shaped shells usually require blanks of irregular shape to provide material where needed. Blanks are made in a number of ways, depending on such factors as gauge, properties of the material, quantity, and size. Sheet metal can be furnished in sheets or in large rolls.

Blanks with straight edges can be produced from sheet, plate, and foil by employing either power- or foot-operated guillotine shears. Plate blanks, which require good edge condition, can be rough, sheared, and then machined or routed to final size.

LOW COST DESIGN

Tolerance	± .010 on centers
	± .005 on hole diameters
Material	16 ga. 1010 sheet
	Std. tolerance .059 ± .006
Bends	Radius inside 1/4 thickness
Hole distortion	Print should note "distortion permissible" if hole is too close to a form
Outside corners	1/16" radius allowed

HIGH COST DESIGN

Tolerance	± .003 on centers
	± .001 on hole diameters
Material	1010 half-hard
	.055 ± .003 thickness
Bends	Inside .005 radius
Hole distortion	± .001 all holes
Outside corners	.001 radius allowed (sharp corners required)

PRICE COMPARISON
BASESD ON TOOLING & RUN OF 1000 PIECES

	LOW COST	HIGH COST	DIFF
MATERIAL COST	$.19	$.38	+$.19
TOOLING	$300	$600	+$300
TOOLING/PIECE (1000 PCS)	.30	.60	.30
RUN LABOR (1000 PCS)	.20/Piece	.35/Piece	+.15
TOTAL	$.69 each	$1.33	$.64

FIGURE 2.57 "Minor" differences in sheet metal design can double part cost. (Courtesy of Dayton Rogers Mfg. Company, Minneapolis, Minn. With permission.)

Blanking with conventional punch and die is usually employed for sheet and thin plate where large quantities are required. Its application can be limited by such factors as press capacity, blank size, and material thickness. Therefore, special shears or other means may more economically cut some blanks.

Blanking action can be summarized as follows. As the blanking load is applied, the cutting edges of the punch and die enter the stock, producing a smooth, sheared surface. As penetration increases, cracks in the stock form at the sharp cutting edges. The remainder of the blanking operation is accomplished by fracturing of the material rather than shearing. This produces the characteristic blanked edge, which is partly shear and partly fracture. A blanked edge produced with the correct tool clearance will exhibit a sheared surface for one third to one half of the metal thickness. Insufficient clearance will increase the amount of sheared surface, and will generally produce a blank with a higher burr. This reduced clearance gives increased blanking load and tool wear. The latter can lead to more rapid deterioration in blank edge quality. Excessive clearance will tend to increase the amount of fractured surface, and the blank will be "broken out" rather than sheared, producing a rough edge.

Figure 2.59, example #1, illustrates results that can be expected from material such as cold-rolled sheet steel, commercial quality, that is less than 1/4 hard temper, or with a shear strength of approximately 45,000 psi. A slight pulldown (A) of the blanked edge and a straight sheared section (B) for about 25 to 30% of the material thickness will occur on the die side. The opposite will occur on piercing, as indicated by (C) and (D) on the punch side. The balance will have breakage. More pulldown and a greater sheared edge will occur for softer material. Example #2 indicates the results of less than the usual clearance. Increased blanking pressure will be required,

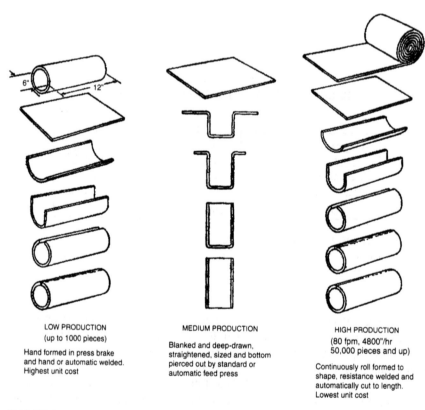

LOW PRODUCTION
(up to 1000 pieces)

Hand formed in press brake
and hand or automatic welded.
Highest unit cost

MEDIUM PRODUCTION

Blanked and deep-drawn,
straightened, sized and bottom
pierced out by standard or
automatic feed press

HIGH PRODUCTION
(80 fpm, 4800"/hr
50,000 pieces and up)

Continuously roll formed to
shape, resistance welded and
automatically cut to length.
Lowest unit cost

FIGURE 2.58 A simple example showing the influence of quantity on process selection for producing a tubular shape of large diameter and thin wall. (From R. A. Bolz, *Production Processes*, Industrial Press, New York, 1963. With permission.)

and double breakage on the blanked edge will occur. At times, double breakage on thick parts might be more desirable than 8 to 10% taper. Example #3 indicates that on very hard materials, the pulldown is negligible and the sheared edge might amount to only 10% of the thickness of the material.

A tolerance of 0.005 in. is normal in the blanking and piercing of aluminum-alloy parts in a punch press. Using a power shear and press brake, it is possible to blank and pierce to a location tolerance of 0.010 in. or less, although tolerances for general press-brake operations usually range from 0.020 to 0.030 in., since the individual punches are usually located manually on the brake. For economy in tool cost, specified tolerance should be no less than is actually necessary for the particular part. A tolerance of 0.005 in. would probably require that the punch and die be jig-ground, adding 30 to 40% to the cost over a "normal" punch press part.

Poor blanks may be produced if there is too little stock material between the blank and the edge of the strip. Typical scrap allowances between the blank and strip edge, and between the blanks, are shown in Figures 2.60 and 2.61.

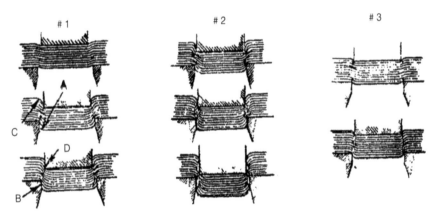

FIGURE 2.59 Typical blanking or punching operations on commercial cold-rolled sheet steel. (Courtesy of Dayton Rogers Mfg. Company, Minneapolis, Minn. With permission.)

Thickness, inch	Blank diameter, inches	Scrap on the side, inch	Scrap between blanks, inch
0.013 to 0.057	0 to 9 10 to 19 20 and over	1/8 3/16 1/4	1/16 3/32 1/8
0.057 to 0.081	0 to 9 10 to 19 20 and over	5/32 7/32 5/16	3/32 1/8 5/32
0.081 to 0.128	0 to 9 10 to 19 20 and over	3/16 1/4 3/8	1/8 5/32 1/16

FIGURE 2.60 Scrap allowances for blanking. (From *Forming Alcoa Aluminum*, Aluminum Company of America, Pittsburgh, Pa., 1981. With permission.)

Piercing

The piercing of sheet, thin plate, and foil is similar to blanking except that the cut blank becomes the scrap and the stock material is kept. When a large number of closely spaced small holes are pierced, the process is described as "perforating." Most techniques used for blanking also apply for piercing. Finish blanking becomes finish piercing, with a radius or chamfer on the punch rather than the die.

Piercing and perforating may either be preparatory operations in the blank development, as in the case of ceiling panels where perforated blanks are made for further forming, or they may be finish operations, as in the case of strainers and colanders, which are formed prior to being perforated.

Piercing is usually performed on a single-action press. The tooling is the same as for the blanking process, except that the die for piercing must match the contour of the part. The punch may be ground at an angle to reduce the load on the press. The dimensions of the pierced hole correspond to the dimensions of the punch.

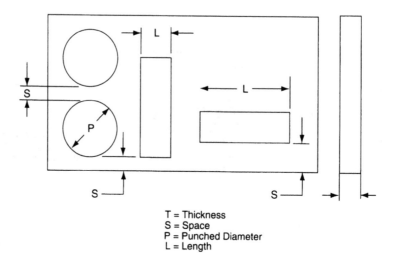

T = Thickness
S = Space
P = Punched Diameter
L = Length

Avoid Minimum Values of "S" Where Possible.

Length "L" is less Than	Punch Dia. "P" is less Than	Space "S" is Minimum
	5T	1.5T
	10T	2T
10T		2T
25T		4T

FIGURE 2.61 Minimum distance between cutouts and sheet edge. (Courtesy of Dayton Rogers Mfg. Company, Minneapolis, Minn. With permission.)

Figure 2.62 shows typical punched holes. Rollover and/or bulge are natural consequences of the punching process, the mechanical properties of the material being punched, and the die application of techniques employed.

A common practice exists in dimensioning punched holes. For general-purpose holes:

Dimension feature size limits only.
Measure feature size only in the burnished land area.
Shape deviations within the feature size limits are permissible.

For special-purpose holes:

Dimension only those elements that affect part function.
When dimensioned, specify the burnished land as minimum only.
Specify burr height as maximum only.

To pierce holes with economical tools and operations, the hole diameter must not be less than the stock thickness (Figure 2.63a). If the hole diameter is less than the material thickness (or less than 0.040 in.), it usually must be drilled and deburred, and each of these operations is slower than punching. Figure 2.63b, indicates a hole diameter with a tolerance of 0.001 in. A hole can be pierced within these limits on the punch side for approximately 25 to 30% of the material thickness, as indicated in Figure 2.63c. The percentage of thickness varies with the shear strength of the materials. On machined holes punched undersize, redrilling and reaming are usually required where the finished hole size is somewhat larger than the stock thickness, as indicated in Figure 2.63d. If a finished hole is required through the stock thickness, as indicated in Figure 2.63e, where the hole size is equal to or less than the material thickness, more costly drilling is required in addition to deburring and reaming, since the part probably cannot be punched.

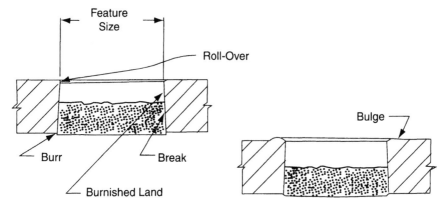

FIGURE 2.62 Typical punched holes showing rollover and bulge. (Courtesy of Dayton Rogers Mfg. Company, Minneapolis, Minn. With permission.)

(a)

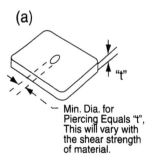

Min. Dia. for
Piercing Equals "t",
This will vary with
the shear strength
of material.

(b)

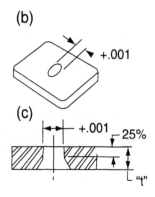

▼ +.001

(c)

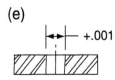

+.001 25%

"t"

(d)

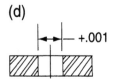

+.001

(e)

+.001

FIGURE 2.63 Tolerances on punched holes in sheet metal. (Courtesy of Dayton Rogers Mfg. Company, Minneapolis, Minn. With permission.)

Figure 2.64 shows that when the web is a minimum of the stock thickness, a hole can be punched, which is less expensive than drilling and deburring (see example e). A web that is less than the stock thickness will result in a bulge on the blank. Bulge conditions increase progressively as the web decreases, until there is a complete breakthrough. However, the bulge is hardly visible until the web is reduced to less than one half the stock thickness. These examples also apply to a web between holes. If a measurable bulge is not permitted, a drilling and deburring operation may be necessary. If the web is too narrow, the profile of the blank might be changed by adding an ear of sufficient dimensions and shape to eliminate the problem, as seen in example (g) in Figure 2.64. Another suggestion might be to change the contour of the blank to include the hole as a notch. The notch could be either be pierced, or be made wide enough in the blank without a piercing or notching operation.

When piercing adjacent to bends, Figure 2.65a indicates that the minimum inside distance required from the edge of a hole to a bend is 1 times the material thickness (T) plus the bend radius (R). Otherwise, distortion will occur, as indicated in Figure 2.65b, or piercing after forming must be considered. Figure 2.65c indicates a similar condition, except for openings with an edge parallel to bend. In this case the following requirements apply for economical tooling and production:

When L is up to 1 in., the distance should be $2T + R$ (minimum).
When L is 1 to 2 in., the distance should be $2\frac{1}{2}T + R$ (minimum).
When L is 2 in. or more, the distance should be $3T$ to $3\frac{1}{2}T + R$ (minimum).

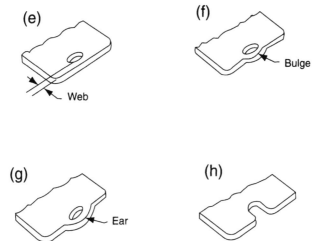

FIGURE 2.64 Bulge conditions when hole is pierced too close to edge. (Courtesy of Dayton Rogers Mfg. Company, Minneapolis, Minn. With permission.)

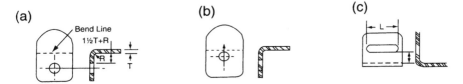

FIGURE 2.65 Problems when piercing too close to bends in sheet metal. (Courtesy of Dayton Rogers Mfg. Company, Minneapolis, Minn. With permission.)

Burrs are ragged, usually sharp, protrusions of edges of metal stampings, as seen in Figure 2.66. Commonly specified permissible burr conditions in order of increasing cost are:

1. Unless otherwise specified, normal acceptable burr can be 10% of stock thickness.
2. A note, "conditioned for handling," is interpreted to mean that normal stamping burrs are to be refined as necessary for average handling.
3. Remove burrs on specified edges.
4. Remove all burrs.
5. Break sharp edges or corners where specified.
6. Break all sharp edges and corners.

2.3.2 Design for Bending

Bending, one of the most common operations in sheet metal working, consists of forming flat metal about a straight axis. Press-brake bending and roll bending are the usual methods, although punch-press tools are often built to produce multiple bends, or bends in conjunction with other operations, such as blanking. The same equipment and tools employed for bending steel and other metals can be used for aluminum, although appropriate changes in bend radii and technique should be considered. Most principles of bending also apply to such complementary operations as flanging, beading, curling, crimping, and lock seaming. Figure 2.67 shows some typical press-brake forming dies.

Factors Influencing Bending

It is important to consider carefully in the design stage the relationships among mechanical properties, dimensions of the part after bending, metal thickness, and

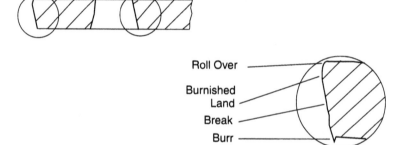

Roll Over
Burnished
Land
Break
Burr

FIGURE 2.66 Section through a pierced hole showing rollover, the burnished land, the break, and burr formation. (Courtesy of Dayton Rogers Mfg. Company, Minneapolis, Minn. With permission.)

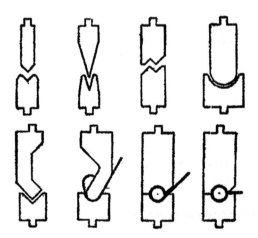

FIGURE 2.67 Typical press-brake forming dies. (Courtesy of Rafter Machine Company, Inc., Belleville, N.J. With permission.)

"formability" of the metal. When a specific alloy is selected for strength or other reasons, the part should be designed in accordance with recommended bend radii to assure producible results (see Figures 2.68 and 2.69). Conversely, when part dimensions are the prime consideration, the material selected should be readily formable to those dimensions. While it is often possible to form sheet to tighter radii than is shown in the figures, it usually costs more due to heating, tool maintenance, greater incidence of failure, etc.

The elongation values of a metal are not valid indicators of bendability, as they are taken over a 2-in. gauge length. The elongation in a very short gauge length controls bending characteristics, and this is frequently not related to the 2-in. value. The empirical data of Figures 2.68 and 2.69 represent the best measure of bendability available. Some other recommendations are given in Figure 2.70. The height of the form of the part has an influence on the bend. Diagram (a) in Figure 2.70 illustrates a 90° bend with insufficient height (h) to form properly. Consequently, stock must be added so the form is high enough (H), and stock is then cut off, which means additional tooling and an additional operation. If h is not high enough, the cutoff die may not have sufficient strength to stand up for a particular material or thickness. This may result in an even higher cost secondary operation, such as milling. Diagram (b) in Figure 2.70 shows how to determine the minimum inside height H, which in this case equals $2\frac{1}{2}$ times the material thickness (T) plus the required bend radius (R). The concept illustrated by diagram (b) is shown in chart form below the diagram. These recommended minimum formed height dimensions are general to cover most variables of design, size, material types, tempers, and thicknesses, but permit economical tooling and production. Proper design, small parts, and easily formed materials, such as aluminum, brass, copper, or mild steel, may be formed with slightly lower inside formed height (roughly 20% less).

Figure 2.71a illustrates a design that is not desirable in terms of either quality or economy. When the bend is inside the blank profile, as shown, the material must be torn through the stock thickness and the bend radius. If the part is under stress, this tear will likely cause fatigue failures. In addition, stock tooling cannot be used, because the flat area adjacent to the bend must be held in position during forming, which means extra tooling expense. Diagram (b) shows a similar condition, but with the bend just outside the blank profile. In this case, the tear proceeds to the tangent of the required bend radius. Diagrams (c) and (d) illustrate a possible solution by changing the blank profile to provide relief for the bend. Besides eliminating the chance of fatigue under stress, there is the possibility of using stock 90° V punches and dies in a press brake. The results are better quality and a less expensive tool price. If the relief notches shown in diagram (d) are wide enough compared to the material thickness and shear strength, or are designed like the relief in diagram (e), they can be included in the blanking operation at very little extra cost for tools and no extra operation.

Distortion and interference of forming is another condition to consider. Figure 2.72a shows a distortion condition that occurs in forming. It is a noticeable distortion when heavy metal is bent with a sharp inside bend radius. On material thicknesses less than 1/16 in., or when the inside forming radius is large in comparison to

Alloys	Tempers	Thickness of Sheet–Inches									
		.016	.025	.032	.040	.050	.063	.090	.125	.190	.250
		Band Radii in 32nds of an inch									
1100	-0	0	0	0	0	0	0	0	0	0	0
	-H12	0	0	0	0	0	0	0	0	3	6
	-H14 ①	0	0	0	0	0	0	0	0	3	6
	-H16	0	0	0	0	1	2	3	4	-	-
	-H18	1	1	2	2	3	3	6	8	-	-
3003 & Alclad 3003 5005 5457 5357 5657	-0	0	0	0	0	0	0	0	0	0	0
	-H12 or -H32	0	0	0	0	0	0	0	0	3	6
	-H14 or -H34 ②	0	0	0	0	0	0	1	2	4	8
	-H16 or -H36	0	0	1	2	2	3	5	6	-	-
	-H18 or -H38	1	2	3	4	5	6	9	12	-	-
3004 & Alclad 3004 5004 5454 5354	-0	0	0	0	0	0	0	0	0	2	4
	-H32	0	0	0	1	1	2	3	4	8	16
	-H34	1	1	1	2	2	3	5	6	12	24
	-H36	1	2	3	4	5	6	9	12	-	-
	-H38	2	3	4	5	7	8	12	16	-	-
5050	-0	0	0	0	0	0	0	0	0	0	0
	-H32	0	0	0	0	0	0	1	2	4	8
	-H34 ③	0	0	0	1	1	2	3	4	8	12
	-H36	1	1	2	2	3	4	6	8	-	-
	-H38	1	2	3	4	5	6	9	12	-	-
5052 5652	-0	0	0	0	0	0	0	0	0	0	0
	-H32	0	0	0	0	0	0	2	3	6	12
	-H34	0	0	0	1	1	2	3	4	8	16
	-H36	1	2	2	2	3	4	6	8	-	-
	-H38	1	2	3	4	5	6	9	12	-	-
5083 ⑤	-0	-	-	-	-	3	4	6	8	12	16
	-H32	-	-	-	-	6	7	10	14	20	28
	-H34	-	-	-	-	8	10	14	20	28	36
	-H113	-	-	-	-	-	-	-	14	20	28
5086 5155	-0	0	0	0	0	0	0	2	3	6	12
	-H32	1	1	1	2	2	3	5	6	12	24
	-H34	1	2	3	4	5	6	9	12	24	32
	-H36	3	4	5	6	7	9	12	18	-	-
5456 ⑤	-0	-	-	-	-	-	-	-	8	12	16
	-H24	-	-	-	-	8	10	14	20	32	40
	-H321	-	-	-	-	-	-	-	-	24	32
2024 Alclad 2024	-0	0	0	0	0	0	0	0	0	2	4
	-T3	1	3	4	5	7	8	12	16	32	48
6061 Alclad 6061	-0 ④	0	0	0	0	0	0	0	0	2	4
	-T6	1	2	2	2	3	4	6	8	18	32
2014	-0	0	0	0	0	0	0	0	0	2	4
	-T6	2	4	6	8	10	12	18	24	40	64
7075 Alclad 7075	-0	0	0	0	1	1	1	3	2	8	16
	-T6	2	4	6	8	10	12	18	24	40	64
7178 Alclad 7178	-0	0	0	0	1	1	1	3	4	8	16
	-T6	3	6	7	9	12	16	24	32	48	96

① Use this data for #1 reflector sheet.
② Use this data for #1, #2, #11 and #12 brazing sheet. #2 reflector sheet and RF10, RF15, RF20 finishing sheet.
③ Use this data for porcelain enameling sheet.
④ Use this data for #21 and #22 brazing sheet.
⑤ These materials may require annealing or stress-relieving after cold forming. If formed at a metal temperature of 425–475°F, a 50 per cent smaller bend radii may be used.

FIGURE 2.68 Minimum recommended inside radii for 90° cold-bending aluminum sheet. (From *Forming Alcoa Aluminum*, Aluminum Company of America, Pittsburgh, Pa., 1981. With permission.)

WORKING FROM THE INSIDE, THE FIGURES SHOWN ARE THE CORRECT BEND ALLOWANCES.
SEE EXAMPLE FOR CLARITY

.06 R. INSIDE

DEVELOPED LENGTH

.750 + .880 + .015 = L645

MATERIAL THICKNESS	INSIDE RADIUS OF BEND													
	.016	.030	.048	.063	.094	.125	.156	.188	.219	.250	.312	.375	.437	.500
.016	.004	.003	.010	.016	.029	.043	.056	.070	.083	.097	.124	.150	.177	.204
.020	.007	.000	.007	.014	.027	.041	.054	.067	.080	.094	.121	.140	175	.201
.025	.010	.004	.004	.010	.023	.037	.051	.064	.077	.091	.116	.144	.171	.198
.030	.013	.007	.001	.007	.020	.034	.047	.061	.074	.088	.115	.141	.168	.195
.032	.014	.008	.000	.006	.019	.033	.046	.059	.073	.086	.113	.140	.167	.194
.036	.017	.010	.003	.003	.016	.030	.044	.057	.070	.084	.111	.137	.164	.191
.040	.020	.013	.006	.001	.014	.027	.040	.054	.067	.081	.108	.135	.161	.188
.045	.022	.015	.009	.002	.011	.025	.038	.052	.065	.078	.105	.132	.159	.186
.048	.025	.018	.011	.004	.008	.022	.035	.049	.062	.076	.103	.130	.156	.183
.050	.027	.020	.013	.007	.007	.020	.033	.047	.060	.074	.101	.127	.154	.181
.060	.032	.026	.019	.013	.001	.014	.028	.041	.054	.068	.095	.122	.148	.175
.063	.036	.029	.022	.015	.003	.011	.025	.038	.052	.065	.092	.119	.145	.172
.075	–	.036	.029	.022	.009	.005	.018	.031	.045	.058	.085	.112	.139	.165
.080	–	.039	.033	.026	.012	.000	.014	.028	.041	.055	.081	.108	.135	.162
.090	–	–	.040	.033	.020	.006	.007	.021	.034	.047	.074	.101	.128	.155
.125	–	–	–	.056	.042	.029	.015	.002	.011	.025	.051	.078	.105	.132
.160	–	–	–	.076	.063	.049	.036	.022	.009	.004	.031	.058	.085	.112
.190	–	–	–	–	.083	.070	.056	.043	.030	.016	.011	.038	.064	.091

NOTE: FIGURES TO THE LEFT OF THE HEAVY LINE ARE TO BE ADDED
FIGURES TO THE RIGHT OF THE HEAVY LINE ARE TO BE SUBTRACTED

FIGURE 2.69 Table of metal-forming bend allowances. (From J. P. Tanner, *Manufacturing Engineering*, Marcel Dekker, New York, 1991. With permission.)

the material thickness, distortion is barely noticeable. The material on the inside of the bend is under compression, which results in this bulge condition on the edges. In addition, the edges on the outside of the bend are under tension and tend to pull inside. This bulge or distorted condition is usually of no concern and is accepted as standard practice unless bulging will cause interference with a mating part. This interference should be referred to on the part drawing so that a secondary operation can be considered to remove it. The extra operation may not require tooling, but it will add to the cost of production. Figure 2.72b illustrates a blank developed to prevent interference resulting from bulge without extra production cost. The upper left-hand form (enlarged section) in Figure 2.72c indicates a fracture condition that occurs when the burr side of the blank is on the outside of the bend. This fracture condition occurs because the burr side of the blank on the outside of the bend is under tension and causes the minute fracture on the sharp edge to open up and in extreme cases to become visible. Blanks should be produced so the burr side will be on the inside of the bend, which is under compression, as in the lower form. However, when blueprint requirements prevent this, or when a bend is in an opposite direction, as in the upper bend, fractures may occur.

Tumbling or deburring well before forming can minimize fracture in most cases. On extra heavy material with a very sharp inside bend radius, or on materials that are

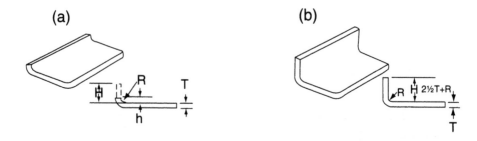

Stock Thickness		INSIDE BEND RADIUS									
		FRAC.	DEC.	FRAC.	DEC.	FRAC.	DEC.	FRAC.	DEC.	FRAC.	DEC.
Frac.	Dec.	Sharp	Sharp	1/32	.031	1/16	.062	3/32	.093	1/8	.125
1/32	.031	5/64	.078	7/64	.109	9/64	.140	11/64	.171	13/64	.203
1/16	.062	5/32	.156	3/16	.187	7/32	.218	1/4	.250	9/32	.281
3/32	.093	15/64	.234	17/64	.265	19/64	.296	21/64	.328	23/64	.359
1/8	.125	5/16	.312	11/32	.343	3/8	.375	13/32	.406	7/16	.437
5/32	.156	25/64	.390	27/64	.421	29/64	.453	31/64	.484	33/64	.515
3/16	.187	15/32	.468	1/2	.500	17/32	.531	9/16	.562	19/32	.593

FIGURE 2.70 Minimum height of form depending on material thickness and inside bend radius. (Courtesy of Dayton Rogers Mfg. Company, Minneapolis, Minn. With permission.)

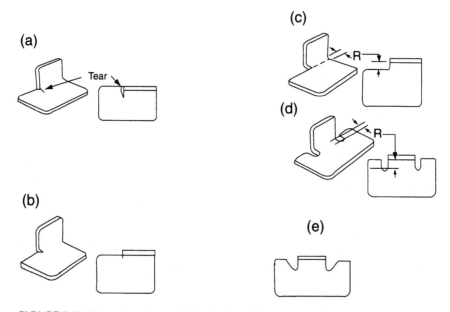

FIGURE 2.71 Examples of poor design for forming, and some alternatives. (Courtesy of Dayton Rogers Mfg. Company, Minneapolis, Minn. With permission.)

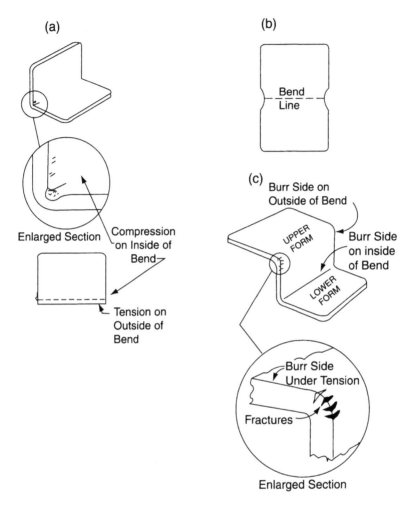

FIGURE 2.72 Examples of distortion and interference in forming. (Courtesy of Dayton Rogers Mfg. Company, Minneapolis, Minn. With permission.)

difficult to form, such as SAE 4130 steel, tumbling well before forming may not be adequate. It may be necessary to hand file or disk sand a radius on the sharp edges. Such secondary operations will add to the cost of production. Therefore, for the most economical production and if design will permit, ample inside bend radii should be permitted for heavy and difficult forming material when the burr side of the blank must be on the outside of the bend. If slight fractures are permissible, this should be indicated on the part drawing.

Springback

Springback is the partial return of the metal to its original shape when bending forces are removed. This is encountered in all bending operations. Springback cannot be eliminated entirely; it must be compensated for by overbending. Varying the equipment settings when using V dies or rolls easily determine the proper amount of overbending. When bending on bottoming dies, the springback must be estimated before the dies are built and compensated for in the design. Analytical methods are available for calculating springback; their accuracy depends on the specific mechanical and physical metal properties, which usually can be obtained only by experiment. Samples should be made to determine the springback of specific bends prior to tool design, and the tools should be built with sufficient stock to allow for rework if close angular tolerances are required.

As a general rule, the springback characteristics of different alloys and tempers will be proportional to their yield strengths, and thin sheet will spring back more than thick sheet in an inverse ratio to the cube of the thickness. Springback is greater for large-radius bends than for tight bends.

Low-Cost Forming

Forming or bending is an easy way to convert flat sheet metal stampings into parts that are useful in three dimensions. However, each forming operation is an added cost. The following suggestions will help to keep forming operations (and costs) to a minimum (see Figure 2.73). Important for this part are the locations of the hole centerlines. Two forming operations, one for each direction of bending, are required to produce this part. An alternative design, which preserves the relationships between the holes, requires only one forming step. The offset (or elongated Z bend) is easy to include in the forming die and is inexpensive.

The hole and boss in the center of the formed part in Figure 2.74 must be extruded. For the first part design, extruding is an additional operation, because the forming press must move in an opposite direction. The alternative design can be formed while at the same time the center hole and boss are extruded in one movement of the forming press.

In the assembly shown in Figure 2.75, two formed parts are nested together. Because R1 = R2, manufacturing variations in the two bend radii may prevent complete contact between the two parts. Either the need for selective assembly or high rejection rates can increase the cost of these parts. When R1 is smaller then R2, the parts do not touch in the bending zone. This design ensures that any two parts, selected at random for assembly, will fit together.

The part shown in Figure 2.76 is made by a process called "channel forming." Severe angles near 90° make die removal difficult, create high stresses in the metal, and require high bending force. The alternative part design, also a formed channel, has slightly open angles and therefore is easy and inexpensive to form.

Slots and holes near the bending zone of a formed part distort with the metal during bending. Often, distortion can be compensated for by stamping a deformed slot in the blank, as seen in Figure 2.77. In this example, the deformed slot (dashed lines) is transformed into the proper shape after forming. Usually, trial-and-error testing is required to determine the exact shape of the slot prior to bending. Twisting is neither a well-known nor a commonly used forming operation, even though it is

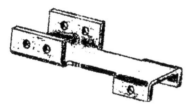 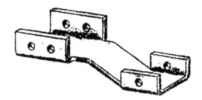

FIGURE 2.73 Redesigning to permit low-cost forming. (Courtesy of Dayton Rogers Mfg. Company, Minneapolis, Minn. With permission.)

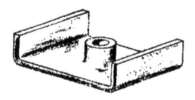 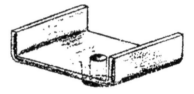

FIGURE 2.74 Redesigning to reduce the cost of the extruded hole. (Courtesy of Dayton Rogers Mfg. Company, Minneapolis, Minn. With permission.)

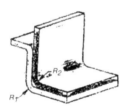 

FIGURE 2.75 Assembly problem caused by nesting when R1 (inside) is larger than R2 (outside). (Courtesy of Dayton Rogers Mfg. Company, Minneapolis, Minn. With permission.)

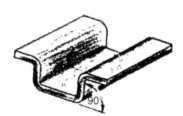 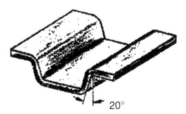

FIGURE 2.76 Angles near 90° make this channel difficult to form. (Courtesy of Dayton Rogers Mfg. Company, Minneapolis, Minn. With permission.)

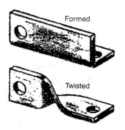

FIGURE 2.77 Example of a distorted slot near a bend, and an example of twisting. (From *Machine Design*, Penton Publications. With permission.)

a simple and inexpensive way to form parts. Twisting can be performed easily with special press tools, jigs, or even by hand. This part, for example, requires less material than an equivalent part bent into an L shape. Twist angle is limited only by the ductility of the material; in some cases, angles over 360° are possible.

Optimizing Bracket Shape

Although a bracket is a rather prosaic component, basically a formed sheet metal part designed to support or mount one component to another, a wide variety of designs are possible. In some cases, unconventional shapes can save material, reduce weight, or add strength. Here are a number of such improvements in bracket design (see Figure 2.78).

Dimples, beads, and other forms of compound curvature improve the strength of a bracket. A bracket shows the simplest application of this technique with two 45° bends at the corners. Ribs and beads provide more strength, however, and can be used as shown to stiffen brackets made from thin-gauge metal. Several of these designs reduce bracket weight by as much as 50% without sacrificing strength.

In many cases, ribs are welded or riveted to brackets for strength. All but two of the examples shown in Figure 2.78 can be made from a single piece of sheet metal and then welded or riveted easily in a second operation. The two- and three-piece assemblies require jigs and fixtures for assembly prior to welding.

"Cutting corners" saves material. For relatively low loading, a bent plate with diagonally cut edges can reduce material requirements by 50%. The example calling for compound bending also saves material but can support higher loads. Shear forming or lancing creates a tab for fastening.

2.3.3 Design for Drawn Parts

Drawing is a process that forms a flat piece of metal into a seamless hollow shape. In a drawing operation, a punch applies force to the center portion of a blank to pull or draw the metal into a die cavity. A blank holder applies a restraining force to the edges of the blank to eliminate wrinkling and to control the movement of metal into the die. Closely related to the process are redrawing, which reduces the cross-section

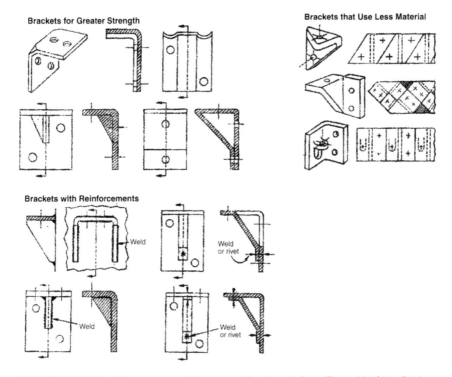

Brackets for Greater Strength

Brackets that Use Less Material

Brackets with Reinforcements

Weld

Weld
or rivet

Weld

Weld
or rivet

FIGURE 2.78 Some improvements in common bracket design. (From *Machine Design*, Penton Publications. With permission.)

dimensions and increases the depth of a drawn part, and reverse drawing, where a draw and a redraw operation are combined in one press stroke. Some drawing operations, such as the forming of shallow shapes, are referred to as pressing or stamping, but the distinction is not consistent.

A blank is the starting form for practically all drawing operations. The surface area of the blank should be only slightly greater than that of the finished product, allowing sufficient metal to hold the blank during drawing and final trimming. Excessive blank metal, aside from incurring unnecessary cost, requires increased drawing pressures that can result in fracture of the metal. See Figure 2.79 for a nomograph for calculating blank diameter.

The word "drawing" is generic and includes the operation in which metal is drawn over a die radius with no attempt to reduce the gauge (pure drawing), the process in which the edges of the blank are gripped tightly by the blank holder and die so the stock cannot slip (pure stretching), and all draw operations that combine drawing and stretching. The metal requirements vary depending on the type of operation. When a part is to be formed primarily by stretching, high elongation in the metal is required. The elongation value, established in a tensile test, is a measure of the relative formability of metals (see Figure 2.80).

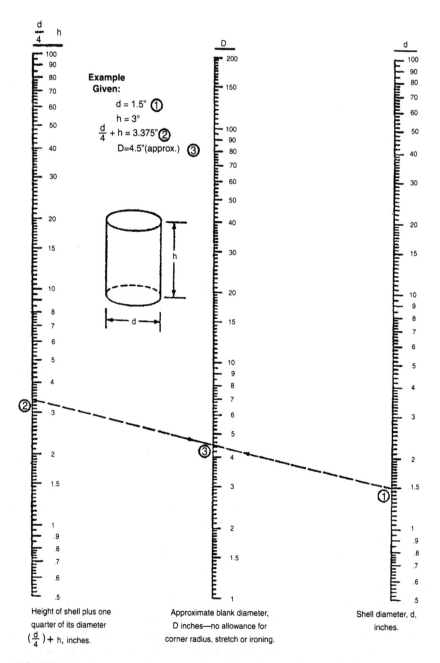

FIGURE 2.79 Nomograph for calculating blank diameter for a cylinder drawn from sheet. (From *Forming Alcoa Aluminum,* Aluminum Company of America, Pittsburgh, Pa.. 1981. With permission.)

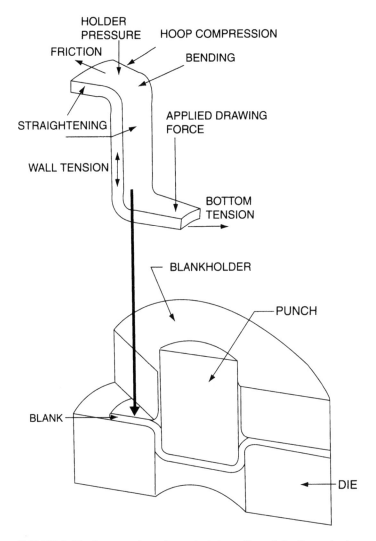

FIGURE 2.80 Cross section of a typical draw die and the forces in the drawn part. (From *Forming Alcoa Aluminum*, Aluminum Company of America, Pittsburgh, Pa., 1981. With permission.)

When the blank is allowed to move between the blank holder and the die as in the case of pure drawing, elongation is less important. Here, the prime factors are tensile strength and the rate of strain hardening. It is not uncommon in producing a round, flat-bottom, straight-sidewall part to take a greater draw reduction with high-strength, low-elongation metal than with annealed stock. The operations are not necessarily accomplished with the same set of tools, however, since higher-strength alloys normally require larger tool radii and greater clearance.

The force exerted by the punch varies with the percentage of reduction, the rate of strain hardening, and the depth of draw. Aluminum, like other metals, strain hardens during draw operations and develops a harder temper with a corresponding increase in tensile and yield strength. If the part is to be drawn successfully, the force exerted by the face of the punch must always be greater than the total loads imposed on that portion of the blank between the blank holder and the die. Also, the metal between the edges of the punch and die must be strong enough to transmit the maximum load without fracturing. This set of conditions establishes a relationship between the blank size and punch size for each alloy-temper combination that dictates the minimum punch size that can be employed in both drawing and redraw operations.

The edge condition of drawn parts can be seen in Figure 2.81, and the surfaces and characteristics of drawn shapes in Figure 2.82.

Edge of blank with earing

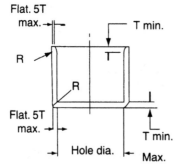

1. Cupped from the blank.

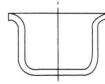

2. partial draw.

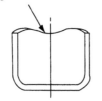

3. Recirawn, struck, punched hole & pinch trimmed.

4. Flat-edge trimmed after drawing.

FIGURE 2.81 Edge conditions of drawn parts. (Courtesy of Dayton Rogers Mfg. Company, Minneapolis, Minn. With permission.)

2.3.4   Design without Fasteners

Tabs

One of the easiest and most economical methods of making permanent and semi-permanent moderate-strength joints in sheet metal parts is by folding or bending of tabs. No screws, rivets, or other loose pieces are required, preparation is simple, and assembly is rapid. Limitations of fastening with tabs are few: the thickness of the deformed member should be between 0.012 and 0.080 in., the method is not recommended for high-strength joints, and the metals should have sufficient ductility so they do not spring back after being bent. Soft steel, aluminum, copper, and brass are the metals most commonly joined by this method. For best results, parts should be designed so that the bending is perpendicular to the direction of sheet rolling, or at least 45E from that direction. Also, the stamping burr should be at the inside of the bend when possible, and the final bend should be made in the same direction as the preliminary bend.

The working principle is relatively straightforward as shown in Figure 2.83. In Figure 2.83a, one or more tabs (also called legs, ears, or tongues) are formed in one of the workpieces, and matching slots slightly longer than the tab widths are punched in the mating part. Both features are made in the press operations that fabricate the parts. In Figure 2.83b, the tab is bent up at an angle that suits the assembly (usually 90°), inserted into the slot, and folded over to form the joint. Where the slot shape need not be a narrow rectangle, preferred shapes are triangular, round, or simple notches, as seen in Figure 2.83c. Punches for these shapes are stronger and last longer. The shape and size of tabs can be varied to suit assembly requirements and appearance, as shown in Figure 2.83d.

Joint strength is increased by using two or more tabs, folded in opposite directions, as seen in Figure 2.84a. Where space is limited, a single tab can be split so that it functions as two tabs, shown in Figure 2.84b. Figure 2.84c shows that joint strength is higher when the tabs are stressed in shear rather than in bending.

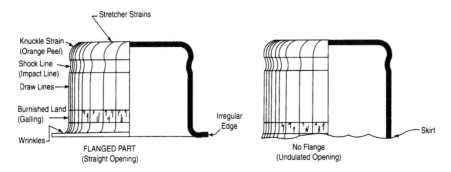

FIGURE 2.82 Surfaces and characteristics of drawn shapes. (Courtesy of Dayton Rogers Mfg. Company, Minneapolis, Minn. With permission.)

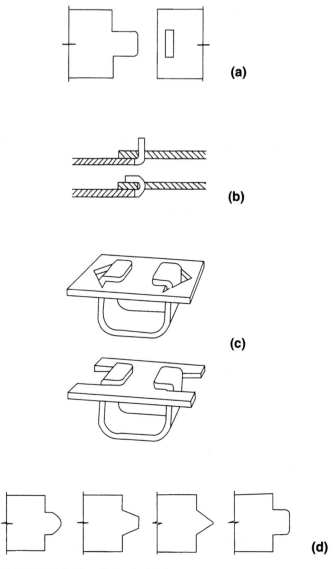

FIGURE 2.83 Use of tabs in joining parts. (From *Machine Design*, Penton Publications. With permission.)

Assembly using tabs need not involve two separate parts; folded tabs can join two ends of the same part, as shown in Figure 2.85. In some cases, tabs, in sandwich fashion, can join three or more parts. Figure 2.86 shows twisting replacing folding. For joints involving relatively thick metal, locking action can be improved by twisting the tabs instead of folding them. One drawback to the twisted tabs in Figure 2.86a is the increased height required; another is that the sharp edges can be a source of injury to personnel. The height of twisted tabs can be reduced using the designs in Figure 2.86b,

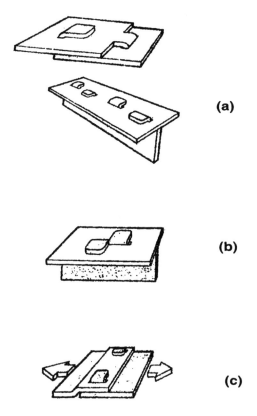

FIGURE 2.84 Joint strength is increased by using two or more tabs. (From *Machine Design*, Penton Publications. With permission.)

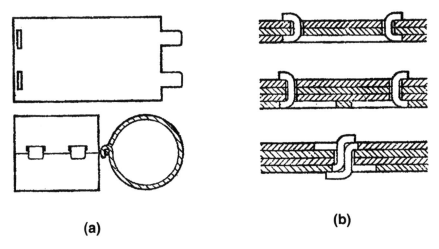

FIGURE 2.85 Joining of a single part or multiple parts with tabs. (From *Machine Design*, Penton Publications. With permission.)

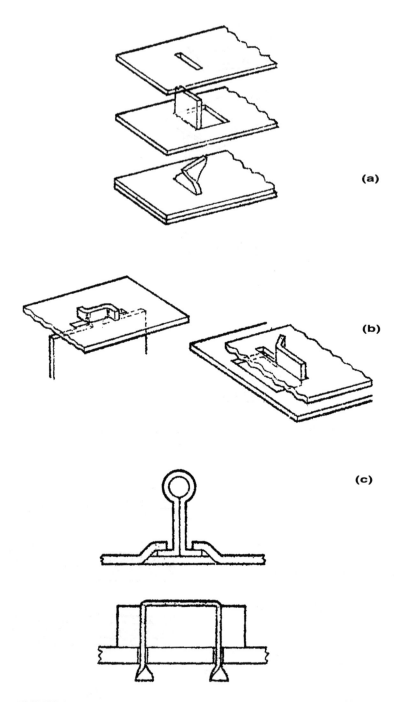

FIGURE 2.86 Use of assembly tabs by twisting rather than folding. (From *Machine Design*, Penton Publications. With permission.)

which are more efficient but also more expensive to stamp. Bulky or non-sheet-metal parts, as shown in Figure 2.86c, can be attached to a sheet metal base with bent tabs or with a metal strap whose ends are twisted to secure the parts together.

Staking

Staking is a simple way to assemble parts in which one part is plastically deformed or spread out around the other. Although staking is most frequently used to fasten lugs, bosses, and covers to sheet metal and plastic parts, the process can be applied to nearly all assemblies where one component is a ductile metal. Here is a review of the many variations of staking, along with several suggestions that can make the procedure more effective.

Figure 2.87a shows that parts can be staked together in one of two ways. To stake this lug into a flat sheet, for example, either the sheet is forced inward (left) or the lug is deformed outward (right). The tool shown here is a circular impact punch with a wedge around the periphery. When a mating part has a small shoulder, as in Figure 2.87b, a washer should be used to distribute the staking impact. In the case of delicate stampings, washers may be required on both sides of the assembly. Figure 2.87c shows that a hollow die can be used when the assembled component must protrude at both sides. For heavy-duty applications, a flat punch shears the metal to produce a larger deformation. Examples so far have shown stepped or shouldered components that require staking from one side only. It is possible, however, to stake inserts from both sides, as shown in Figure 2.87d.

Parts are not always staked all around their periphery. As shown in Figure 2.88a, parts can be "prick punched" in specified locations, especially for thin or weak materials and when low joint strengths can be tolerated. Staking points are selected according to strength or impact requirements, but usually they are symmetrical around the insert.

The staked assemblies illustrated so far will not resist rotational load on the insert. One way to prevent rotation under light loads is to notch the surface of the plate, and stake the insert into the notches (Figure 2.88b). For high-rotational loading, however, noncircular inserts and matching holes should be specified. Two stampings, such as a cover and a base plate, can also be staked together. In Figure 2.88c, for example, an ear on one stamping is assembled and staked at right angles to the plate. Figure 2.88d shows two variations for heavy-duty assembly by staking using rivets (left) and longitudinal spreading of a sheet metal tab (right). Long assemblies may require several staking points or ears for rigidity. In addition, washers (right) can be used to prevent damage while staking.

"Progressive" staking, shown in Figure 2.88e, eliminates the sudden blow of a punch or die and is often used for light load-bearing assemblies. A toothed wheel (left) pushed along an ear or tab and a roller or wheel forced against a rotating insert or rivet (right) are two progressive staking methods.

Lock Seams

Seams provide a simple way to join sheet metal parts such as panels, tubes, housings, containers, and ducts. Basically, a seam is a shape formed in the two mating

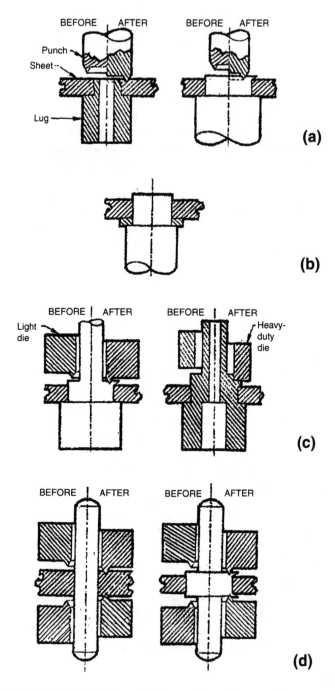

FIGURE 2.87 Recipe for well-done staking. (From *Machine Design*, Penton Publications. With permission.)

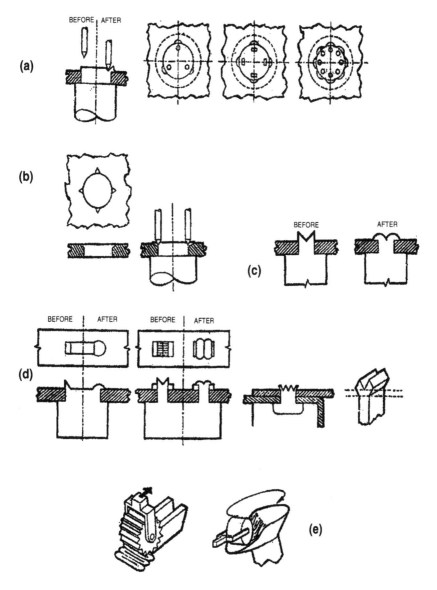

FIGURE 2.88 Additional staking methods. (From *Machine Design*, Penton Publications. With permission.)

edges of sheet metal parts, so that when the parts are assembled, they lock together by pressure or mechanical "flattening." Seams are economical and are often used instead of mechanical fasteners, especially when holes for fasteners would create stress concentrations, or when welding or brazing heat would deform thin metal. Seams are often stronger than other fastening alternatives and are as heat-resistant as the sheet metal. In addition, the curved, overlapping column of a metal seam can help stiffen the assembled part. (See Figure 2.89.)

Seams require a great amount of plastic flow in a metal. Generally, seams are limited to 0.011- to 0.050-in.-thick sheet; the formability of the alloy determines whether the sheet is too thick to be seamed without cracking. A rule of thumb for seams in carbon steel is to design the seam width between 3/16 and 1/2 in. for sheets up to 1/32 in. thick. Thicker sheets require wider seams.

There are four types of seams. Straight seams are used to make tubelike parts out of flat sheets or to join parts such as formed sheet metal panels. Closed-bottom seams can join a tubular body to a flat top or bottom (for example, sealing a top to

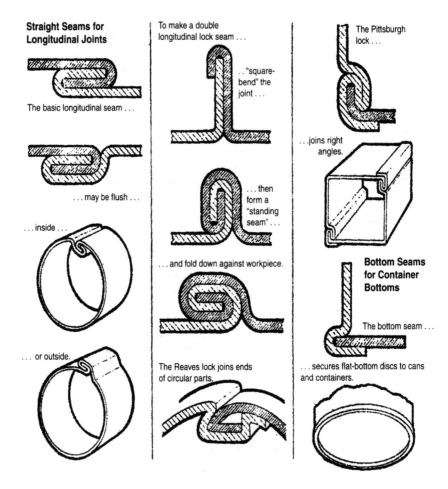

Straight Seams for Longitudinal Joints

The basic longitudinal seam . . .

. . . may be flush . . .

. . . inside . . .

. . . or outside.

To make a double longitudinal lock seam . . .

. . "square-bend" the joint . . .

. . . then form a "standing seam" . . .

. . . and fold down against workpiece.

The Reaves lock joins ends of circular parts.

The Pittsburgh lock . . .

. . . joins right angles.

Bottom Seams for Container Bottoms

The bottom seam . . .

. . . secures flat-bottom discs to cans and containers.

FIGURE 2.89 Additional staking methods. (From *Machine Design*, Penton Publications. With permission.)

a food can). Closed middle seams connect circular or rectangular tubing or similar parts end to end (as in connecting ductwork). A fourth type of seam is locked with a strap, clip, or band between the mating parts. Of course, this seam requires a third component for each joint. The illustrations here review most of the common variations of the four types of seams. (See Figure 2.90.)

2.3.5 Bibliography

Alting, L., *Production Processes,* Marcel Dekker, New York, 1982.
Bolz, R.A., *Production Processes,* Industrial Press, New York, 1963.
Forming Alcoa Aluminum, Aluminum Company of America, Pittsburgh, Pa., 1981.
Machine Design, Penton Publications.
Tanner, J. P., *Manufacturing Engineering,* Marcel Dekker, New York, 1991.

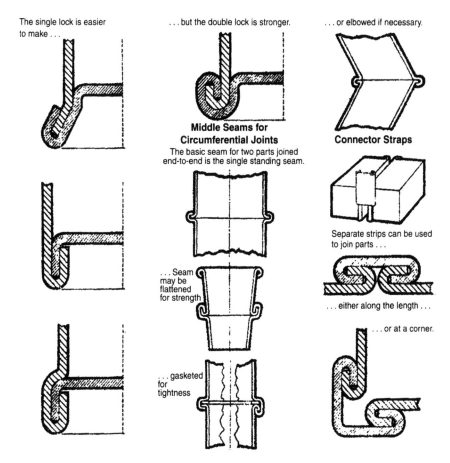

FIGURE 2.90 Variations of lock seams in sheet metal fabrication. (From *Machine Design,* Penton Publications. With permission.)

2.4 PROCESS SELECTION

In Chapters 1 and 2, a large amount of material was covered in a few pages. The selection of the optimum product design and production process is not a simple task; there is no magic formula to make the decisions automatic. However, there are some good guidelines and techniques that will assist the responsible manufacturing engineer and product development professional in separating out the very bad designs and processes and help us in selecting the better ones.

Worldwide competition is real, and is *now*. The market in general (and the customer in particular) is very demanding. It will take a team effort with a degree of cooperation that was unheard of a few years ago to design, produce, and market high-quality products at a low price.

One must understand:

- Who the customers are
- What they want
- What they expect the life of the product to be
- Whether they expect to be able to repair it, or have it repaired
- When they want it
- How much they are willing to pay for it
- What their actions will be if the product does not meet all their stated and unstated requirements

Some customers want a product at the lowest possible price that will just meet their minimum requirements for a short period of time, and then they plan to dispose of it. Other customers will want a good-quality product that will always perform its intended function, and that will have a reasonable life, for which they expect to pay a "fair" price.

The best possible product at the lowest possible price should be our goal in every case. Good design does not automatically translate to a high price. Someone once said, "Good, fast, and cheap—choose any two." Although the design effort is not without cost, it is usually only a very small percentage of the overall product price; however, it does have a major influence on product manufacturing cost and quality. Poor-quality workmanship in the shop is unforgivable, as is scrap and rework. A lot of our poor quality is due to a poor design; the product is not "producible." This is the kind of reasoning that leads to the belief that the functions of design, manufacturing, quality, marketing, purchasing, and others must work together during the design and development phases of any product to develop the best product configuration, production processes, and production sources. The best course of action often involves choosing to build those parts of the products that are most critical or most costly, and to purchase the others from those who manufacture them best. The core competency of the manufacturing group should be emphasized, and the things that can be managed well should be kept in the fabrication portfolio. Working in cooperation with qualified suppliers during the early product development phase before moving into production is a wise decision.

One should never attempt to put a product into production that has not had sufficient environmental testing, in addition to functional testing. Someone once said wisely, "One test is worth a thousand expert opinions." This applies to children's toys as surely as to aerospace products. One of the authors (Jack M. Walker) had the opportunity to work with the head of R&D of a major toy manufacturer as a consultant while we were in the development phase of a sophisticated weapon system for the U.S. military. It was surprising to learn that the toys were tested in about the same harsh environments as the military weapon, and that the toy "customer" was more demanding of the product. It was not enough that a toy function the day the child took it home; it was expected to lie out in the snow all winter and still work when the child picked it up the next spring.

The responsible manufacturing professional will insist that any manufacturing process continuously produce a good-quality product. Discussions of quality with a manufacturer of camera flashbulbs yielded the following valuable information that is applicable to all manufacturing. Their goal was never to produce a flashbulb that did not function every time, even after rough handling and under low-voltage conditions. The company once recalled 20 million flashbulbs and replaced them, at no charge to the dealers, because they thought that a few of them might be defective due to a variation discovered in the manufacturing process. Their reasoning was that if someone used their flashbulbs to take a once-in-a-lifetime photo (and aren't they all?) and the bulb did not fire, the customer would never again buy their flashbulbs, and probably none of their other name-brand products!

Manufacturers in the United States are capable of producing very high-quality products at very low cost. Many have become careless, or shortsighted, in past decades. Products have been built successfully and efforts have been made to sell those products. For a long time, this succeeded. Now the design process has to be reversed: first, the manufacturing engineer and product developer must learn what the customer wants; second, they must decide how the product can be made; and finally, they must figure out how the product can be designed so it can be built with high quality at low cost. (Don't forget the worldwide market—of both competitors and customers.)

In summary, Chapter 1 points out how to select the optimum product and process design. Subchapter 1.1 discusses the concurrent engineering process, which may be a new way of thinking for some (perhaps for company administration as well), in order to be able to combine our collective talents and experience and put them to better use. The "tribal knowledge" that all manufacturing companies possess is usually a manufacturer's most valuable asset. The important dos and don'ts as well as the how-tos are usually not written; they are scattered in the brains of people throughout the company. Concurrent engineering (by whatever name it is called) is a recognition of this, and an attempt to change from after-the-fact negative attitudes and statements like "I knew that wouldn't work" to positive attitudes and activities before the fact and during the design and development process. The use of this tribal knowledge in a positive, cooperative environment is a *must* for future success in our business. There are some excellent tools available, as well as talented people that can help our companies in this effort. An approach such as QFD (see Subchapter 1.2), where the customer's wants are reviewed and prioritized, and the technical requirements to achieve them defined in a matrix, is the best way to start any product development program. The product requirements are better defined, as a target, in this manner. How competition does in comparison with the approach taken to customer satisfaction is also revealed.

Once the big picture of product design is viewed, some concepts of actual hardware design that will meet the technical requirements can be studied. This is the goal to provide what the customer wants in a product. Processes such as DFA or DFMA may be necessary to be successful (see Subchapter 1.3). The ultimate product design should not be the result of "the loudest voice," which may be very persuasive in supporting his or her favorite design concept. The processes that are inherent in each design concept must be analyzed and the cost elements reviewed in a formal manner. This takes considerable skill, since seemingly minor or unimportant details may become significant cost drivers. However, until they are brought out during the review and analysis process, we really do not know (quantitatively) what the impact of each element of the design will be on the product production. The quality of the final product is also largely inherent in the detail design. The demonstrated manufacturing tolerances must be smaller than the product demands.

The importance of trying to do things correctly the first time cannot be overstated. Carpenters repeat the old adage "Measure twice, cut once." This is true in manufacturing products as well. If there is any question about some part of the product design, consider models, or preferably a prototype part that is truly representative of the final product. By utilizing some of today's rapid prototyping techniques (see Subchapter 1.4), and the companies that specialize in short-lead-time deliveries, we can have a lot more confidence in our proposed design. We should insist on sufficient functional and environmental testing, to prove that the design works under the anticipated conditions. While this sometimes may seem like a waste of time and money in getting a product to market, it is usually not. It may save the commitment of funds for the wrong production tooling, which requires expensive redesign and retooling after start of production, the worst of all situations!

This chapter has introduced some of the more common manufacturing processes. There are advantages and disadvantages to all of them. Again, a formal estimating technique is the only approach that is quantitative rather than intuitive. Checklists re valuable triggers, since no one can remember everything all the time. Think of the space shuttle: even though it is fully automated and essentially controlled by computers, the pilots go through extensive checklists to prevent oversight and error. The remaining chapters in this handbook will also be of assistance in influencing the process selection, and in describing the influence of the process on the ultimate product design. They include plastics, composites, nontraditional processing, assembly, factory requirements, factory design, and the human factors of the people involved. Things such as ergonomics, anthropometry, safety, industrial hygiene, and training are important to consider today and tomorrow.

3 The Deductive Machine Troubleshooting Process

Richard D. Crowson

3.0 SYSTEMATIC DESIGN ENGINEERING

Some areas of mechanical engineering that have the potential to be dangerous or fatal if they fail are pressure vessel design, aeronautical design, aerospace design, and marine design. In these specific high-risk design areas, regulatory agencies require very strict procedures for the design tasks, but troubleshooting is not always emphasized equally.

The practice of medicine also has the potential to be dangerous or fatal if errors occur. Both the engineer and physician must rely on deductive reasoning and systematic approaches to problem solving.

The engineering process of problem solving is like the diagnostic process used by a physician in that both use existing scientific knowledge and do not seek to discover basic knowledge.

The physician and engineer employ empirical information and professional judgment in solving immediate problems. Additionally, the problems in medicine and mechanical engineering may not have just one correct answer. The answers to the problem may have an answer that is better than other answers (Juvinall and Marshek, 2000).

3.0.1 Ground Rules for Use of Deductive Reasoning in Data Analysis

When employing deductive reasoning, one must proceed from general to specific statements utilizing the specific rules of logic. The system for organization of known facts to reach a conclusion is called syllogism.

A syllogism contains: (1) the major premise, (2) the minor premise, and (3) the conclusion. An example of a syllogism is as follows:

1. All human beings have a finite life span (major premise).
2. John Smith is a human being (minor premise).

3. Therefore, John Smith has a finite life span (conclusion) (Ary, Jacobs, and Razavieh, 1979).

The first ground rule of deductive reasoning is to insure that all premises of the syllogism are true. If the premises are true, the conclusion is also true.

A second ground rule of deductive reasoning is to recognize the limitation of deductive reasoning. The premise must be true in order to arrive at a true conclusion. When attempting to establish true premises, scientific inquiry cannot depend on deductive reasoning alone. It is necessary to establish the truth of statements dealing with scientific phenomena through experimentation.

Deductive reasoning can organize information that is already known and can illustrate new relations of that information. However, it is not a source of new truth.

A third ground rule of deductive reasoning is that the researcher must have a clear understanding of the theory guiding the investigation. Deductive reasoning is helpful in troubleshooting problems encountered in debugging machines. It links theory and observation, allowing the troubleshooter to deduce from existing theory what phenomena should be observed.

When troubleshooting mechanical designs, the theory is influenced by design goals originated by the design engineer.

A troubleshooter must first understand the desired operational specification and design goals of the machine in order to solve problems related to the initial start up of that machine. If this specification and goal is not achieved, the troubleshooter then uses observation and deductive reasoning to provide a hypotheses to be tested. This hypothesis ideally provides reasons for the fact that the design goals cannot be achieved. It may also assist the technicians in making changes to the machine to bring the machine in line with the design engineer's goals.

3.0.2   Inductive Reasoning

Francis Bacon (1561–1626) introduced an approach of reasoning that encouraged the investigator to establish general conclusions from facts derived by direct observation. In this system of reasoning, observations were made on particular events in a class, and from these observations inferences were made about the whole class.

An ancient anecdotal statement attributed to Bacon illustrates the importance of observation. That statement is as follows:

In the year of our Lord 1432, there arose a grievous quarrel among the brethren over the number of teeth in the mouth of a horse. For 13 days, the disputation raged without ceasing. All the ancient books and chronicles were fetched out, and wonderful and ponderous erudition, such as was never before heard of in this region, was made manifest. At the beginning of the 14th day, a youthful friar of goodly bearing asked his learned superiors for permission to add a word, and straightway, to the wonderment of the disputants, whose deep wisdom he sore vexed, he beseeched them to unbend in a manner coarse and un-heard-of, and to look in the open mouth of a horse and find an answer to their questionings. At this, their dignity being grievously hurt, they waxed exceedingly wroth; and, joining in a mighty uproar, they flew upon him and smote him hip and thigh, and cast him out forthwith. For, said they, surely Satan hath tempted this bold neophyte to declare unholy days more of grievous strife the dove of peace sat on the assembly, and they as one man, declaring the problem to be an everlasting mystery because of a grievous dearth of historical and theological evidence thereof, so ordered the same writ down. (Ary, Jacobs, and Razavieh, 1979, p. 6)

This story explains that the young man was attempting to apply a new way of seeking truth. He wanted to seek facts instead of depending upon authority or speculation alone. This is the fundamental principal of modern scientific investigation. The scientific method follows prescribed steps.

3.0.3   Scientific Method Steps

Inductive reasoning alone is an inadequate method of solving problems. The isolated knowledge and information that is found by induction does not advance scientific knowledge.

Induction and deduction best serve the investigator when they are integrated into one single technique called the scientific method. This method involves moving inductively from observations to hypothesis and then deductively from hypotheses to the logical implications of the hypothesis.

One way the scientific method and inductive reasoning differ is in the use of a hypothesis. The scientific method requires the formation of a hypothesis by the investigator. The steps of inductive reasoning include first making an observation, then organizing the information discovered by inquiry. The scientific approach to problem solving differs in that the investigator logically predicts the result of a series of events and sets out to prove or disprove the prediction by systematic investigation.

The steps of the scientific method vary from author to author, but the systematic process of inquiry is the important part of the method.

The scientific method can be defined as consisting of the following five steps. In step one, definition of the problem, the investigator begins with a problem or question. The question is formatted in such a way that observation or experimentation can provide the solution. Step two consists of formulating a hypothesis, which is thought by the investigator to be the answer to the question. Step three is defined as the deductive reasoning step. This is the point in the process where the implications of the hypothesis are taken to a theoretical conclusion and plans are developed which will guide the observations that will prove or disprove the hypothesis. Step four is the step to collect and analyze data. Observation, testing, and experimentation are used to analyze the data collected. The fifth and final step is confirmation and or rejection of the hypothesis.

After collecting the data, the results are analyzed, and the evidence presented either supports or disproves the hypothesis.

3.0.4   Balancing Science and Profit

When considering a problem to be solved at the debug phase of an initial start-up of a new machine, economic factors must be considered. The machine being debugged represents a financial investment to the principals involved in manufacturing it.

If the machine is a prototype or first article that is to become a product, then the investment in development of that machine may be considerable.

Time is represented in terms of financial opportunity; therefore, pressure to complete the debug process on or before schedule is exerted by administrative personnel. This pressure may sway technicians and result in a compromise of design goals in order to meet a delivery schedule. The pressure of finances and schedule will complicate the problem-solving equation by introducing human emotion and stress. These factors will be counterproductive in terms of keeping open communication and in following logical deductive steps to a solution.

Machines are an essential part of our lives. They affect our health, well being, and pleasure. In the third century B.C.E., simple machines consisted of levers, wedges, wheels, and screws. Archimedes claimed he could move the earth if given a place to stand and a lever long enough.

The design of simple machines has been refined for centuries. As technology develops, methods of efficiently manufacturing machines that are more complex must be developed.

Systems engineering techniques are used to design machines for manufacture. Machine design is a well-documented, mature process. It occurs at the time of the initial start-up of a machine and precedes release of the machine for repetitive manufacturing. Most aspects of machine debug have not been addressed in the texts reviewed for this research.

Financial pressures exerted by administrative personnel and customers to ship a product early cause it to be shipped before it has reached its fullest design potential.

Pressure of this kind is exerted because engineers want to perfect a machine beyond the cost-effective practical limit. The desire to ship too quickly and a lack of systematic methods of debug may motivate technicians to shorten the debug phase of machine manufacture. However, shipping a machine prematurely does not make long-term financial sense. Focusing on short-term gains from fast assembly of a machine before proper debug techniques are applied may cause the engineering design hours devoted to innovation to be lost. Additionally, resulting field repairs and warranty costs due to inadequately debugged machines cause a negative effect on profits.

Software and computers are now included in the definition of machine, because the term "machine" includes mechanically, electrically, or electronically operated devices that perform desired tasks (*Merriam-Webster's Collegiate Dictionary,* 10th ed.).

Software design and debug has recently moved into the forefront of engineering cost budgeting and analysis. Software design and debug has caused many projects to overrun budget and schedule estimates. This has attracted attention from both the government and industry. The U.S. Defense Department estimated that $36 billion was spent on mission-critical software in 1995. One software researcher stated that this estimate might be off by 50 to 150%. Studies are being conducted at universities, government institutions, and in industry to determine the best methods to debug software.

The software debug process is greatly impacted by the skill level, knowledge, and training of the individual programmer. Accurate software debug time estimates are almost impossible to do with current methods. A study indicated that any given software engineer may be 28 times more productive than any other software developer of equal experience and training. The time required to debug software has also been found to vary as much as 28:1 from person to person. This variability generated interest in finding standard methods to design, encode, and debug software. Studies have been conducted to find ways to equalize the time required for these tasks.

The U.S. government issued a challenge to improve software debug techniques. A joint industry, university, and government group met this challenge. Their findings are documented at http://www.bmpcoe.org./guideline/books/bestprac/002fc7fd.html. Since that challenge was undertaken, software debug methods have been vastly improved. Government contractors now claim a two-thirds reduction in debug time as a direct result of site design for test strategies for Surface-Mounted Technology modules.

The need to understand and systemize the software debug process is valid. However, this chapter asserts that a similar problem in the debug of machines exists in the manufacturing arena today, due in part to increased complexity in technology. It is the contention of this research that efforts should be made to discover new machine debug processes that equal those formalized for software. A disproportionate amount of time and money seems to have been spent on software debug as compared to the machine debug processes. When comparing the potential savings from improved machine debug methods to improved software debug methods, improvements in machined debug may result in an equal or greater cost savings to manufactures and end users of machines.

The U.S. Census Bureau's *Statistical Abstract of the United States: 2000* for the period from 1990 to 1998 reported that $118.2 billion in gross domestic product was spent for industrial machinery alone. Electronic, electrical, motor-vehicle, instrument, and miscellaneous manufacturing industries were reported to have produced $418.6 billion. The manufacturing revenue for these machines is $571.9 billion. This amount represents 6.5% of the gross domestic product for the reporting period. The $36 billion reported for mission-critical software contrasts dramatically to the machine manufacturing total. Applying a factor of 150% for estimation error, $36 billion increases to $54 billion. This would be equivalent to only 0.6% of the gross domestic product and only 9.4% of the mechanical manufacturing product. No amount of money designated for machine debug research has been reported in any of the resources selected for this research. This data provides strong evidence that more emphasis is being placed on software debug research as compared to machine debug research, and that more money is spent on machine manufacture than on software development. It is the contention of this research that if manufacturing debug techniques have not kept pace with increases in technology, then the machines being manufactured today do not achieve design goals efficiently. Therefore, manufacturers of machines will improve profits and productivity if machine debug processes are improved. A company will improve its profits if the machine debug process effort is patterned after the design process effort. Therefore, more effort and more money should be put into making the machine debug process more effective.

3.0.5   Sources of Data Used in Deductive Reasoning

Data is accumulated from several sources in order to balance the perspective of this study. For example, debug facts that are taken from case studies reveal interaction between engineers and technicians. This interaction is an important factor in the attempt to understand the dynamics of communication during the debug phase of machine development.

Data derived from literature includes information taken from many sources, such as engineering texts, medical texts, crime-scene investigation texts, legal reasoning texts, and other sources that discuss problem solving and logical thinking. Much of the data compiled for this reason is empirically derived and should represent practical data.

The information and data taken from interviews should be used to highlight the troubleshooting mindset. When interviewing those involved in the use of or assembly of a machine, the data that is learned is similar to processes common to engineers, physicians, and other professionals.

Personal experience gained and shared by the investigator is fundamental to the direction the research takes.

Although this chapter proposes an inductive study be used in the troubleshooting process, the many years of the investigator's practical experience should influence the process of developing a hypothesis for investigation. Textbooks devoted to research used for education were also consulted, as were diagnostic texts covering medical, clinical, and psychological diagnosis techniques.

3.1 MACHINE DEBUG PROCESS

"Professional engineering is concerned with obtaining solutions to practical problems" (Juvinall and Marshek, 2000). Many product development processes have been defined in terms of phases. Koller's method breaks down the design process into a 20 steps. These steps are organized into three phases. The first is the product planning phase; next is the product development or design phase. Finally, he defines the finished product phase as the last phase, out of which the product is delivered. According to Koller, the transition from the development phase to the finished product phase is accomplished by moving from the production documents to the finished product.

An additional phase must be added to a development process in order to be successful in today's highly technical manufacturing environment. This additional phase to be added is the machine debug phase. This phase may also be thought of as a process.

Successfully debugging a machine involves technical problem solving. *Debug* is defined in this study as a methodical procedure of defining, understanding, and eliminating problems encountered when initially starting up a machine. The problems encountered during the debug phase have many components. They are the application of deductive problem-solving techniques and the succinct communication of design goals by engineers to technicians.

The cost of manufacturing has increased because new technology has made product development more complex than in previous decades. For example, picosecond and femtosecond lasers placed upon nanometer linear motors are used in micromachining applications such as photoresist mask repair. A machine such as this is used in semiconductor manufacture and costs as much as $2.2 million to purchase. One reason machine debug has become more problematic is the increased technical complexity in machines being manufactured today. Current global economic pressures have forced manufacturers to research ways to develop products more efficiently to help offset increased costs. These economic pressures are related to new technology development and high export tariffs.

The problem is that present mechanical debug techniques taught in universities and practiced in industry today are not efficient enough for the increased complexity of technologically advanced machines. Current machine debug relies on intuition and does not encourage team problem solving and brainstorming. Debug practices need to be based upon formal deductive reasoning and scientific investigative processes. The techniques described in this chapter aim to teach the engineer to understand problems of debug faced by the technician and likewise to teach the technician the problems facing the engineer in the debug phase. Another aim of the solution to this problem is to empower both the engineer and technician with debug resolution techniques and checklists that are designed to facilitate the debug process. The subproblems are: (1) the need to improve communication of design goals by the design engineer to the machine builder, and (2) the need of improved intercultural communication techniques to be used by both engineer and technician. These intercultural techniques encompass both ethnic and corporate cultures.

A machine design process that does not include adequate debug plans wastes time and money. Many professions other than engineering employ scientifically based problem-solving methods. Checklists and systems derived from a broad spectrum of problem diagnosis and resolution techniques are needed for successful machine debug. This research proposes to identify and evaluate problem-solving methods for use during the initial start-up of a new machine.

Most engineering schools do not teach intercultural communication in their curriculum. Some schools do not adequately prepare their graduates in project management skills. This research proposes to incorporate both into debug methodology.

3.2   ERROR BUDGETING

The fundamentals of economic analysis, project management, the process of design, and two design case studies are all at the heart of error budgeting. A foundation of theory for precision must be laid and understood in any error budget. This is done by defining the principles of accuracy, repeatability, and resolution. Dr. Alexander Slocum, a professor of mechanical engineering at MIT, introduces the concept of error budgeting in his book *Precision Machine Design* (1992). This book is rich with practical examples. His error budget is a spreadsheet-based system for defining error-contributing elements of any machine. His book illustrates design methods used to integrate precision components into machines. Many chapters discuss sensors, error mapping, system design consideration, bearings, power generation, and transmission. The book contains a floppy computer disk with the spreadsheets defined in the book.

This is an extremely helpful book. Dr. Slocum teaches a two-day seminar in which the text and the spreadsheets are explained. This text is a milestone in engineering technological educational achievement.

Dr. Slocum teaches senior-level or graduate engineers from this text. The book is also written for use by practicing professionals. The Precision Engineering Research Group, a member of the Laboratory for Manufacturing and Productivity at MIT, was established in 1977 to conduct research in manufacturing with the purpose of development of fundamentals in manufacturing science. This group also uses this book.

Dr. Slocum mixes design analysis and creative thought with practical examples. The book contains methods of design written for the student or professional. Those are applicable as design examples widely found in industry. No other book reviewed in this research combines theory, innovative technique, and design methodology as well as this book does.

The strength of this book lies in the orderly writing style used in the presentation of facts and fundamentals before outlines of principles. The purpose and application of principles are easily understood. Spreadsheets provide mathematical modeling tools necessary to achieve design goals. Several corporate contributors were involved in the development of this book. Their product-specific examples make it practical. Dr. Slocum's method is the best example of a design checklist reviewed in this research. The error budgeting approach is useful in designing, troubleshooting, or debugging a machine.

Advances in technology influence the development of greater complexity of consumer products. For example, Quantum Well infrared sensors coupled with a fiber-optic delivery system and nanometer positioning stages may provide advanced medical imaging devices that operate from within the body. These advanced products will require high levels of technical skill to reliably manufacture and service. The product development process as defined by K. Roth (1968) follows these steps: (1) task formulation phase, (2) functional phase, (3) form design phase, and (4) result. For Roth, and most product development theorists, the result is production drawings; then the product appears. This research proposes that an additional phase must be considered to exist between the production of the drawings and the successful completion of a product.

This additional phase is the debug phase. After a machine is designed and the first prototype is built, a team of technicians will start the machine for the first time. When the initial start-up of a machine is performed, a specialized task must occur to achieve the desired result from that machine. This specialized task is called the debug process.

Debug in this chapter is defined as the methodical procedure of understanding and eliminating problems encountered when initially starting up a machine. Oftentimes, design limitations are discovered that require immediate attention from engineers and technicians to perform the machine debug task. If the debug process is not performed efficiently, the machine may not perform to the designer's expectations. An improperly performed debug process may also result in higher product cost and may cause the product to reach the market later than expected. This may negatively affect the competitiveness of the product by affording the competition an advantage of being first to market with an innovation.

This main purpose of this chapter is to suggest efficient deductive and systems approaches to machine debugging. A second aim of this chapter is to discuss models of efficient communication between engineers and machine-building technicians during the machine debug process. Both techniques may result in higher productivity and a decrease in the time required for production ramp-up. These recommendations are intended to promote lower product development costs through improved efficiency in machine debugging.

These recommended steps to improve technical problem solving, if followed, may assist the design engineer and the machine builder in the successful execution of their tasks. Specific guidelines for improving communication techniques are also recommended.

The deductive reasoning approach to problem solving discussed in this chapter has two parts. Part 1 is the discussion of methods for the design engineer to consider when guiding machine building technicians and assemblers in the implementation of design goals through deductive debug techniques during the debug phase of product development. Part 2 is the discussion of ways to succinctly communicate the design engineer's goals for machine debug to the machine builder. The solution to both parts of this problem incorporates an understanding of deductive reasoning, systematic engineering, human interaction, psychology, and corporate cultural influences.

Communication models are presented that will help the reader visualize the essential elements of communication. The concept of cognitive models of thought was explained to be the way humans think when communicating goals or ideas. David I. Cleland presents ideas related to cultural considerations in project management. In his text, he gives examples of corporate culture and the culture of successful product development teams.

Paul G. Hebert, in a text related to ethnic intercultural communication, states that words have implicit as well as explicit meanings. His definitions discuss denotative and connotative meanings of words.

Sherwood G. Lingenfelter and Marvin K. Mayers explain the difference between time-oriented and event-oriented people. They developed a grid that contrasts American, Latin American, Korean, and Japense cultural priorities assigned to time and events. An understanding of differences in culture-based time orientation will help the engineer and technician from different cultures more efficiently achieve common goals.

David J. Hesselgrave defines concrete relational thinking, the importance of learning the language of the receptor culture, linguistic mirrors of culture, and the traditional approach to communication behavior. These cognitive processes define a way of thinking that is culturally specific. Understanding these points helps the engineer and technician to bridge corporate and ethnic gaps that impede technology transfer.

Studies produced by Ilya Grinbery define the basic principles for system engineering when used to develop advanced technology for any engineering system. These principles are: (1) functional analysis, (2) requirements or criteria, (3) synthesis, and (4) testing.

G. Seliger describes engineering synthesis and natural science analysis techniques. Included in this paper were discussions of: (1) inductive and deductive innovative paths of thought, (2) trial-and-error methods for solving inventive problems in engineering, (3) TRIZ—the theory of inventive problem solving, (4) general design and development methods, (5) axiomatic design, (6) design for specific goals, called "design for ease of" (DFX), and (7) utilizations of design methods and other modeling and design methods. Principles of problem solving for product innovation were combined with principles of system engineering to develop systematic and process approaches to troubleshooting. These principles were developed as a result of this study and are presented in the appendix.

John A. Stratton, G. Seliger, K. Ueda, A. Markus, L. Monostori, H.J.J. Kals, and T. Aria define emergent synthesis methodologies for manufacturing. The text covered the problem with synthesis, problem classification, concepts of emergence, and emergent synthetic approaches to problem solving. Problems were classified as (1) complete problems, (2) incomplete environment problems, and (3) incomplete specification problems, with definitions of each type of problem classification. This text was combined with other material to build the framework of a discussion presented in the appendix, outlining methods designed to improve technical communication.

3.3 ENGINEERING CASE STUDIES USED IN THE DEBUG PROCESS

Information gathered from case studies of new product debug processes is used to develop baseline time values to compare various methods for debug. Secondary causes of ineffective debug, such as poor communication, are revealed in the case histories. Conflicts that affect troubleshooting are detailed and discussed as well. Environmental factors affecting debug are discussed.

The troubleshooting problem is viewed from multiple viewpoints in the case histories.

3.3.1 Abbreviated Case Study of High-Frequency Ultrasonic Medical Scanner

An anecdotal example of machine debug follows for illustrative purposes to describe the mechanical debug process. This is provided at this point in the research in order to help the reader understand the need for systematic debug methods.

Setting

An elderly medical doctor and researcher in England wanted to develop a high-frequency handheld ultrasonic scanning device that could be applied to any part of the body.

Early experiments showed that the device could be used to treat burn patients and could be used in geriatric pressure wound treatment. It was hypothesized that physicians could judge the progressive growth of new skin without removing gauze from the burns by using the ultrasonic scanner. This would minimize an extremely painful process that burn patients must endure.

The scanner was also believed to allow physicians to diagnose and treat pressure wounds, called bedsores, which occur in bedridden elderly patients. This treatment would occur before the wounds surfaced on the top layer of skin. A pressure wound typically lies beneath the skin and does not receive proper attention until it is visually observed, which is generally too late in the treatment cycle to relieve the symptoms.

Personalities

The physician who originated the idea was a renowned anatomist and one of the editors of the publication *Gray's Anatomy*. This elderly female physician had been married to another physician that had recently died. His dream had been to produce this device. The surviving spouse undertook the project as a memorial to the deceased husband.

The original engineer and developer of the device was a British engineer who was experienced in software and embedded processor design. This engineer was skilled in electronic design and considered himself a mechanical engineer as well. His formal training was in electrical engineering, but his experience in mechanical devices was obtained as a garage hobbyist. He was very confident in his mechanical abilities and knowledge. This engineer developed a proof-of-concept device.

A financial partner in the United States was engaged to secure Federal Drug Administration (FDA) certification and to market the device in the United States. This partner had developed companies in the past and was good at generating interest in new development projects. He was also good at convincing investors to put capital up for new projects. This partner secured the aid of an electrical engineer to build nine prototypes patterned from the original development device.

The electrical engineer from the United States oversaw the fabrication of the nine prototypes. This electrical engineer was given large quantities of stock in the newly formed company as payment for the manufacture of the prototypes and for production device manufacture. This electrical engineer ran a small company that manufactured an electronic device used as an alcohol breath analyzer. Because of his success with electronic devices, he felt confident that the development of this new scanner would be no problem even though it was a combined mechanical and electrical product. He was also contracted to do product development and manufacture of the device in the United States.

Sequence of Events

The first nine prototypes were fabricated as copies of the original device with minor changes to improve some functionality of the unit. The fabrication of these devices was very difficult. The devices were eventually made to work, sporadically. An FDA certification application was filed. Some of the devices were shipped to physicians for evaluation. A product development mechanical engineer was hired to correct the problems with the device 18 months after the first nine prototypes were fabricated and to release tooling in order to product large quantities of the device.

Problems

The problems were classified into the following groups: (1) mechanical, (2) manufacturing, (3) process, (4) reliability, (5) thermal, (6) precision, (7) structural, (8) ethnic-cultural, (9) corporate-cultural, (10) financial, (11) regulatory, and others. Some of the major problems are as follows. The nine units were completed, but they required thousands of hours to manufacture. When they were completed, they did not function properly or consistently.

The units leaked water, they jammed mechanically, and the various internal components were sensitive to the internal heat generated by the stepper motor driving a screw mechanism that provided the translation action of the transducer.

The cost of the devices was more than ten times the expected cost because the electrical engineers had the housings machined out of blocks of virgin acrylonitrile-butadiene-styrene (ABS) plastic in lieu of molding them from glass or talc-filled ABS. The nonfilled ABS was not strong enough to resist flexure induced by holding the scanner.

Additionally, the internal mechanisms were very sensitive to the heat generated from the stepper motor installed to drive the transducer. The heat generated from the stepper motor was trapped inside the housing because the unit had to be sealed. The sealing was required to allow the units to be washed and sterilized. The aluminum parts expanded at a rate of three times that of the stainless steel components. The ABS housing grew at a rate of six times that of the stainless steel parts.

The initial design of this mechanism anchored the steel parts to the plastic housing. When the plastic expanded due to heat or external heat, the mechanisms became misaligned and bound.

The transducer was designed to reciprocate 15 mm by a linear screw driven by a screw attached to a stepper motor. The driving mechanism often jammed and would not free itself. This required the unit to be disassembled to free the screw from the nut. The first unit and the copied nine units did not actually translate the required 15 mm due to dimensional errors in the housing and tolerance stack-up errors.

The transducers were manufactured from a silver-filled epoxy and gold and polymer laminated foil. The manufacture of the transducers was not predictable. The yield of these parts was less than 1%—this is to say that out of 100 parts manufactured, only 1 worked within specification.

Because the device had a cell filled with water to facilitate the ultrasonic signal, water would occasionally leak from the water-filled cell and damage the electronic parts.

Conflicts

The conflicts in this project included the pressure to manufacture the product induced by the electrical engineer and financial partner. The electrical engineer applied pressure to the mechanical engineer to complete the task quickly because he had potential financial gains to be realized. This pressure was problematic, because no systematic debug process was followed with the initial unit or with the nine prototype units. Repetitive manufacture of the prototypes was not possible until the design problems were corrected.

The conflict also occurred because the electrical engineers did not understand the theories of the mechanical devices that were not yielding reliable results. The lack of understanding of repetitive manufacturing issues was not understood and also caused conflicts. The corporate culture of a small electronic-device manufacturing group attempting to manufacture a device that required micron precision required an entire work cell modification and mechanical assembly training that did not exist in the core technical training of the assemblers.

The ethnic cultural conflicts existed in the fact that the British engineer did not understand mechanical issues or U.S. manufacturing processes, and did not accept the idea of systematic debug or product development processes.

The electrical engineer agreed that the mechanism needed redesign, but wanted the redesign to be done in a way that would not be obvious externally so as not to alert the FDA of any changes.

McGraw-Hill published *Product Design and Development* in 1995. The authors, Karl T. Ulrich and Steven D. Eppinger, are associate professors at the Massachusetts Institute of Technology (MIT), Sloan School of Management.

3.4 DEBUGGING MACHINES THROUGH BURN-IN ANALYSIS

Burn-In: An Engineering Approach to the Design and Analysis of Burn-In Procedures by Finn Jensen and Niels Erik Petersen (1982) is a classic book describing engineering science related to the burn-in process.

The book begins with a description of the burn-in problem in Chapter 1. Then, in Chapter 2, the mathematical approach and design philosophy of the entire book is discussed. Chapters 3 through 6 cover various aspects of system failure patterns and procedures. Actual research data, charts, graphs, and case study references accompany each method discussed. Chapter 7 explains accelerated burn-in testing procedures. Chapter 8 deals with the economy of burn-in. Moreover, the final chapter, Chapter 9, explains steps used to plan and control the production burn-in of systems. The book also contains four helpful appendixes that have tables useful for statistical analysis of burn-in. The bibliography contains a wealth of references for additional research on the subject.

The book is the result of a research project initiated at the Academy of Denmark and sponsored jointly by the Danish Council of Technology and a number of Danish manufacturing companies. It was written for the professional reliability engineer to assist with techniques of burn-in. It also describes benefits gained from following the burn-in process.

The reader is given detailed descriptions of a systematic method of burn-in and reasons for burn-in awareness. The book is successful in describing the burn-in practice. The book contains a wealth of data for use in reliability engineering and establishing burn-in procedures. Much of the data is presented in graphical form. Some of the formulas presented have direct application to any component, subassembly, or complete system. This is its strength.

A weakness is the lack of discussion of machine debug. The authors of *Burn-In* presume that mechanical and electronic systems will be available for lengthy testing immediately upon first assembly. Application of burn-in procedures described in this book is possible after debug. The book describes a key part of successful manufacturing process. It is therefore a very valuable tool for design and reliability engineers. The information it contains can be used for sustaining engineering and continued development of new products.

3.5 REFERENCES

Al-Assaf, A. F. and Schmele, J. A., *The Textbook of Total Quality Healthcare*, St. Lucie Press, 1994.

Ary, D., Jacobs, L. C., and Razavieh, A., *Introduction to Research in Education*, 2nd ed., Holt, Rinehart and Winston, 1979.

Barness, L. A., *Manual of Pediatric Physical Diagnosis*, Year Book Medical Publishers, 1972.

Bass, B. M., *Stodghill's Handbook of Leadership*, Free Press, 1981.

Grinberg, L. and Stratton, J. A., *A Systems Engineering Approach to Engineering Design Methodology*.

Jensen, F. and Petersen, N. E., *Burn-In: An Engineering Approach to the Design and Analysis of Burn-In Procedures*, John Wiley & Sons, 1982.

Juvinall and Marshek, *Fundamentals of Machine Component Design*, John Wiley & Sons, 2000.

Kardos, G. and Smith, C. O., On writing engineering cases, in *Proceedings of ASEE National Conference on Engineering Case Studies*, 1979, p. 8, http://www.civeng.carleton.ca/ECL/cwrtng.html (last accessed July 2002).

Koller, R., *Eine algorithmisch-physikalisch orientierte Konstruktionsmethodik*, Z. VDI 115, 1973 (in German).

Koller, R., *Konstruktionsmethode für den Maschinen—Gerate-und Apparatebau*, Spriner, 1976 (in German).

Lauer, J. M., Montague, G., Lunsford, A., and Emig, J., *Four Worlds of Writing*, 2nd ed., Harper & Row, 1985.

Leedy, P. D., *Practical Research Planning and Design*, Macmillan, 1977.

Lochner, R. H. and Matar, J. E., *Designing for Quality: An Introduction to the Best of Taguchi and Western Methods of Statistical Experimental Design*, ASQC Quality Press, 1990.

Moran, E., Managing the minefields of global product development, *QUIRK's Marketing Research Review*, 3, http://www.ams-inc.com/readings/eileen_article3.htm (last accessed July 2001).

O'Hara, C. E. and O'Hara, G. L., *Fundamentals of Criminal Investigation*, 5th ed., Charles C. Thomas, 1980.

Pahl, G., Beitz, W., and Wallace, K., *Fundamentals of the Systematic Approach*, Springer-Verlag, 1993.

Roth, K., *Gliederung und Rahmen einer neuen Maschinen-Gaerate-Konstruck-stionslehre*, Feinwerktechnik 72, 1968 (in German).

Roth, K., Franke, H. J., and Simonek, R., *Algorithmisches Auswahlverfahren zur Konstruktion mit Katalogen*, Feinwerktechnik 75, 1971 (in German).

Seoger, G., *Product Innovation—Industrial Approach,* Technical University Berlin, Germany, Department of Assembly Technology and Factory Management, 2002, http://www.sme. org/cgi-bin/libhtml.pl?/library/libhome.htm&&&SME& (last accessed 2002).

Slocum, A. H., *Precision Machine Design,* Prentice Hall, 1992.

Ueda, K., Markus, A., Monostori, L., Kals, K., and Arai, T., *Emergent Synthesis Methodologies for Manufacturing,* Society of Manufacturing Engineers, 2002, http://www.sme.org/ cgi-bin/libhtml.pl?/library/libhome.htm&&&SME& (last accessed March 2002).

U.S. Census Bureau, *Statistical Abstract of the United States: 2000.*

4 Factory Requirements

Vijay S. Sheth

and

Richard D. Crowson

4.0 INTRODUCTION TO FACILITIES REQUIREMENTS

It is all about cost! In the past a factory was one of the largest fixed assets that a manufacturing firm owned. Many exciting changes in manufacturing are occurring as the global marketplace has opened up. According to the Internal Revenue Service, a building has a normal life of 40 years, and therefore 1/40th of its cost becomes part of the manufacturer's depreciation expense every year.

Since offshore manufacturing has demonstrated itself in most domestic markets, the size and role of the modern manufacturer has been a source of change. The equipment that is required inside the building probably costs more than the building itself, and although it is depreciated over a shorter period of time, it is also a big part of manufacturing overhead cost. Buildings and equipment must be maintained in good repair. The best product design and the best production line in the world alone do not insure success in business. The facility itself is a large part of the bottom-line equation. This might indicate that the smaller the facility, the less expensive and simpler it could be—the better for the total enterprise! New trends in manufacturing require partnerships with manufactures all over the world. Asia has become a major player in the world's manufacturing industry. For manufacturers in the United States to compete, a new

Figure 4.1 shows such an "efficient" factory in Poland in the early 1900s. There was little concern for the employees; there were so few jobs that people would work hard under the worst of conditions for whatever wages were offered. The competition for jobs was such that workers developed very good individual skills in order to keep their jobs. The quality of the product produced was not really much of a concern, since it was probably as good as anyone else around was producing. In fact, the products were usually consumed locally, and sold at whatever price was necessary in order for the factory to make money. There was little in the way of competition for the product, and not much concern about inventory levels. The raw materials were quite simple, and low in cost. There was no need to be too concerned about reducing costs, as long as sufficient product came out the door to satisfy the market demand. Since labor was so cheap, there was no need to spend a lot of money for exotic tooling, and inspection as such was unheard of. One supervisor, usually the largest and "most positive" of the experienced workers, was sufficient to make sure that everyone was working. There

FIGURE 4.1

was almost no government regulation to worry about. Waste was dumped out back, or in the river, and forgotten. Cost control and accounting procedures were minimal. Even then, however, there were two kinds of costs: those that added direct value to the product, and those that were necessary but did not add directly to the product value.

Manufacturers who consider the changes in the world we live in today, and try to imagine the one we will live in tomorrow, will find they do not have a historical precedent to rely on in making decisions. Life is more complicated than ever before, and running a successful factory today, and tomorrow, is expected to be a lot more complex. Worldwide competition must be considered, both for domestic products and for similar competing products. The customer demands production of high-quality products, with short delivery times, utilizing exotic materials and processes, all at competitive prices. In this environment, the manufacturer needs to improve existing products constantly, and to develop and produce new products with a short lead time. Rules and regulations on every imaginable subject, with fines and other penalties, can influence the end product cost. The majority of the domestic workforce is not highly skilled. Trends in the United States seem to indicate the employee is not especially motivated, or loyal, to the company. Employees demand and expect benefits such as high wages, longer vacations, excellent working conditions, good tools to work with, training and instructions, health insurance, retirement benefits, and so on.

So, what is the manufacturer to do in this climate of competition with lower-cost manufacturing workforces from other countries? A good product line and a well-designed product that is capable of being produced with high quality and at low cost may not be possible. Let us assume for the moment that we have accomplished this important step (see Chapters 1 and 2 on product development and design). We now need the facilities to produce this product effectively, at low cost and with high quality. We must furnish our workers with a safe, environmentally correct workplace, and furnish them the best tools and equipment that are available (within reason, after some accurate trade-off studies). We need to establish goals and standards for all operations on the product production floor, and have real-time feedback on our status—and especially any problems that might come up. We need to know what our material requirements are, and where the material is. It may be in the warehouse, in stores, in transit, at a supplier, on the production floor waiting—or actually in the factory being processed as planned. We need a way to move the material, and the product in process, through the factory. We must keep our inventories of raw materials, work in process, and completed product to the very minimum. This says that we must have a good factory layout, the correct equipment and tooling, and a good flow through the factory for both the product and the materials. Our information and control systems must provide up-to-the-minute information about our total enterprise operation. Our systems and procedures for operating the factory must be correct and disciplined. The same goes for all the other "non-value-added" effort and costs.

What are the factory requirements that establish this "world-class factory"?
What kind of buildings do we need? (Or can we use the ones we have?)
How much floor space do we need for the production of our products?
What special features are required in the production buildings?
> Floor space
> Ceiling height
> Structure
> Floor strength and rigidity
> Environmental controls (ventilation, temperature, lighting, etc.)
> Safety considerations

What do we need to consider for material?
> Receiving and shipping (single point, or deliveries direct to various places in the production line?)
> Space for storage (warehouses, stores, on the production line, or at suppliers?)
> Transportation of materials to the workplace (conveyors, AGVs, hand trucks, fork lifts?)
> Movement of material in process (Hand pass, conveyors, hand trucks, forklifts, cranes?)
> Total accountability for all material (what, where, how much—versus need)

What space do we need for other supporting functions?
> Offices (how many, where located, how fancy?)
> Meeting rooms
> Communications support (central mainframe computer, network distribution, telephones, video conferencing, etc.)
> Laboratories (quality, materials and processes, testing, etc.)

Fortunately, this complex task can be broken down into manageable pieces, and tools and techniques are available to assist us in the analysis of each of the elements of the factory. We will ensure that the facility and equipment are not only adequate to compete in the marketplace today, but will be flexible enough to permit us to continue in the future. This is the subject of Chapter 4. The emphasis is on the planning approach to facility requirements, layouts, and implementation in general, not on the mechanics of site selection, plant design, contracting, and installations, which are covered in Chapter 4.

4.1 GENERAL FACTORY REQUIREMENTS

Many people think that a manufacturing building is sort of like a warehouse. It has a floor, walls, and a roof. If we add some lighting, and perhaps a restroom or two, we will be ready for production. This makes one visualize a rectangular steel frame sitting on a concrete slab floor. Across the top are standard steel bar joists and a simple roof. The walls are prefabricated steel or aluminum panels, attached to stringers or purloins on the standard post-and-beam wall construction. This may not be a bad way to start our thinking. After all, we just want to do some work and make money—with the lowest possible fixed investment and operating costs!

Now back to the real world of today—and tomorrow. In our original thinking about the factory, we would probably need to do some site preparation (after having selected the site), including roads, drainage, parking lots, a loading dock, and perhaps some security fencing, regardless of any other special conditions known at the time. We could continue this study for a while, and specify some landscaping out front; an office for the boss; water, telephone, and power lines coming in; and a sewer going out. However, we have only set the stage to ask some pretty basic general questions.

How strong should the floor be, and where should it be stronger or have a better base prepared underneath it?

How high should the walls be (what ceiling height is needed) and where?

What kind of loads should the roof support? Perhaps you had in mind an overhead crane system, or air-conditioning units mounted on the roof.

How about the walls—will we need insulation, or a finished interior?

Will there be a mezzanine, either now or later?

How much square footage, and cubic volume, are needed?

How many people will work here? How many in the shop, and how many in offices?

What kind of lighting levels are needed in the shop, and in the offices?

Does it look like a long, narrow building, or is it more square in shape?

Will we need multiple loading and unloading docks? How many need adjustable height provisions, or dock levelers of some sort?

Is there a limit on how close together the internal supporting columns for the roof can be?

How fancy should the offices be? How many will have full-height walls and ceilings, and how many can just have privacy partitions?

You can see that there are going to be a great number of general specifications needed, whether we remodel an existing building or build a new one. In any case, there will need to be some architectural and engineering (A&E) work, even at this early planning stage. The type of questions asked above, although very basic, will have a great influence on the cost of construction, and the time required to do the work. For example, how much are we willing to spend to develop an image for our company? This includes the facade and landscaping, among other factors.

4.1.1 Special Requirements

In addition to the "simple" questions above, there will be some more specific general requirements that must also be defined. The question of exhaust vents into the atmosphere is becoming a big design problem today. The exhaust may come from a paint shop, or a chemical processing line, or dust from a grinding operation, or whatever. The same goes for all types of waste disposal, especially anything considered hazardous by one of the environmental agencies. Even such things as a chemical sink in the materials laboratory needs consideration. If there are any types of cryogenics involved—and the use of CO_2 for cooling is quite common—they require special lines, etc. The problem of how clean your compressed air needs to be—or how dry—can make a difference in the plumbing and piping, as well as the compressors, dryers, etc. Your communication systems may require a backbone down the entire factory, with specified drops or closets which are much more cost-effective if installed or provided for in the initial building plans. Some existing plants have actually installed radio-frequency transmitters at each computer workstation, as being more cost-effective than installing hard lines in an existing building. Fire protection may turn out to be a problem, or at least worth consideration in the general planning of the factory. Storm sewers and retention ponds on site are a requirement in some localities.

These types of decisions are part of the overall facility requirements, in addition to providing for the production process and material handling, which are usually thought of as the main drivers in factory planning. In today's world they are not the only drivers.

4.2 REQUIREMENTS OF THE MANUFACTURING PROCESS

The manufacturing process is the primary basis for the factory requirements. It is essential that the process be thoroughly understood and revised if necessary for ease of material handling and manufacturability. For instance, it may be possible to combine operations, to modify the process to suit some existing equipment, or to alter the sequences so that a common line can be set up for multiple products.

The material going through the manufacturing facility is either waiting, in motion, or being worked on. The only productive time is when it is being worked on, and this may be only a fraction of the total time it spends in the plant. (See Figure 4.2.)

The process can be visualized by constructing a process flow chart which shows each operation and move and any delays in the process. By analyzing various elements of the process flow chart, it is possible to lay out a plant to minimize unproductive

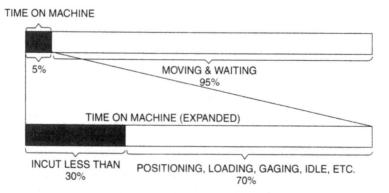

FIGURE 4.2   In batch-type shops, the average part spends most of its life in the shop just waiting to be cut, or waiting to be moved. The results: great work-in-process inventories, very slow through-put rates. (Courtesy of Cincinnati Milacron. Cincinnati, Ohio. With Permission.)

elements. Once requirements of the plant layout are clearly understood, whether for a new facility or for a modification of an existing plant, it is necessary to work on the details which help in defining the factory requirements, including any buildings and the preliminary layout of the factory.

In many instances, the process starts by understanding the manufacturing process for the product manufactured. Sometimes, depending on the existing plant and equipment, it may be beneficial to change the process before further work is done on the facility planning. Many of these decisions involve capital investment considerations and a commitment to the process which is economical in the long range. This activity, the manufacturability of product, is often accomplished concurrently with the product design to save time and cost (see Chapters 1 and 2). However, it is essential that the manufacturing engineer understand the importance of the process required by the product as designed, and the available resources. Many decisions, such as regarding storage, handling of materials, equipment availability, manpower, and support services are necessary during the process of laying out the facility.

4.2.1   Block Plan

Imagine designing and setting up a 50,000 ft.^{2} plant involving over 200 pieces of different equipment linked to produce an optimum flow. This would be a horrifying task if one were to start at the beginning with a piece-by-piece placement of equipment. However, there is a simpler and more logical approach to this problem in combining tasks or equipment into major functional groups. Simplification can also be achieved by dividing plant activities into smaller functional groups. In order to do this one must have a knowledge of process details, material flow, and the number of support activities required. Once this information is obtained, it is possible to locate the different groups in the layout with their relationship to each other.

Block Layout Approach

One approach to layout which shows the outlines of activities or departments is known as a block layout, or block plan. For instance, a block layout for a department store has products assigned to different areas, such as men's, ladies', cosmetics, kitchenware, appliances, and toys. It is usual practice to prepare a preliminary *block layout* before the detailed factory layout is prepared. See Figure 4.3 for an example of a preliminary block layout.

Block layouts are used to provide preliminary information to other groups in the factory and to the A&Es involved in the construction of a new facility. A preliminary assembly flow chart as shown in Figure 4.4 may be about the level of detail available at the time the preliminary block plan is prepared. It will be based heavily on previous experience, and will be the baseline for preliminary work by the A&Es. A block layout showing special requirements such as processed steam, chemical processing, storage, and overhead equipment will be required to address the key issues of facility planning. The preliminary block layout will change as more detailed analyses are performed. Early considerations must include the following:

Size of the facilities: Minimum or maximum practical size for operating cost, organizational effectiveness, and risk exposure.

Lease, buy, or build: A review of the alternatives available from architectural designs based on the block layout provided. This will also provide cost estimates for management review and perhaps a major shift in direction from original goals, based on financial commitment.

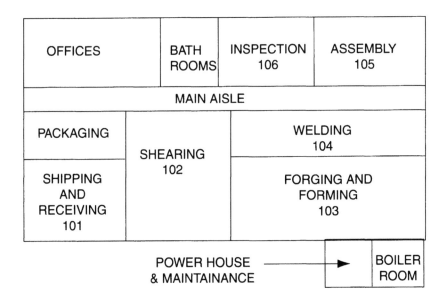

FIGURE 4.3 Preliminary block plan based on experience with similar product production.

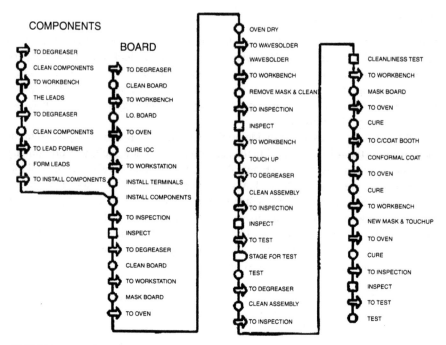

FIGURE 4.4   Preliminary assembly flow chart for a circuit card assembly.

Operating requirements: This consists of details regarding flexibility and expansion possibilities. It provides alternatives for material handling and storage requirements.

Requirements for preparing a more complete block layout include the following:

Square feet required by each block
Relationship of blocks to each other
Shipping and receiving dock locations
Department requiring the use of hazardous material
Special utility requirements
Plant service requirements, e.g., tool cribs, stores, maintenance, engineering support
Emergency exits
Main aisle locations

Block Layout Elements

It is important that a balance be achieved between the total square footage required by blocks, service areas, and aisles and the available square footage in the building. This is the first step in determining that all the activities to be housed in the building are provided for. The following are typical details that are summarized early in the facility planning process:

Department/function square footage required

Shipping and receiving	3,600
Storage	10,800
Square shear	9,800
Roll and swage	22,000
Flame cut and weld	15,000
Assemble	22,900
Clean and pickle	17,800
Paint and plating	15,000
Packaging	6,000
Support offices	20,500
Main aisles—10% assumed	14,400
Total	157,800

If the required square footage is more than is available in the existing facility, a decision must be made as to how to provide the additional square footage. This might mean an extension to the existing building for some of the services, construction of a mezzanine to take advantage of the building cube, or a new building or buildings.

Notice that the square footage values are rounded to the nearest hundred square feet and the aisle requirement is established as an approximation based on experience. Questions still remain as to how to establish a more accurate square footage for each department and how to arrange the blocks. For this, we must have more knowledge of the process and production requirements. Some of the specific elements needed are shown below, followed by a more detailed description:

Process routing sheet
Flow process chart
Relationship chart
Trip frequency chart

Process Routing Sheet

One important document used in the plant by production workers, supervisors, and schedulers is the process routing document. Figure 4.5 shows a basic form of routing as an assembly process flow. This document, prepared by manufacturing engineers, depicts the best and the most practical method of processing incoming material to obtain the desired product. In the case of a new product manufactured in an existing facility, it is generally prepared to take advantage of existing resources which include equipment, the material handling system, storage, and the skill of the personnel. In the case of new facilities, the manufacturing process and layout require a thoughtful approach and heavy involvement of management. For instance, a part can be machined on a low-cost mill or on a high-speed, computerized, numerical-controlled machine with an automatic tool changer, costing several hundred thousand dollars.

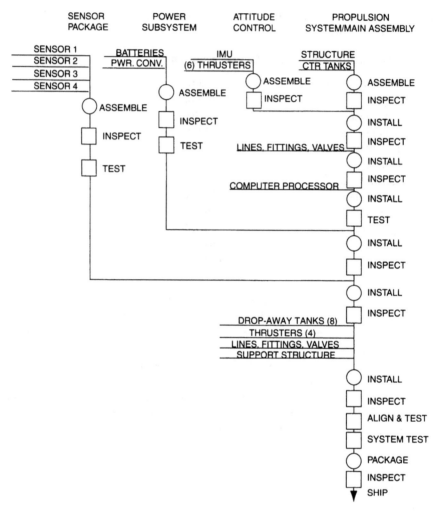

FIGURE 4.5   Assembly flow chart.

Flow Process Chart

A flow process chart (Figure 4.6) is developed for each component and subassembly in the manufacturing of the final product. It is a descriptive method in chart format to show the flow of material through various process activities. These activities have been coded by the American Society of Mechanical Engineers (ASME). An explanation of these activities, which are also represented by symbols in the flow chart, follows:

   Operations (to change form): An operation is performed when material is intentionally changed in its physical or chemical form to obtain a desired result. This is preceded and followed by one or more of the other four classifications as described below.

NO. __2__
PAGE _1_ OF _1_

SUMMARY						
	PRESENT		PROPOSED		DIFFERENCE	
	NO.	TIME	NO.	TIME	NO.	TIME
○ OPERATIONS	13		10		3	
⇨ TRANSPORTATIONS	4		2		2	
☐ INSPECTIONS	0		0		0	
D DELAYS	5		3		2	
▽ STORAGES	0		0		0	
DISTANCE TRAVELED	120 FT.		42 FT.		78 FT.	

JOB ___Pricing and Posting Orders___

☐ MAN OR ☑ MATERIAL ___Unpriced Orders___
CHART BEGINS ___On A's desk___
CHART ENDS ___On C's desk___
CHARTED BY ___John Smith___ DATE ___4/15/83___

	DETAILS OF (PROPOSED) METHOD		NOTES
1	Placed on A's desk	○⇨☐D▽	
2	Time Stamped and Stored	○⇨☐D▽	
3	Placed in OUT baskets	○⇨☐D▽	2 wire baskets
4	Waits	○⇨☐D▽	Shorter wait
5	Picked up by B	○⇨☐D▽	
6	To desk	○⇨☐D▽ 14	
7	Placed on desk	○⇨☐D▽	
8	Priced	●⇨☐D▽	New rotary price file
9	To C's desk	○⇨☐D▽ 28	By B
10	Placed on desk	○⇨☐D▽	
11	Waits	○⇨☐D▽	
12	Posted and Sorted	●⇨☐D▽	
13	Placed in Envelopes	○⇨☐D▽	
14	Placed in OUT basket	○⇨☐D▽	
15	Waits	○⇨☐D▽	Picked up by messenger

FIGURE 4.6   Flow process chart.

Transportations (to move material): When the purpose is to move an object for a specific reason, transportation occurs. Many times transportation is combined with operation for savings in time and space. In this case, it is considered to be an operation activity. Transportation is identified separately in the flow process since it signifies: (1) the need for handling and storage methods and equipment, and (2), a non-value-added function which should be minimized or preferably eliminated.

Inspections (to verify): An inspection occurs when the objective is to verify quality or quantity separately from operation activities. Here, the only objective is to inspect. Frequent inspections performed by the operator during machining operations are not separate inspection classifications, but rather are a part of the operation.

Delays (to wait): A delay occurs to an object when it is not possible to perform the operation, move, inspection, or store function intended. Here the object is simply waiting for the next action to process it further toward its completion.

Storages (to hold): Here an object is protected and held until further action is warranted. The material is kept ready for the move condition.

The productive time usually occurs only under the operation and inspection classifications. The others are contributors to material handling and storage costs, which should be kept to a minimum. This method is a very useful tool in determining handling requirements, process equipment, and possible process improvements based on the ratio of productive time to total time required to complete the process. The column marked "notes" can be used to estimate the time required to perform the corresponding activity (except storage), and also to identify any special requirements such as equipment, tooling, storage, etc. This information is vital to the layout, which in turn will help establish the size of the building requirements.

Reasons for Flow Process Chart

To understand the overall nature of the system being studied
To eliminate costly errors by properly analyzing the material flow
To allow adequate space to avoid safety problems
To eliminate flow patterns that are not suitable, resulting in additional handling
To allow storage space adequate to support the production rate
To locate and size aisles appropriate for product handled
To show the possibility of combining operations by grouping machines or operations to avoid extra handling, storage, and delays
To avoid cross traffic and backtracking of the material in process
To decide whether product flow or process flow concepts of factory flow will be adopted

A flow process chart can be constructed by using a process routing sheet (see Figure 4.7), which provides details of processing the product. If the factory requirements must be established prior to the final design of the product, as often is necessary, preliminary process routing sheets can be prepared during the product design process. They are usually needed in any case during design evolution, in order to help select the optimum product design.

Relationship Chart

A relationship chart is extremely effective where product or traffic patterns cannot be established easily. Figure 4.8 shows a relationship chart using the following steps:

1. List all departments or activity centers in the blocks shown on the left.
2. Using appropriate closeness ratings from a value box, assign closeness ratings in the top half of the blocks intersecting two departments or activity centers.
3. Assign an appropriate code from the reason for the closeness block in the bottom half of the block corresponding to the closeness rating.

ROUTING

Part Name _____ Blade Pieton _____ Module _____ AD7562-A2 _____ Part No. _____ 6510777

Material   Bronze Forging   A29   Date   No. of Sheets   1   Sheet No.   2   track No.   9

OPERATION	EQUIPMENT	Oper. No.	Dept & Group	STANDARD TIME			No. Max	Group Standard Hours
				Reg'd No.	Minimum	Max'm		
Face, turn, chamfer & form groove in small O. D.	(#5 W & S)	27A	44A3	2776		.1530		.2202
Face, turn, large end	(#5 W & S)	27B	44A3	2806		.0672		
Inspect & credit Group 44A3		24	33	Ind.	Labor			
Drill (2) 21/64 moles & (2) 7/32" holes & drill burr (4) holes	(4 Spell. L. G.)	13	44B1	3539		.0730		.0130
Inspect & credit Group 4481		24	33	Ind	Labor			
Mill slot 3/32 & 7/64" deep	(Kent Owens)	31A	44B2	2883		.0237		.0237
Inspect & credit Group 4482		24	33	Ind	Labor			
Millthreads	(Leed Bradner)	31B	44B5	2897		.0721 (1 oper. 2 Mach)		.0721
Inspect & credit Group 4485		24	33	Ind	Labor			
Burr slots, groove & threads	(Burr Room)	7A	44D	3255		.0330		.0467
Scratch brush slotted end	(Burr Room)	7B	44D	3389		.0090		
Mesh	(Burr Room)	10	44D	3390		.0047		
Inspect & credit Group 44D		24	30	Ind	Labor			
MOVE TO BOND ROOM								
(Used on Piston Assembly - 6500386)								
	Totals							

FIGURE 4.7   Process routing sheet, showing standard times, that can be used to prepare a flow process chart.

It is important to understand the functions of all departments and appropriately assign frequencies and closeness values between any two departments. Many times, inconsistency arises when two departments rate their relationship with each other differently. One way to avoid this inconsistency is to have one person who is knowledgeable about functions provide ratings for each pair of departments. Additional charts can be prepared between the factory floor departments, or within a department.

Trip Frequency Chart

The travel pattern between departments or activity centers can be shown by a chart such as the one in Figure 4.9, which can help in providing valuable data for location of the blocks. (Refer back to the original block plan in Figure 4.3.) The following steps will aid in constructing this chart:

1. Construct a matrix showing departments on both the left and the top of the matrix. Each department represents a block to be shown on the block plan.
2. Estimate the number of moves required by the material handlers to support production in an average shift or day between the two departments—e. g.,

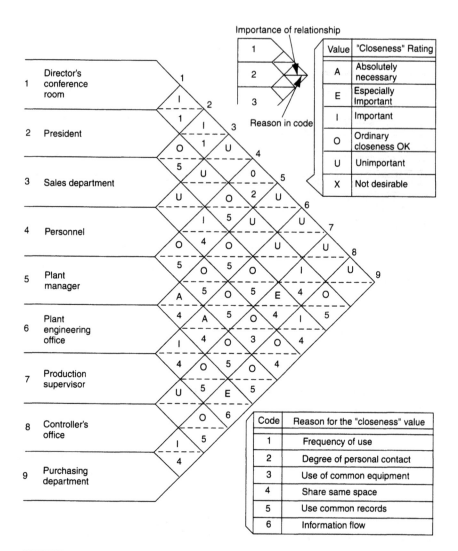

FIGURE 4.8   Relationship chart.

eight trips are required to deliver material from Department 103 to Department 104.
3. The higher the number, the closer the departments (blocks) should be located to reduce traveling distance, time, and downtime. The other advantage is that the material traffic is not going through unrelated areas, thereby eliminating safety problems and congestion.

The number shown in each block represents the relationship of one specific department to another for the movement of goods. The blocks are filled by analyzing

To From	101	102	103	104	105	106
101	–	15	4	3	4	2
102	9	–	12	3	4	7
103	0	7	–	8	1	0
104	0	2	1	–	12	2
105	0	0	1	9	–	10
106	1	0	0	2	8	–

FIGURE 4.9 Trip frequency chart used to establish travel patterns between departments.

product flow and tallying the number of times material movement is required over a period of one shift or one day. The greater the frequency, the more significant is the requirement to locate them closer to each other. A carefully planned block plan can save operation costs in handling equipment and overhead support labor. There are also other advantages—for example, savings in aisle space by providing wide aisles only between blocks where necessary.

Other Elements in a Block Layout

Flexibility in facility planning refers to allowances required for uncertainties. For instance, a specific department cannot be laid out due to unresolved questions on the method of material handling or not knowing the stage of production that the material is received from the supplier. Sometimes, there is flexibility in routing a product in more than one way between departments. If these variables are unanswered during the block layout, it is necessary to take precautions and allow for these uncertainties. Product rework should be allowed for.

The expansion is always difficult to plan unless one has a definite idea of what, how much, and when it is required. For an existing building which has more area available than the sum total of all blocks, it is possible to prorate additional square footage to the blocks which are apt to be affected by the expansion. A later version of the block plan is shown in Figure 4.10.

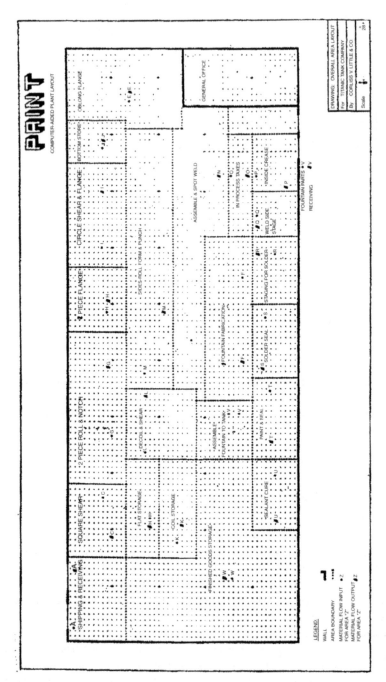

FIGURE 4.10 Later version of a block layout with pickup (*) and set-down (#) points, developed using the cursor on an Exidy Sorcerer computer. Distances are computed automatically and provided to the scoring routine. (Courtesy of Corliss V. Little & Co., Raytown, Mo. With Permission.)

4.2.2   Requirements for Preparing a Plant Layout

Plant layout is one of the most important phases in any facility-planning and material handling project related to either changes in existing products or new product manufacturing. It depicts the thoughtful, well-planned process of integrating equipment, material, and manpower for processing the product through the plant in an efficient manner. This means that the material moves from receiving (as raw materials and parts) to shipping (as finished product) in the shortest time and with the least amount of handling. This is important, since the more time the material spends in the plant, the more costs it collects in terms of inventory, obsolescence, overhead, and labor charges.

Plant layout is the beginning of a footprint of the actual plant arrangement to obtain the most efficient flow of products. See Figure 4.11 for some examples of desirable and undesirable flow patterns. Plant layout is also the basis for determining the cost of the facility in the case of new construction or alterations. Sometimes it is useful to work the problem of facility requirements in reverse. For example, if the final output of the studies is a layout of the factory (new or modified existing), how do we collect the information needed in order to complete the layout? The following tools, techniques, and other elements will help in our analysis and factory requirements definition:

1. Assembly flow chart
2. Flow pattern of product
3. Production rate and number of shifts
4. Equipment configuration

A: DESIRABLE FLOW PATTERNS

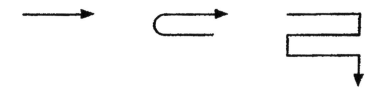

B: UNDESIRABLE FLOW PATTERNS

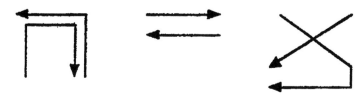

FIGURE 4.11 Examples of desirable and undesirable flow patterns in a factory.

5. Flooring preparation
6. Utilities
7. Store arrangement
8. Shipping and receiving
9. Material handling methodology
10. Support functions
11. Work-in-process storage
12. Safety and emergency evacuation

Following are brief descriptions of some of these factors.

Flow Pattern of Product

Plant layout is not simply an arrangement of machines, departments, and services best suited to the physical dimensions of the plant. It is a carefully thought-out plan for installing equipment as the product smoothly follows the process as determined by the process flow chart. It is often difficult to obtain an ideal condition in an existing plant in which a new product is introduced, for several reasons. Some of them are:

Cost to rearrange equipment is prohibitive.
Plant configuration does not allow for installation of the most effective arrangement.
More than one product line shares common equipment.
Incremental addition of machinery is not strategically located due to lack of space.
Process sequence changes after the installation is completed.

Process Flow or Product Flow?

One important consideration in layout is whether the plant arrangement will be based on a process or a product flow. This decision is based mainly on the characteristics of the product and its production rate. The basic difference is in how the product flows through the plant during manufacturing. It has implications for the initial investment in equipment, the size of the facility, and future requirements. Figure 4.12 shows some of the differences in these arrangements. In process flow, similar functions to be performed on the product are grouped together. The functions can be by either machines or tasks. For example, all drill presses may be grouped together, or all manual packaging performed at one location. With this type of arrangement, various products manufactured in the facility travel haphazardly depending on which function is to be performed on it next.

In product flow, one product is processed independently of another in the same plant. In this case, all functions required to be performed on the product are so arranged as to obtain a smooth flow of material with the least amount of downtime and handling.

Process or product flow can be arranged either for machines or for departments. Figure 4.13 shows the department arrangements. In process flow, all similar machines are arranged together, which is not the case in product flow. It is not

OPERATION SEQUENCE

Part	Oper. 1	Oper. 2	Oper. 3
A	lathe	drill	lathe
B	drill	mill	
C	lathe	mill	drill

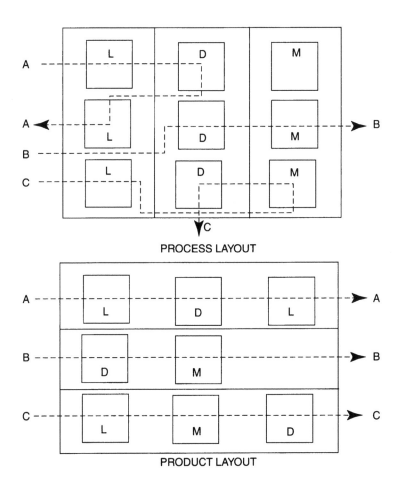

PROCESS LAYOUT

PRODUCT LAYOUT

FIGURE 4.12 Influence on plant layout based on process flow or product flow through the plant.

PRODUCT	FIRST SEQUENCE	SECOND SEQUENCE	THIRD SEQUENCE	FOURTH SEQUENCE
X	Forming (F)	Machining (M)	Chem. Process (CP)	Assembly (A)
Y	Machining	Assembly	Chem. Process	Forming
Z	Forming	Machining	Assembly	Chem. Process

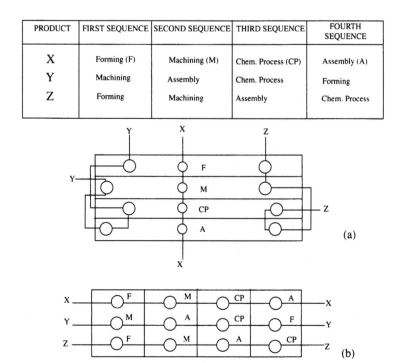

FIGURE 4.13 Flow patterns within a department for the same product: (a) process flow; (b) product flow.

uncommon in a job shop to find all drill presses standing side by side, and then all the lathes, milling machines, grinders, and so on. Process flow plants usually have single-function departments, such as packaging, inspection, painting, or chemical processing, for all products.

In product flow plants, where more than one product may be manufactured, there will be some duplications of functions. This type of arrangement is most effective when the product is mass-produced. Common examples are manufacturing facilities for appliances, automobiles, garden equipment, and housewares.

Advantages of Product Flow Arrangement

Material moves progressively in a definite pattern from machine to machine or from one department to another, causing less damage to material.

Workers involved with the handling of the product experience less confusion, since only one product is involved. They attain a degree of expertise in processing the product.

Material is handled efficiently. There is less possibility of a mix-up with parts from other products.

Processing time is greatly reduced, since equipment and workers are assigned to a particular product.

Less work-in-process inventory is required.

Inspection can be reduced, since workers become proficient in one part or a function. Further, the part is frequently tried out on the very next operation, thereby eliminating the need for inspection.

Delays are reduced, since there is no schedule conflict with other products.

Production control tasks become simpler, since only one product is made in each production area. This greatly reduces paperwork such as move tickets at every stage of production.

Material staging areas between machines can be reduced, since there are fewer delays and the product can be moved as a continuous flow instead of a batch.

Investment in material handling equipment can be reduced, since the product does not have to be moved in batches over a long distance, thereby eliminating the need for bigger or faster equipment.

It is easy to train workers, since only one product is involved. Supervisors can be trained easily for the processing of just one product.

Disadvantages of Product Flow Arrangement

Some duplication of machines is necessary, which requires high capital outlays.

It is difficult to justify high-speed, automatic machines if product volume is low.

Machines are not used to their fullest potential—in terms of both capabilities and utilization.

If specialized machines are used, they cost more and become obsolete if production of the product ceases.

Balance of the production line is difficult. As a result, labor utilization may drop.

Any changes in processing of a part may cause major changes in the line layout, balancing, and equipment.

Major breakdown of a machine may idle the entire line.

Specialized material handling methods are used. They are costly and of little value if the product is discontinued.

Workers exhibit special skills and have to be retrained for different assignments.

Manufacturing cost is higher if the product line is not moving at full planned capacity.

Advantages of Process Flow Arrangement

Flexibility in scheduling is available, since a number of machines are available in the same group.

Product flows in batches. This simplifies handling and reduces the frequency of moves.

Machines are of a generic type and are better utilized. This reduces initial capital investment.

A single machine breakdown does not affect the schedule much, since the product can be routed to other, similar machines in the same area.

If specialized handling equipment is required, it can be shared with other products with some modification.

Workers are skilled to run any equipment within the group. This provides flexibility if an assigned worker is not available or quits. This eliminates delays waiting for a worker.

Supervisors become proficient in the process functions they supervise. They can help troubleshoot problems related to setup and quality without much outside help.

When a new product is introduced in the plant, it can be processed more efficiently initially, since most of the equipment is already in use.

Overtime schedules can be avoided, since flexibility exists with similar equipment. Breakdown of one machine does not necessarily hold up successive operations if part can be routed differently.

Layout is less susceptible to changes in product mix over the long term.

Generally suitable for small production runs.

Disadvantages of Process Flow Arrangement

Material takes a long route and may be subjected to loss or damage.

Production control has to closely follow up and coordinate the required date of the product at each function.

Main aisles and aisles between the machines have to be wide for transportation of material in bulk.

A large number of rejections of parts results if the error is not caught until a subsequent assembly operation.

Delays can occur when parts obtained from previous operations have to be scrapped and replaced by new ones from the very beginning of the line.

Production Rate and Number of Shifts

It is usually not possible to plan facilities without knowing the required rate of production. This requirement is established initially by management in order to facilitate not only the planning of the facility but also the commitment to capital outlays.

Equipment Configuration

Before plotting a layout, it is necessary to have all the physical characteristics available for the equipment to be housed in the plant. The following are the details required:

1. Size of the equipment: length, width, and in some cases height, to make sure that equipment can be placed under a low ceiling height or under the mezzanine.
2. Operator working area in relation to the equipment.
3. Access for material handling and storage equipment, such as the forklift, containers, and racks.

4. Storage for parts which are completed and waiting to be processed.
5. Storage for fixtures and tooling which are stored near the equipment for ease of handling.
6. Inspection table by the equipment if required.
7. Maintenance access from the sides and from behind the equipment.
8. Special characteristics of machines. A large press may require a foundation separated from the rest of the flooring to make sure that the vibration generated while operating the machine does not affect other areas where precision operations such as inspection, spot welding, or automated electronic assemblies are performed. If this is not taken into consideration, then quality of production may be affected in other areas when such tasks are performed simultaneously with press operations.
9. Pit requirements for equipment or for services such as an underground chip conveyor. It becomes expensive to construct a pit after the flooring is poured, because rerouting of underground utilities and drains may be necessary. In the case of existing facilities, if the details of underground utilities are known, it is advisable to consider alternative plant arrangements that do not require costly rerouting of the utilities.
10. Location of the controller and machine controls in case of CNC equipment or other special equipment when a separate operator control panel is provided with a flexible conduit. This provides flexibility in locating the panel to obtain smooth material flow from and to the machine. See Figure 4.14

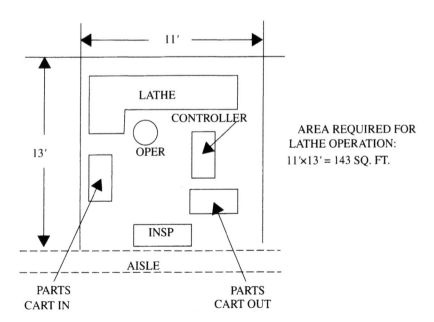

FIGURE 4.14 Computation of functional area requirement for a cell or equipment which includes materials staging and secondary operations.

for an example of the actual floor space required for a lathe and the supporting elements in a cell arrangement.

11. The material handling method is of concern in the case of heavy product
or tooling. If a bridge crane is necessary, then obviously the equipment
must be located in the area covered by the bridge crane. If a floor-mounted
jib crane is required, then it is necessary to locate a jib crane column in
the right relationship to the equipment and parts containers. Think of the
situation in which a jib crane requirement is not considered in the initial
layout. If there is not enough room, then it becomes expensive to relocate
the equipment once the requirement surfaces. In the case of new facilities,
it is also necessary to specify the load requirement of the bridge crane as
part of facility planning so that the building structure is designed to support the bridge crane load in addition to the building and the roof.

12. The upgrading plan for the equipment and material handling system or the
production rate increase in the foreseeable future must also be considered.
This is usually the case when a new product is in development, financial resources are limited, and there is uncertainty in the business. Many
government contractors in the defense industry delay large investments in
productivity improvements until a new product is developed and tried out.
In many instances, the layout is revised when it is decided to go from the
developmental phase of a new product to the production phase. However,
a good estimate must be made when designing the new building.

13. Weight of equipment. Some processing equipment must be massive in
order to handle heavy products such as steel fabrication. Not only must
the flooring be able to support the equipment load without buckling or
cracking, but also the other floor should be able to support the weight of
equipment being moved inside the plant for replacement or relocation.

Machine Capabilities

The production rate desired in a plant is a function of the total output required from
the plant. Suppose that the production schedule is set for 320 units per day. In order
to determine the floor space requirement, one must know whether the plant is scheduled for one shift or two shifts. Further, what is the capacity of the equipment through
which the product is processed? If a machine's capacity is 25 units per hour, then
only 200 units can be produced in a given shift. There is a choice of either scheduling
a second shift or installing a second, similar machine to obtain the same output.

On the surface, this situation may seem to have a minor effect since the labor cost
is not affected. However, there are other concerns:

Investment in the second machine
Floor space occupied by the second machine
Maintenance of the second machine
Supervision for the second shift
Support services for the second shift
Future capacity requirements

Floor Requirements

In a new facility, it is necessary that some of the known requirements of the utilities be planned to be underground to avoid overhead exposure and high maintenance costs. For example, if the process water usage is known, underground sewer lines can be provided that can eliminate expenses for holding tanks and pump stations. Similarly, if the departmental process layout shows heavy usage of electrical power in a certain area, it is possible to lay cables underground. Well-planned utilities underground may result in savings in maintenance cost, valuable floor space, and expensive service equipment requirements, although the A&E design work will assist in making the final decisions.

Flooring requirements also include ramps for ingress and egress to the facility with fork trucks. The shipping and receiving dock must also be planned with cavities for a dock leveler to be installed.

Sometimes it is a challenge to accommodate new requirements in an existing plant. Simply knowing the size of the equipment without supporting details can lead to disaster when the actual facility work begins. The new equipment requirement must be fully supported by the layout and the various details to help the facility engineer in designing and planning the installation work.

Utilities

The basic utilities, such as water and electric power, are usually available in plentiful quantities. However, there are processes which demand high usage of both. For example, a totally air-conditioned facility in Florida dealing in missile production requires the heavy use of electric power for heat treatment and curing ovens. Another example is a chemical treatment plant that uses large amounts of water. It is necessary to estimate requirements at the planning stage so that the A&E firm and contractors are aware of it, and the additional cost to bring the requirements to the plant is also known. A facility may require an electrical substation or a holding tank for water. It may also require a separate distribution center and a waste disposal system. All these requirements are shown on the plant layout or accompanying documents.

Other utilities of concern are compressed air for air-powered tools, fiber-optic data lines for computers, telephone lines, and process steam requirements. A knowledge of these requirements is useful since the activities involving these services can be located strategically in order to control the overhead expense and to maintain safety in the plant.

Cellular Flow

When one studies a process flow chart, one will find that the process moves smoothly along flow lines. On the other hand, with a variety of products, each varying a little bit, there is another approach to processing in a plant; that is, as cellular flow (Figure 4.15). In this method, different products which take the same processing equipment are grouped as a family. A cell consists of a group of machines arranged such that a product can be processed progressively from one workstation to another without having to wait for a batch to be completed or requiring additional handling between the operations. In essence, this arrangement is a small production flow line within a plant.

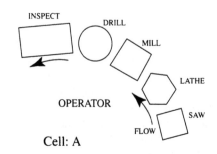

Cell: A

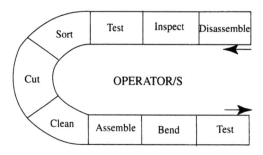

Cell: B

FIGURE 4.15 Cellular flow, consisting of a group of machines arranged such that a product can be processed from one workstation to another progressively.

Cellular flow is not an entirely new concept, but has received wide exposure recently in large companies manufacturing a high volume of a variety of products. One automobile exhaust-system manufacturing plant has set up its equipment in a pattern so that an exhaust pipe can be bent on a pipe bender, the pipe ends faced on a facing machine, one end sized to the required diameter, and another one formed to accept a flange. All four operations are performed on four different machines by the same operator without laying the pipes down between operations. Further, equipment is built on casters so it can be wheeled in and out to suit cell requirements. The immediate benefits realized with this arrangement are reduction in the material handling, elimination of in-process inventory between operations, and reduction in floor space.

Essentials of a Cellular Layout

Multiple products can be grouped into families with similar process requirements.

Parts do not require storage between operations. They are transferred to the next operation either manually or via a simple transfer device such as a chute or a small conveyor section.

The majority of parts through the same cell are processed in the same way. No
special setup is required, although some parts may skip a machine or two
in the setup.

The operator is capable of operating all machines in a cell.

Equipment is arranged close together and there is no staging between
equipment.

Equipment in cells are usually single-function, basic machines.

Concerns of a Cellular Layout

Equipment utilization is low.

A breakdown in one piece of equipment may paralyze the cell operation.

Tight quality control is necessary in order to reduce scrap.

In a block layout or plant layout, one must weigh each alternative and then decide
between cellular and flow approaches. Based on each method, the areas required are
calculated to provide the total requirement for that block. However, it must be noted
that the length of a block is critical to a straight-line flow within the block.

Progressive Assembly Operations

Many assembly operations cannot be economically automated, and are performed by
setting up progressive assembly lines, as shown in Figure 4.16. Subassemblies and
components requiring heavy equipment are set up away from the assembly lines, and
the parts are generally fed to the line as required. This method provides the advantage
of minimum movement of the material in a steady flow. Operations are also synchronized on the line, which gives the benefit of steady output provided all operations are
performed as planned.

It is necessary to study in detail the operations affected by this approach as to space
and equipment requirements before making a layout. This approach is most suitable
for light assembly work requiring bench operations, such as assembly of pressure
gauges, small kitchen appliances, electrical switches, and controls. However, mass
production of automobiles uses a similar approach, whereby there is a central facility setup with progressive parallel assembly lines for assembly of automobiles. The
major subassemblies, such as frame, body, transmission, engine, axles, and doors, are
merged into these lines with the help of automated and material handling equipment.
This equipment may consist of overhead, underground, and floor conveyors, AGVs,
and industrial trucks. The feeder lines are either connected to separate facilities nearby
or are serviced through receiving areas for parts received from outside the plant.

Line Balancing for Detail Layout

"Line balancing" refers to the requirements of equal time for operations performed
on a progressive assembly line. In order to achieve a smooth flow of product through
the line and gain high productivity, it is necessary that the manufacturing engineer
distribute the total work in equal amounts to various stations on the line.

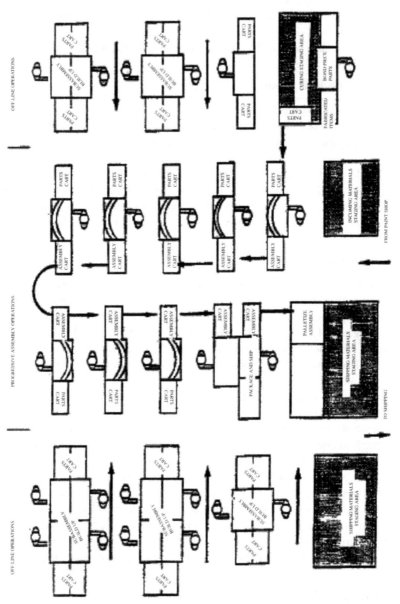

FIGURE 4.16 Progressive assembly lines supplemented by off-line operations.

Operations that are well suited for this type of application are manual and bench type, using small equipment. With large and expensive equipment it is necessary to perform detailed analyses to make sure that the balancing requirements do not end up with too much idle time on the equipment, or else result in a cumbersome line layout. In many situations, it is preferred that operations requiring the use of heavy equipment are performed separately and away from the line. The parts are then introduced to the progressive assembly line at the workstation on the line where they are assembled to the product moving line. This is also referred to as a feeder line concept, in which parts are either brought into the line as batches, or via a conveyor system that is synchronized with the main conveyor speed or operation time of the main line.

Even with manual operations, it is sometimes difficult to split the operations to even out the time. This is particularly true with light assembly-type operations. In these cases, it is necessary to introduce such activities as general cleanup, servicing other stations with supplies, inspection, testing, documentation, and labeling at the stations that have idle time left from the assigned task.

In the progressive line concept, the major elements that determine the layout are the production rate (pieces per hour), the clean breakdown of the entire process into separate operations, and the time required to perform each operation (also referred to as cycle time). Following are the steps in balancing a production line.

1. Determine whether a manufacturing process is conducive to forming a progressive assembly line.
2. Follow the process and itemize operations in chronological order.
3. Determine the time required to perform each operation by an average worker. Use stopwatch study or predetermined time standards.
4. Determine the production rate desired (units per hour) based on daily output expected.
5. Determine the total time required to manufacture one unit (labor hours per unit) by adding each operational time.
6. Determine the number of workstations to meet the production requirement by multiplying the production rate by labor hours per unit.
7. Determine the time allowed to perform the task on each unit (hours per unit) at each workstation (cycle time) by finding the reciprocal of units per hour.
8. Arrange the tasks on each workstation so as to progressively assemble the unit by either combining the operations, dividing into subassemblies, or duplicating the operations on two or more workstations so there is an even distribution of work. In cases where the operations are to be divided into more than one workstation, it may be necessary to review the time studies of operations to determine the break in the operation. Essentially, the product moves progressively from one workstation to the next at the end of the cycle time.

For process type operations that take longer than the cycle time determined, it may be necessary to duplicate workstations for that operation. Again, there may be some idle time left over. This may be filled by splitting other operations and redistributing the task. However, line balancing requires a paper exercise in balancing the time for each workstation and often a rough layout to make sure that the area assigned to each workstation is adequate for smooth flow, safety to the worker, and efficient materials handling.

One method of overcoming the limitations of line balancing manufacturing is by utilizing a transporter as a material handling medium. A transporter does not require balancing of operations and yet allows flexibility to the extent that even unrelated products can be run on the workstations served by the line. In essence, the line can be run as a progressive assembly line just like line balancing manufacturing, or as a multiproduct operations line served by common powered material handling equipment. With the transporter it is necessary to deliver parts to the workstations served in small batches and to return the completed product to the dispatcher at the head of the transporter for rerouting to the next operation workstation.

Batch Processing versus Continuous Flow

As discussed above, there will be some operations that are not a part of a fabrication center, nor part of an assembly line. The example in Figure 4.17 shows parts processed through a curing oven in batches. This type of process flow is easily adaptable to existing equipment or standard equipment. Although it is easy to set up and requires the minimum in equipment purchase, it will require more units in flow due to the batch process (perhaps 36 units cured at one time). It is less productive, and requires handling between operations. Figure 4.18 shows the same task setup with a continuous flow arrangement. The higher production rate capability must be weighed against the required production rate, to make certain that the higher cost of the conveyor system and flow-through oven, the longer lead time for the equipment, and the larger floor space requirement will be offset by the cost savings in labor. While there are advantages and disadvantages to both of these processes, one should be selected prior to making the final layout or ordering any capital equipment.

4.3   REQUIREMENTS FOR THE FLOW OF MATERIALS

4.3.0   Introduction

One of the three main ingredients of plant activities is material. The other two are labor and machines. It is necessary that raw materials and parts be available when required. Since delays are unproductive and expensive, they must be eliminated whenever possible.

For the plant layout to be effective, it must consider the movement and the handling of the production materials, supplies, and tooling. Many times the layout is prepared around the existing material handling equipment and methods presently in use in the plant. The manufacturing work centers must be adequately supplied with the material so there is a smooth flow of production and no downtime waiting for the materials, handling equipment, tote boxes, containers, and carts. An effective layout must take into consideration the production planning and control requirements, such as batch

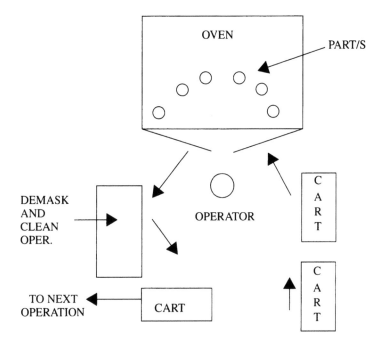

FIGURE 4.17 Parts being processed through a curing oven in batches.

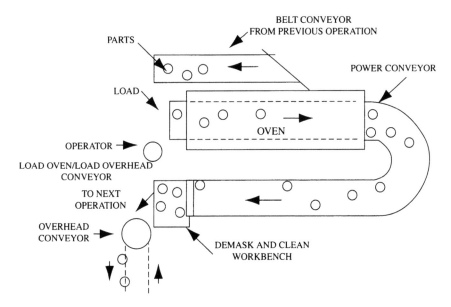

FIGURE 4.18 Continuous flow arrangement showing parts loaded from one end and flowing continuously on a conveyor which carries them in and out of the oven for processing.

sizes, mode of transportation, storage space by the workstation, and follow-through with shop work orders for effective control of the work-in-process material.

On the other hand, an improper layout that does not allow for the systematic flow of the product may result in parts shortages, quality problems, missing operations on some parts, and damage to the product due to handling.

With the modern concepts of minimum inventory and just-in-time manufacturing, the emphasis has been on making the material available when needed. It makes sense to carefully plan out the store requirements, since the cost of production in most industries includes over 50% material costs. In paper-converting facilities, the material cost may be as high as 80%. Large companies emphasizing low overhead costs have realized the value of subcontracting unskilled and semiskilled operations to smaller companies with less overhead, lower wages, and smaller benefit packages. This in turn increases material costs while reducing in-plant labor content.

On the other hand, companies have realized the value of high inventory turnover to keep inventory shrinkage and costs to the minimum. This has become more important than ever before, since the advances in technology and global competition have made product life cycles shorter. Inventory shrinkage includes material obsolescence, damage, and loss. For continuous flow production, it is necessary that material flow be smooth. The material handling department must be well equipped to serve all production activities in the plant efficiently.

4.3.1   Materials Handling in Support of Production

In a manufacturing plant, the overall process is broken into various small operations connected by numerous transportation and staging steps. Materials handling should be considered an integral part of the total manufacturing operation. It is a catalyst that provides the reactions of people, materials, and machines. The material inside the plant is in one of two physical states in support of production: (1) properly stored and ready for its next move, and (2) properly transported. The goal is to keep the cost of these functions to a minimum, since neither of these add value to the product but both are an essential part of the total product cost. The material handling system in the factory operation must integrate the following functions:

Transportation of raw material and parts to the factory
Receiving, storage, and retrieval
Movement of material to production centers
Transportation between operations and staging
Packaging and shipping

A proper flow pattern of material will help the manufacturing process by moving materials in the shortest possible time through the plant. It will also help in reducing rework, congestion, and safety problems. In order to obtain proper flow, one should consider the factors affecting the flow:

Plan to minimize storage of material at the workstation.
Minimize frequency of material handling.

Move material the shortest distance and in a straight line wherever possible.

Standardize containers and material handling equipment.

Minimize paperwork accompanying material movement.

Decide on the type of flow process best suited for manufacturing—batch, continuous, or cell.

Standardize production equipment.

Combine operations.

Reduce distances between departments.

Remove bulky outer packing from incoming materials.

Avoid using specialized handling equipment.

Avoid large variations in production output.

Plan for common use of special requirements.

Familiarize workers with system requirements.

Locate support services nearby to reduce production downtime.

Figure 4.19 shows various flow patterns in manufacturing plants. The beginning and the end of the flows depend to a certain extent on the locations of both the receiving and shipping docks. Other factors that determine the flow are the type of the products manufactured and the locations of the key equipment in the plant. The key equipment here is defined as the equipment that must be placed only at certain locations inside the plant. In this case, the flow pattern is worked around the equipment.

Figures 4.20a–4.20c and 4.20f show parallel manufacturing line flows usually adaptable to a high-volume, light to medium products manufacturing plant, e.g., electrical motors of different sizes, lawn mowers, small kitchen appliances. Figures 4.20d and 4.20e show a looped flow for a processing plant with little variations in the product, e.g., corrugated containers, automobile bumpers, dishwashers, and clothes washers and dryers. Figures 4.20g and 4.20h are for setups where more than one product is manufactured with subassemblies during processing, e.g., building hardware such as aluminum doors and windows.

4.3.2 Stores

Centralized Stores

Referring to Figure 4.21, method 1, the material is received at the receiving area, then inspected and routed for storage in a central store. From there it is retrieved for delivery to the workstations. However, provision has to be made to provide enough storage area for material received from the store. Frequently the workstations may end up being mini-stockrooms for items that are brought in as unit loads containing a large quantity of items whose usage is minimal.

Benefits

Materials can be bought in bulk for use on more than one product.

Avoids duplications of storage functions such as inventory control, supervision, and space in the storeroom.

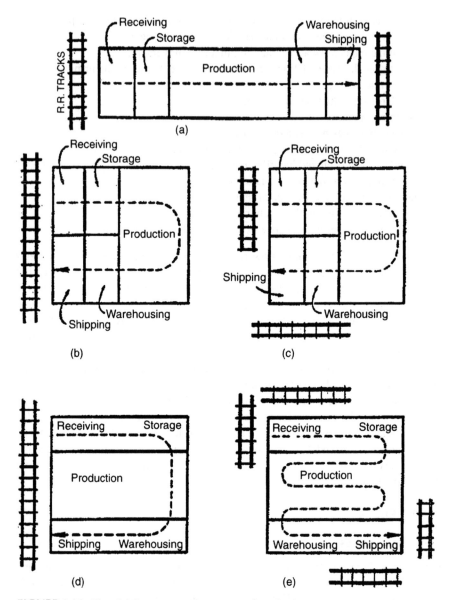

FIGURE 4.19 Material flow patterns in support of production.

Increased utilization of storage equipment such as automatic storage and retrieval systems (AS/RS) and vertical and horizontal carousels.

Increased utilization of material handling equipment.

Material handling system can be automated, since there is economic justification due to the large volume of products handled.

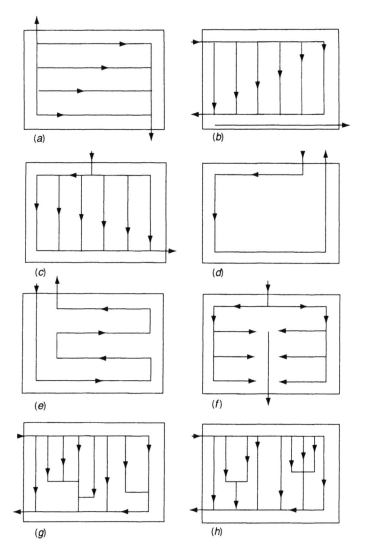

FIGURE 4.20 Types of manufacturing flow patterns: (a–c, f) parallel line flows; (d, e) a looped flow; (g, h) more than a single product.

Better control of material, since access to the stores can be limited to only a few authorized employees.

Flexibility due to cross-training of personnel and greater familiarity with a larger variety of parts.

Less prone to parts shortages for common materials, since they can be borrowed from materials allocated to other product lines.

Less investment in inventory, since duplicate storage is avoided of common items such as hardware, chemicals, and shop supply items.

METHOD 1   CENTRALIZED STORES

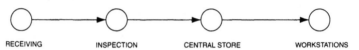

| RECEIVING | INSPECTION | CENTRAL STORE | WORKSTATIONS |

METHOD 2   DE-CENTRALIZED STORES

| RECEIVING | INSPECTION | DE-CENTRALIZED STORE | WORKSTATIONS |

METHOD 3   CONTINUOUS FLOW MANUFACTURING (CFM)

| RECEIVING (ON DOCK) | WORKSTATIONS |

METHOD 4   CENTRALIZED DISTRIBUTION STORE

| RECEIVING | INSPECTION | CENTRAL STORE | PRODUCTION CONTROL CRIBS | WORKSTATIONS |

METHOD 5   LIMITED CONTINUOUS FLOW MANUFACTURING

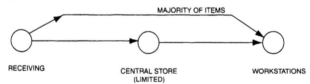

MAJORITY OF ITEMS

| RECEIVING | CENTRAL STORE (LIMITED) | WORKSTATIONS |

FIGURE 4.21 Material storage arrangements.

Drawbacks

Not all materials can be centralized, such as hazardous materials like oil, chemicals, explosives, and some packaging.

Excessive control on low-cost items such as nuts, bolts, or rivets can add cost burden and may cause production delays if not obtained in time from central store.

Breakdown of storage and handling equipment may paralyze entire plant operation.

Conflict in priority of servicing the needs of each production center by stores personnel.

Kitting function must be addressed properly to make materials available at proper workstation.

Production control has less flexibility and must properly coordinate production schedule and material requirement to avoid disruption and material jam-ups.

Emergency replacement and last-minute schedule changes may disrupt store's routine and affect other areas served.

Decentralized Stores

As shown in Figure 4.21, method 2, material is received at one receiving center, inspected, and routed to one of the many decentralized stores serving the corresponding production lines. These stores are naturally smaller in size than the centralized store, since each of them deals only with the materials required by the production line it serves. For instance, a central store might be storing 6000 items, whereas distributive stores might keep only 800 items. It is necessary that the receiving department have proper information on which store to route the material to when it is received. This can be done in a number of ways, such as coding receiving papers, identifying packaging material, or cross-referencing part numbers to store location on computer. However, with this method, the material is segregated at a common receiving center and distributed to properly identified decentralized stores.

Benefits

Store personnel work in close relationship with the production line they support and with material handlers to avoid delays.

More familiarity with products, and therefore less susceptibility to error in filling orders.

Less investment is required in automatic storage and handling equipment.

Store location is usually closer to the point of use, therefore avoiding delays in filling emergency requirements.

Store can easily be moved or rearranged in case of product change or reconfiguration of product line.

It is easier to implement in-store kitting requirements.

There is less damage to material since smaller quantities are handled.

There is better control of reordering, due to familiarity with parts and rate of usage.

Less storage space is required by workstations due to reduced lead time and improved response time.

Provides better support to continuous flow manufacturing concept.

Drawbacks

Duplication of storage for common parts results in higher overall inventory cost.

Labor utilization may suffer due to low store activity level.

Duplication of clerical and accounting functions and computer equipment.

Unauthorized and undocumented withdrawal of materials by unauthorized personnel.

Coordination and control by central authority on all decentralized stores to maintain uniformity can be a challenge to management.

4.3.3   Continuous Flow Manufacturing

What is continuous flow manufacturing, and how does it affect material supply?

In recent years, with so much emphasis on the Japanese production technique of just-in-time, American industries have started seeing the benefits of reducing inventory cost. A way of achieving this is to maintain a continuous flow of products through the manufacturing floor by eliminating batch-type production orders. Batches create a stop-and-go environment with increased setup, inspection, and delays. Often operators have to be retrained to perform the same job, due to the time between batches. The tooling condition, material readiness, and support services have to be lined up when a new production batch is begun. All these create inefficiency in the operation and increase the overall cost of manufacturing.

Automobile assembly is one example of continuous flow production which has been in use for decades. However, over the years improvements have been made to allow for more flexibility and productivity.

In continuous flow, the batch size is reduced almost to one unit, with the flexibility of allowing necessary changes to be made in a particular production unit. This allows for production of a multitude of models or variations every day. The material is stored directly at the point of use. (See Figure 4.21, method 3.) Receiving inspection is also eliminated. It is the responsibility of the supplier to assure that 100% acceptable parts are supplied and stored at the workstation at the right time.

However, this system requires total control over labor, material, and machines. Labor must be skilled to adapt to new variations. The right material in the exact quantity must be available at the proper work center. Every operator is his own inspector and should be able to perform the job requiring no delays or further inspection. Equipment should be well maintained and thoroughly checked periodically to avoid downtime, since the effectiveness of a continuous production line depends on achieving and maintaining high reliability of all its components.

Continuous production is not a new concept. Chrysler, and others, have extended it by involving major suppliers in delivering products on dock as required. It may be necessary for the supplier to deliver parts and materials two or three times daily, properly inspected and arranged so that they can be brought right on to the production line for use the same day. This may call for several loading ramps to be built into the building, along the production line.

Benefits

Quality awareness reduces product cost in the long run.

Inventory cost is greatly reduced.

Better service from suppliers, since they are also part of the continuous production team.

Materials handling can be simplified and automated due to steady demand in small volumes.

Operators are involved in keeping production line running for maximum output.

Drawbacks

Needs a great deal of planning and effort for start-up.

Many variables make it difficult to keep the line running smoothly.

May initially discourage management due to too many breakdowns, stoppages, and high costs.

As flow is depicted in Figure 4.21, method 4, the materials are received at the dock, inspected, and put away in a central store for future use. From there, they are retrieved by various production control cribs when required to support the production floor. In this case, the production control cribs usually act as temporary staging areas, so that materials are readily made available to production. The is usually the case in a large plant manufacturing several products. Production control cribs are usually situated near the production line to facilitate effective service. In many cases, it is the function of the production control crib to bring material from the store, unpack, and sort it out in kit form for deliveries directly to workstations.

Figure 4.21 method 5, shows yet another variation of a continuous flow system which is either adapted as a step toward continuous flow or used as a compromise in case of a product which is not fully adaptable to continuous flow. However, a properly designed system can yield some benefits to improve the efficiency for other systems used concurrently.

4.3.4 Establishing Stores Requirements

In arriving at the facility requirements for a manufacturing plant, the material handling and control system plays a large part. This system must be selected early in the process of developing a production block plan and layout. The number of parts, the quantity of each expected to be in the plant at one time, and the location, weight, and cube should be estimated. See Figure 4.22 for a materials storage analysis. Adequate floor space and volume can then be established, the material handling system defined, and equipment placed on order. Some of the equipment may be included in the cost of construction and provided by the building contractor. Other equipment may be handled by the company, and installed upon building completion. Some of the equipment that has "roots" is often assigned to the building contractor. This gives single-point responsibility for the coordination of structure, piping, and electrical requirements, and it is accepted upon demonstrated performance of the equipment.

MATERIALS STORAGE ANALYSIS

Approved By: GLP   By: VS   Page 2 of 9   Date: 4.5.89

No.	Item	Cont. Type	Size (Ft)	Weight (Lbs)	Quant. per Unit Load	Space Req'd (CuFt)	Handl Eqpmt Req'd	Handl and Storg Limit.	Prefer Storg Eqpmt	Storg Method	Pick Method	Storg Capac (Unit Load)	Total Space Req'd (CuFt)	Storg Cata-gory	Comments
1	VINYL FLOORING	WOOD	2X2 X12	1100	4 ROLLS	52	SIDE LOADER OR FORK TRUCK	WIDE AISLE CANT. BACKS	SIDE LOADER	RACKS TRUCK SIDE LOADER	FORK LOADS	12	624	A	
2	LAWN MOWER	CARTON PALLET	1¼X 2½X4	450	4 PIECES	15	FORK TRUCK	-	FORK TRUCK	PALLET BACKS	FORK TRUCK CARTON	8	120	B	
3	CAMPING FUEL (LIQUID)	CANS CARTON	1½X 2X4	15	10 GAL	12	PALLET JACK	HAZ. SUB.	FORK TRUCK	MARKED AREAS	MANUAL	10	120	D	HAZARDOUS SUBSTANCE
4	ANCHOR BOLTS-A	FIBER BOX	0.8X 1X1	60	50 PIECES	0.8	HAND CART	-	FORK TRUCK-L	CARTONS	MANUAL	5	4	C	
5	FLOOR SEALANT KIT	CAN CARTON	1.5X 2X3	125	12 QUARTS	9	FORK TRUCK	-	-	-	MANUAL	4	36	D,E	SHELF LIFE CONTROL KEEP IN COOL PLACE

CATAGORIES: A - Extra Long Items, B - Bulk Storage, C - Medium Shelf Storage, D - Special Storage, E - Rotate Stock.

FIGURE 4.22 Material storage analysis worksheet.

This often includes the air-conditioning equipment; compressed air; special cryogenic equipment, controls, and lines; communication and data lines; pumps; and other items that have complex interfaces or that could delay construction if they are not properly coordinated and available on schedule.

The completion of the material handling system requirements then feeds into a revised block plan and factory flow and layout.

4.4 ANALYSIS OF FACTORY EQUIPMENT NEEDS

4.4.0 Introduction and Overview

Equipment engineering encompasses many functions. Large companies have a person or department assigned to equipment engineering. Responsibilities may vary from company to company, but essentially the function is to select and procure the right equipment for the given set of data. It is essential that the responsible person or group involved in equipment engineering be business-oriented and have an engineering background.

Methodology of equipment engineering is an important and integral part of an organization, since it often represents one of the most vital functions of facilities planning. Proper selection of equipment provides smooth operation and lowest operating costs. On the other hand, improper selection results in cost overruns, downtime, quality defects, and management headaches. Equipment engineering is also involved in environmental and safety issues, which may have a large effect on the corporate bottom line.

4.4.1 Business Considerations of Equipment Acquisition

Equipment is a piece of machinery to support production in the plant. Equipment is distinguished generally from tooling in the sense that equipment is general-purpose and can be used for more than one particular application. It is a standard product for manufacturers. In contrast, tooling is especially made for a particular application, and is useless for other applications. Equipment is usually depreciated over its lifetime under Internal Revenue Service guidelines, except in the case of certain research and development projects where it could be expensed earlier. However, tooling is considered an up-front expense to produce an item. The yearly depreciation value provided for equipment is a bookkeeping entry which adds to the overhead expenses of doing business, but provides cash flow in terms of tax credits. The equipment in this category is commonly known as "capital equipment," since its book value at any time is considered to be a part of the capital assets of a company.

There is a rather large gray area between tooling and capital equipment. For example, an off-the-shelf machine can have some special features added to make it unique to the support of only one product. This may be done by the equipment manufacturer, or it may be designed and added to the machine in-house after receipt. Depending on your accounting department interpretation (subject to agreement with the IRS), the above example could be classified as capital equipment, since the modification is considered minor and perhaps usable on other products at some time in the future. Or it could be considered tooling, since the equipment is made special-purpose by your design, and the machine is only a part of the special-purpose tool. Or perhaps the most conservative approach would be to classify the standard part of the machine as capital, subject to depreciation, with the special features classified as tooling, charged to the product and expensed at the outset. This "color of the money" is worthy of some serious discussion at times, and should be well understood by the manufacturing engineer. The other choice is to make the equipment a part of the basic building, and treat it totally differently. One of the reasons that this discussion becomes important is that you may not want the equipment to become a part of your continuing overhead rate, which you would have to consider in your product cost every year. It might be wise to lease the equipment, which would be part of your annual cost, but you could terminate the lease, or exchange the equipment and lease a more modern machine as one becomes available in the future. This would reduce the amount of cash expenditure by your company required to set up a new facility.

4.4.2 The Process of Equipment Acquisition

The following sequence of events describes in general the process of acquiring equipment:

1. Plant layout showing the process flow and equipment in general
2. Flow charts, process flows, and any other documents that describe in greater detail the production process that requires the piece of equipment
3. Review of equipment literature from various suppliers
4. Preparation of equipment specifications

5. Solicitation of quotes for qualified sources
6. Selection of equipment
7. Justification and procurement package
8. Installation, start-up, training, and acceptance
9. Preventive maintenance, record keeping, guarantees, and follow-up

Plant Layout

The function of the plant layout in acquiring equipment is to help in preparing specifications and visualizing how the proposed equipment fits in the product flow of the plant. The selection of supply voltage and current, location of the breaker box, location of control panels in relation to building walls, access for maintenance, and other factors may be influenced by the equipment selection. It would be embarrassing to discover that the power drive for a machine cannot be replaced without moving the machine away from a wall, since there are no removable panels to provide access for maintenance from the front. It may be valuable for the equipment supplier to review the plant layout with you in terms of performance and accessibility of the equipment.

Certain equipment requires a special environment. A jig borer requires a clean, vibration-free, temperature-controlled room with a special foundation. Welding equipment requires protection from excessive air movement, which would otherwise blow off the arc or shield. An ammunition assembly machine may require a separation wall of reinforced concrete with a blow-off roof.

Manufacturers' Literature

Once the requirement for a particular type of machine is known, it is necessary to contact various manufacturers of this equipment and obtain their descriptive literature. A phone call, after checking the *Thomas Register* for prospective suppliers, will usually save correspondence time. Narrow down your selection to four or five manufacturers to consider asking for formal bids to your specification.

Equipment Specification

Equipment specification may be the most important part of the equipment engineering function. The preparation of equipment specifications is often a long, tedious process requiring help from other disciplines in the plant. It is the single common document to which all bidders respond. It spells out sufficient detail to ensure that the installed equipment will perform as planned. It establishes the responsibilities of the seller and the buyer during acquisition and after installation. It becomes a legal document binding both parties if there is a disagreement between buyer and seller in the future, since it becomes a part of the contract.

The suppliers' specifications and product descriptions will aid in preparation of the specification that you create, since you would like to buy "off the shelf" if possible, and not create an increased price by forcing the supplier to make a special model just to meet your request. However, in certain situations it is necessary for the supplier to modify the product to meet specification. One example is a forklift truck with a 3500-lb. capacity requirement. Since models are available in 3000- or 4000-lb. ratings, the manufacturer quotes on the 4000-lb. model, which uses a beefed-up chassis and frame assembly. The other alternative is to modify a 3000-lb. truck by adding the appropriate counterbalance to support a 3500-lb. weight. This alternative is sometimes less costly, and is acceptable in many situations.

Some of the important features to be defined in the specification include:

Intended use

Performance requirement (size, capacity, etc.)

Design and construction (rigidity, tolerances, hookup requirements, controls, etc.)

Data (maintenance procedures, schematics, etc.)

Installation (can vary from turnkey to technical assistance to none)

Warranty

Training (could be at your plant or supplier's plant; may be included in price or at an additional charge)

Acceptance test (important in high-speed or special-purpose equipment—may be witnessed at supplier's plant or at yours)

Maintenance service (may be needed if you do not have in-house capability)

Spare parts (should obtain the critical ones, at supplier's recommendation, as part of equipment purchase)

Selection of Equipment

Prior to placing a purchase order, it is necessary to make sure the supplier is capable, both financially and technologically, of providing the product to your satisfaction. The most common problems are:

Lack of working capital

Creditworthiness of the supplier

Unavailability of technical support

Uncontrollable growth at supplier's plant

Work delays due to safety and labor problems

Problems in meeting technical performance specification during construction

In any case it is necessary that the supplier selected be not only economical, but capable of supporting the quoted commitment to the equipment. A customer should pursue the following before placing the purchase order:

Obtain a credit reference from the financial institution with which the supplier is connected.

Call past customers who have obtained similar equipment regarding their experience.

Check sales, service, and warranty support by the supplier.

Visit the supplier's facility to review the manufacturing process and safety records.

Obtain a surety bond for any down payment made with the purchase order.

When delivery of key equipment in a system installation is critical, one approach is to attach a penalty clause to the purchase order for late delivery by the vendor.

Once the bids are received, the review process begins. The first step is to review the bidders' responses to the specification to make sure that the offers meet all the requirements. In certain cases, the bidder may have proposed an alternative solution, which should also be reviewed. Conversations between the purchasing agent and the bidders must be documented to avoid any possible future misunderstanding. (See Figure 4.23 for an example of a telephone log.) If a compromise is given to one bidder, then the other bidders should have an opportunity to revise their bids as well. Once all the bids have been reviewed and contacts made with the bidders' customers, the next step is to select the supplier. Award of the purchase order should not be based solely on the lowest bid received, but should also include reliability, experience of others, delivery lead time, installation, and operating costs. An example of a bid review sheet for four suppliers of a vertical carousel is shown in Figure 4.24. It provides salient features and other details of the equipment offered by the bidders. It may bring out irregularities, which may trigger further inquiries and uncover hidden traits. For instance, one brand may be 30% less costly than the others. On checking, you may find that the base casting is machined in a foreign country by an unproven source. A table travel may be manual in one machine but powered in another. One

DATE	DETAILS
	By: ___MPN___
	TELECOM LOG
	SUPPLIER/EQUIPMENT : ___ACME #3 CNC MILL___
5-6-92	JW called to find out if we can accept single phase motor at a savings of $ 750. I agreed to accept it.
5-15-92	Called to discuss with S. Roberts (Chief Engineer. ACME) regarding reliability of Mars Computer System. The equipment has 11/4" dismeter drive shaft instead of 1" - industry standard.
5-18-92	Called to find out users in this area: They are: 1. Reliance Technology 407-268-9532. 2. Rocket Science Corp.. Miami 305-212-7359.
5-20-92	Called Reliance Mr. Moocham. Chief Engineer. He experienced no major breakdown with the similar equipment - just a few oil leaks. Prompt service and parts inventory well kept at regional warehouse.

FIGURE 4.23 Telephone log used in the equipment selection process.

VERTICAL CAROUSEL FOR TOOL CRIB (DEPT. F203)

VENDOR:	1	2	3	4
MANUFACTURER:				
MODEL:	2412-8/28	140-1100-15	501 31/14	2400
DATA BASE:	MMS-2/5400-II INTERFACE	CIN/PL	F/P 2000	ACS10
DIMENSIONS: (HXHXD)	22.2 X 10'11" X 6' 2"	21.6' X 10.1'X 4-1/2'	21.1' X 10' X 5.6'	23' X 10' X 5 1/2'
PAINT:	TAN	BEIGE	BEIGE	PLANTIMUN
SAFETY FEATURES:	SAFETY BARS & INFARED PHTOEYE	TOUGH BAR & PHOTOCELL	HICROSWITCH	INFRARED
UNEVEN LOAD:	20% IMBALANCE	20% IMBALANCE	IMBALANCE INDICATOR	UP TO 1800#
MOTOR HP:	7.5	6.5	5.9	5
CARRIER: PITCH (IN.)	14"	18"	14"	13 1/2"
DIMENSIONS (HXWXD):	14" X 24 3/8" X 102"	16.5" X 15.9" X 96.5"	12" X 24.4" X 98.4"	12" X 24" X 98"
NUMBER:	28	26	31	30
WEIGHT CAPACITY LBS	1,200	1,100	1,100	1,150
CU. FT. STORAGE (TOTAL)	564	381	517	490
HARRANTY:	1 YEAR (ON SITE P+L)	1 YEAR (PARTS ONLY)	1 YEAR (PARTS ONLY)	1 YEAR
DELIVERY:	12 WEEKS	16 WEEKS	16 WEEKS	12 WEEKS
FOB	WISCONSIN	BOSTON	BSOTON	J.J.
TURNKEY INST.	INCLUDED	FORKTRUCK & DRIVER REQUIRED	FORKTRUCK & DRIVER REQUIRED	INCLUDED
NOISE LEVEL: DbA.	N/A	65	60	
SPEED IN/SECONDS:	5.1	6.0	5.3	-
PRICE:	$33,760.25	$28,839.00	$33,080.00	$48,468.00
INSTALLATION	2,400.00	2,500.00	3,576.00	3,400.00
BARCODE CAPABILITY:	3,622	INCLUDED	INCLUDED	INCLUDED
FIFO ACCESS:	872	INCLUDED	INCLUDED	INCLUDED
DATA BASE:	4,284.60	INCLUDED	INCLUDED	4,555.00
HAIN FAME INTERFACE:	1,850.00	INCLUDED	INCL. BASE PRICE	INCLUDED
TRAINING SOFTWARE:	$400/DAT+T&E ($1,400.)	$450/DAY+T&E ($1,500.)	1,275.00	INCLUDED
INTERFACE BOX (NETWORK):	INCLUDED	INCLUDED	6,495 (NOT REQ'D)	1,960.00
OVERLIGHTS:	INCLUDED	INCLUDED	150	316
POSITION LIGHTS:	INCLUDED	825	500	741
KIT PROCESSING:	INCLUDED	INCLUDED	995	-
19 DIGIT CAPACITY:	INCLUDED	INCLUDED	$1,877.	INCLUDED
EMERGENCY OPERATION:	HAND CRANK	WAND CRANK	HAND CRANK	HAND CRANK
MANUFACTURED IN:	USA	WEST GERMANY	WEST GERMANY	USA
SLOTTED CARRIERS:	INCLUDED	390 (STYLE e)	INCLUDED	INCLUDED
TOTAL COST:	$46,789.	$34,054.	$45,942.	$59,440.

FIGURE 4.24 Equipment bid review summary sheet.

supplier may have just switched to using ball bearings from heavy-duty roller bearings in order to reduce costs. Many construction details are generally not mentioned in the manufacturers' literature and are made available only by questioning the suppliers and their customers.

4.4.3 Justification of Equipment Purchases

Even though the piece of equipment finally selected is shown on the preliminary plant layout, it will probably require separate purchasing approval at the management level. An important part of factory equipment engineering is to quantitatively justify the purchase of each piece of equipment. Rather than making a long list of advantages and disadvantages, it is better to present to the decision makers your reasoning in monetary terms. Also, any alternative solutions to the purchase of this specific piece of equipment should be presented. The following is an example cost justification for the installation of a bridge crane for a Cincinnati machining center:

A. Savings Calculation

1. Savings in space and overhead expense because wide aisles and large staging areas will not be necessary, which otherwise is required for present use of fork truck for handling:

Estimated square footage required with fork truck handling   1,627

Estimated square footage required with bridge crane   1,147
Yearly savings: 480 ft.^{2} @ $10/ft.^{2}/yr. = $4,800/yr.

2. Elimination of production delays, as operator will not have to wait for fork truck which is also used in other areas of the plant:

   Estimated at 1 hr./shift $\times$ $12/hr. $\times$ 2 shifts/day $\times$ 240 days/yr. = $5,760/yr.

3. Savings from elimination of damage due to improper handling by fork trucks. Machined parts and fixtures are costly, and if not handled properly while positioning, the machine may cause damage and dents that require expensive rework and replacement. Based on a minimum of one incident per year:

Material and controls	$8,430
Labor (190 hrs. $\times$ $12/hr.)	$2,280
Total	$10,710/yr.

4. Delays and rework, which may require overtime operation to meet schedule requirements:

   10 hrs./wk. $\times$ $6/hr. (OT prem.) $\times$ 50 wks./yr. = $6,000

5. Handling of fixtures and parts without bridge crane, which may require additional time of 1 hr./shift:

   1 hr./shift $\times$ 2 shifts/day $\times$ 240 days/yr. $\times$ $12/hr. = $5,760/yr.

6. Total cost savings estimated:

   $4,800 + $5,760 + $10,710 + $6,000 + $5,760 = $33,030/yr.

B. Equipment Cost (Capital Investment)

Two-ton bridge crane with controls	$62,500
Installation, testing, and certification	$9,000
Freight and taxes	$8,500
Total	$80,000

C. Prospective Rate of Return on Investment

Original cost	$80,000
Salvage value	$2,000
Expected life	20 years
Straight-line depreciation	$3,900/yr.

Return on investment: ($33,030 $-$ $3,900 $\div$ $80,000) $\times$ 100 = 36.4%

This method of calculating the prospective rate of return is also referred to as the "original book" method. The rate of return on the investment is the interest rate at which the present value of the net cash flow is zero. In this case, 36.4% is considered the interest rate of money invested and shows the relative attractiveness of buying the bridge crane in comparison with other projects showing different returns on the investment.

If the investment money is borrowed from a financial institution at 12% interest to purchase the bridge crane, then the purchase in effect provides additional cash flow (profit) of $19,520 per year to the facility (36.4% − 12% = 24.4% × $80,000 = $19,520).

However, if the total savings generated is only $12,000 per year instead of $33,030, then the return on investment would be 10%. In this case, the manufacturing engineer would not recommend the purchase of the bridge crane, since it creates a negative cash flow of $1,600 per year—meaning it would cost more to operate the bridge crane than continue the present method.

4.4.4 Equipment Purchase Priority

Earlier in this chapter, we discussed the possibility (likelihood!) of not being able to get all the equipment to optimize our operations in a "world-class manufacturing" mode—at one time. We therefore suggested that we build the optimum building required, even if we could not afford all the equipment desired prior to starting production. The decision then becomes one of deciding which equipment, machine, or system to proceed with first, and which ones could come online as funding was available.

Visualize a complete list of equipment prepared by reviewing the entire production and materials handling processes, and adding the equipment items that were omitted from the building program, and finally adding the other capital expenditure items from the remainder of the factory. These could include computer systems, networks, software packages, additional terminals, and a host of other pieces of equipment needed to properly outfit the new product facility.

Priority Assessment

The facility floor, walls, roof, lighting, etc. plus the equipment that is absolutely necessary to be able to operate the plant comes first.

There will be some machines and other pieces of equipment that are absolutely required in order to start production. Without a certain minimum amount of the necessary machine tools, etc., we could not produce the product. These pieces of equipment are next on our priority list.

From this point on, the decision as to which expenditure to make next will be based on its value to the operation by making financial analyses like we addressed in Subchapter 4.4.3. Although all companies seem to have different criteria for justifying equipment purchases, the bottom line is always the same—which one makes the most money?

4.5   IMPLEMENTATION PLAN FOR FACTORY AND EQUIPMENT

4.5.0   Introduction and Overview

We talked in Chapters 1 and 2 about the requirements for a good product design, and the need to pay close attention to the needs and wants of our customers. We analyzed and revised our existing or proposed product design, with a team of experts from all functions in the factory. We introduced concurrent engineering, QFD, and DFA as tools to guide us to assure that we did not overlook anything, that we had an organized approach to our analyses, and that our product was the best that we could develop. We discussed the advantage of rapid prototyping techniques to obtain some of the parts. This allows visual and "feel" evaluation, and permits testing in some cases to prove to ourselves that our product is as we had visualized it. We looked at the dimensions, limits, and tolerances of various processes, to help establish the details of the product design that could be produced. We considered the tolerances that the various processes and machines are capable of holding. We made cost studies of the various design and process alternatives, and understood that we could meet the cost and quality requirements of the product in production.

Now, in this chapter, we started thinking about the factory required to produce our product. In Subchapter 4.1 we looked at some of the general factory requirements, and outlined information that would be needed before we could define the physical building arrangement that would best suit our product production. In Subchapter 4.2 we looked at specific factory requirements from the product point of view: what would be needed to support the direct manufacturing operations that we had outlined? In Subchapter 4.3 we looked at the factory requirements from the materials and materials handling point of view. How should we receive, store, and transport the materials to the work stations when needed, and move the product progressively through the production process? More tools and techniques were explored that would help us evaluate our factory requirements from the product and materials point of view. Subchapter 4.4 talked about the selection and justification of capital equipment needed to support both production and material flow, since they are both part of the factory production operation.

The final "missing link" in the factory requirements process is to determine how we want to operate the plant. What does our top management have in mind for our future world-class factory goals? What systems will we set up, and how will we actually run the plant? This must be done prior to the final factory requirements definition, or selection of equipment, or making a final layout, or designing a factory (Chapter 4). There are a lot of options at this point, and perhaps we should look again at the world-class manufacturer characteristics that we introduced in Chapter 1. They are as follows:

Costs to produce down 20–50%
Manufacturing lead time decreased by 50–90%
Overall cycle time decreased by 50%
Inventory down 50%+

Cost of quality reduced by 50%+ (less than 5% of sales)
Factory floor space reduced by 30–70%
Purchasing costs down 5–10% every year
Inventory turns of raw material a minimum of 25%
Manufacturing cycle times with less than 25% queue time
On-time delivery 98%
Quality defects from any cause less than 200 per million

While we can be certain that no one would agree that *all* of the above performance improvements are required, there is a strong message that no one can dispute. We must make a paradigm shift in the way we operate to achieve these "future state" conditions, and perhaps make some major changes to achieve any one of them!

If we believe (as your authors do!) that the above are not just goals, but rather essential elements for success in manufacturing, perhaps we should let these elements guide us in establishing the final requirements for the factory required to produce our product. It would be hard to believe that we could achieve *all* the above goals, and implement *all* the requirements in setting up our factory (existing or new construction) to produce our new product that we have so carefully developed—at least not prior to producing the first product for delivery to our customers. However, we must *consider* all of them in developing the requirements for our facility layout to build the new product—and future products. We need a plan to get there! Most companies will have yearly goals, and 5-year or 10-year operating plans. We can assume that the success of our new product will play a role in these future plans. The questions that come to mind include the following:

How much money for capital expenditures is available this year for the new factory (or new arrangement)?

How much money is available for capital expenditures on this product (or product line) in future years?

What do the cash flow analyses justify in the way of company investment for this product?

What will be the projected cost for the design and build of the new facility?

What will be the cost for the optimum machines and equipment to support production?

What would be the minimum investment required to support starting production—even if our direct labor cost goals could not be met in this configuration?

What would be the cost of the optimum material storage and handling system and equipment?

What would be the cost of the minimum equipment for storage and handling to support production—even if the support cost was not the lowest?

What are the risks involved in the success of this project?

These kinds of questions (and others) will be asked prior to the start of the project and during the design development phase of the product. They will be asked again after the studies are more definite, prior to giving approval to proceed with the optimum factory plan. The bottom line in factory requirements is going to be a compromise. The maximum expenditures required for buildings, production equipment, and material handling equipment to obtain the optimum product and material flow which gives us the highest-quality and lowest-cost product is one extreme. As manufacturing engineers we should define this configuration. However, we should be prepared to offer alternative plans and plans for progressive implementation.

There are many unique circumstances in all companies that will give different answers to the above questions, however we must proceed to support the product production:

> We should establish the factory building design for the optimum system— even though it may not have *all* the desired features and equipment in place the day we start production.
> It will be a progressive activity, as the product production process matures— as the market is proven—and more funding is available.
> We must never forget that the objective of running a factory is to make money for the investors, and this real world can only stand so much negative cash flow—or only so much time before the investment begins to actually pay off.

Having established this as a background check on the cost and feasibility of our factory, we can proceed with a summary of the requirements for its design and construction.

4.5.1   Implementing a Plan for Revisions to an Existing Factory

If rearrangement of the factory to meet the production needs is fairly minor, perhaps the work can be done with your in-house staff. This usually requires some sort of factory order, usually initiated by the manufacturing engineer, to start the actual layout and rearrangement process. After review by the appropriate group responsible for performing the work, a cost estimate should be prepared and approved. The actual layout can now be prepared in the form that is common with your plant procedures, and the necessary engineering calculations for power and other utilities made. This is followed by making more detailed drawings of the modifications to power, water, compressed air, air conditioning, lighting, and other systems. A master schedule of all these activities should then be prepared, and a detailed move schedule for all the major and minor pieces of equipment prepared. In some cases, it may be necessary to contact local trades for proposals to perform some of this work, in conjunction with your in-house maintenance workforce.

The coordination of this "minor" rearrangement is usually more difficult and time-consuming than originally visualized. People get very emotional about moving their offices, making certain that they have the required computer hookups, that the floor machines are accessible for maintenance, and, most important, that production of other products does not suffer. Sometimes part of the work may have to be scheduled at night or on weekends. An example of such a problem would be in cleaning and sealing a concrete floor. This requires the use of strong chemicals, which could cause damage to electronic products and creates strong odors. Another example would be welding of brackets, etc., to the overhead structure, where the danger of fire, or damage to work being performed under the overhead activities, must be considered.

In other words, the process is about the same as in designing and constructing a new building, only on a much smaller scale. Quite often consultants will be needed, and in many cases an A&E firm will be hired to do some of the design and coordination. In many cases, a new building was lower in cost than the same area available by adding a mezzanine to an existing facility. This is one of the reasons for making in-house plans and estimates, followed by the appropriate management approvals.

4.5.2 Implementing a Plan for a New Factory Building

When it has been decided that a major add-on to an existing building, or an entirely new building, is required, a more complex set of circumstances is created. It becomes the task of the manufacturing company to provide to the A&E firm, or the construction management firm, sufficient information to assure that the new facility will meet your requirements. Chapter 4 outlines this program is considerable detail. The manufacturing company should set up a building committee and appoint a project manager (PM). The PM becomes the funnel for all information transfer in both directions. The building committee is his or her sounding board, and in most cases approval authority, during all phases from this point on. All of the flow charts, analyses, preliminary block plans, layouts, and studies that have been prepared by the manufacturing engineer and others should be provided to the design agent, through the PM. The company activities do not stop at this point, since there will be continual discussions and joint decisions made during design and construction. The company will want to check most of the work done as it progresses. They will help in obtaining permits, coordination, and approvals from the local regulatory agencies, and in general making certain that all factors internal and external to the construction proceed smoothly. This usually involves some local political contacts, and working with other influential and interested parties in the community and state. There may be state road funds available to help finance the project, or special agreements made that will ease some of the regulatory problems.

5 Factory Design

William L. Walker

5.0 INTRODUCTION TO FACTORY DESIGN AND CONSTRUCTION

This chapter introduces the processes of facility design and construction in general, and some of the specific options that can significantly effect the outcome. We will start with a look at opportunities for the management of an existing manufacturing firm to shape the resulting designs. Then we will address the two principal design methodologies and explore some of their differences. Earlier chapters provided many of the technical considerations related to manufacturing. This chapter is intended to build upon that information in order to best support the real needs of the factory.

A successful factory design is much more than the sum of the electrical and mechanical requirements of the equipment nameplates. Design is more intuitive, based on experience with codes and with people. There are thousands of decisions that must be made during design and construction. One way to deal with the complexities is to break the process of design into smaller, more manageable groups. A good building program will help define and reinforce the expectations of management.[1] One important benefit is that it can lead the architect or engineer to ask the right questions and make the right decisions during the process of design.

While every element of the new factory is important, some elements are more important than others. That hierarchy must be determined through careful evaluation. The building committee is responsible for determining such issues for the designers. The "building program" as discussed in this chapter is the best opportunity for the manufacturing company to define the requirements and expectations of the new factory.

You might think of the building program as a recipe for a $100-million five-course meal. If you are going to pay that kind of money for something, you want to be confident of success before you put it in the oven. It must provide enough specific information to guide a complete stranger through a very complicated series of decisions. There are high expectations and high stakes. Plans and specifications developed from a good

---

1. The term "program" is not just a general plan for building the new facility. Used by A&E firms, it becomes the specific document that ultimately guides each architect, engineer, and designer in the preparation of construction drawings, specifications, and contracts for work to be done. It translates the users' requirements into the language of the building trades.

program will yield superior results through cooperative teamwork. Construction costs are notoriously volatile, and time is always an important factor. Construction may become one of the highest-risk challenges many companies will face in the next century. Furthermore, in many cases factory refurbishing, additions, and new construction may be a requirement for survival in the near future.

The last section of this chapter deals with construction. In order to survive recent economic changes, many building construction companies have modified the way they do business. There are many advantages to each type of company, whether it be a general contractor or construction management firm. Subchapter 5.5 discusses some of their benefits and strengths. The language of the building designer is used on occasion to discuss some terms and processes.

5.1   THE TWENTY-FIRST-CENTURY MANUFACTURING FACILITY

Designing the twenty-first-century manufacturing facility will be more challenging than ever before. No longer can we expect to draw all of our inferences from what has worked in the past. We must now have a process of design that draws from the solutions used in past decades and develops appropriate responses for the future. Changes in management and society have mandated new performance criteria as well as new functions that must be planned for. Changes in materials and equipment are only beginning to affect the design and construction industries. With the current rapid market and technology changes, we are faced with different evaluation criteria than in the past. The profit potentials of tomorrow, coupled with modern corporate philosophies, have created new facilities challenges.

One flawed decision can result in millions of dollars lost and potential human and business disasters. With this added pressure comes another series of changes in the way businesses make decisions. Many large companies distribute responsibility to a group of area managers such as fiscal, safety, engineering, manufacturing, support staff, legal, etc. This can make for a lot of lengthy meetings, difficulty in reaching decisions, and even major disagreements. While it does provide many ways of looking at the outcome of the decisions, it can also create delays in making one.

While there are always work-arounds to obstacles, some of them create their own negative results and can snowball into big problems. What, then, is the best way to deal with the challenge? Is it better to have one senior-level manager in charge of all the decisions, or maybe hire an outside guru with 40 years' experience? The decisions must be made for every firm based on its specific needs, resources, and situations. The method for determining future decisions comes from management's goals and objectives.

5.1.1   Accounting Systems

Years ago most firms utilized annual profit statements in evaluating the performance of the directors responsible for factory performance. As an example, accounting was usually based on a fairly narrow scope, seldom accounting for waste removal or treatment.

Wastes could be dumped or stored in drums for indeterminate periods to avoid quarterly accounting for storage, disposal, and byproducts in product cost analysis. This practice resulted in skewed data and large waste dumps.

Responsible firms are moving toward full-cycle cost accounting, from source to disposal, including energy and transportation costs. Some firms have begun to carry all expense and income figures as they are incurred, tracked against each product to which they apply. In this way management is provided a more complete and accurate indication of the real costs of doing business. Regardless of the way a firm tracks expenses, it is imperative that the decision makers understand the complete impact of their decisions in the case of new construction. Understanding the real meaning of the data available helps their decisions to yield more consistent success.

5.1.2 The Building Committee

The first thing that must be done by the manufacturing firm is to form a building committee. Upper-level management must be represented, preferably with full decision-making and budgetary authority. This representative may not be the company's eventual project manager (PM) for design and construction, but it must be a person who is strong-willed, experienced, and capable of maintaining focus on the big picture. His or her primary responsibilities are to ensure that the design feedback supplied to the design firm is the best possible, and that the overall project cost targets are met. All area representatives offer their insights and desires. It is up to this senior manager to resolve conflicts and guide the committee to clear, enforceable agreements and directives. Often this person is responsible for the ultimate building program, whether an outside consultant is hired for the actual preparation of the document or it is done in-house. The size and makeup of the committee will depend on the structure of the firm, but experience indicates that it should not exceed nine people.

5.1.3 The Project Manager

The project manager is the single most important player throughout the design and construction process. This person must have experience in similar situations, common sense, and commitment to the project goals. Whether the goal is the cheapest initial-cost space or the most vibration-free work platform achievable, the day-to-day control of the project is the responsibility of the PM. Once the committee decides the issues and objectives and assigns them significance factors (hierarchy), the PM will be the primary point of interface. It may be the PM's responsibility to originate and then implement the hierarchy chart with help from the building committee. The PM will also be responsible for paperwork throughout the project. From the selection of the design firm through construction, this person will represent the firm in meetings with outside agents and agencies. See Figure 5.1 for a diagram showing the relationships among all parties.

Meetings in which decisions are made relative to the project may be held as often as six times a day for 2 years. The PM will be called upon to evaluate options or to make recommendations for changes in the plans and specifications. At each phase of

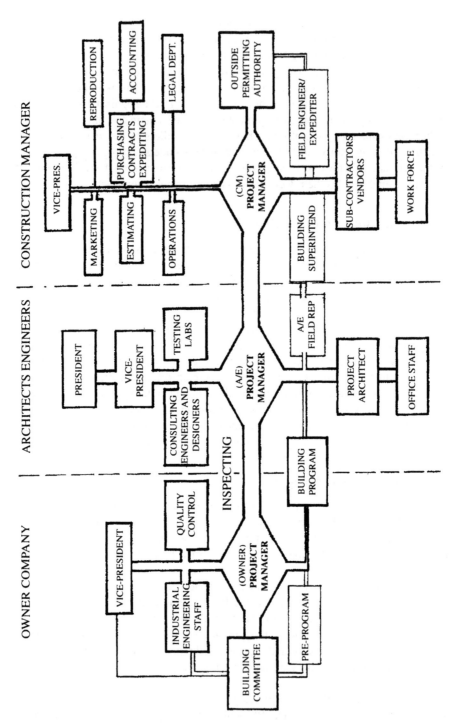

FIGURE 5.1   Relationships among participants in a typical construction program.

design, the PM will be responsible for review and approval of design documents. One of the most important duties is to make sure that other members of the team perform. Milestones must be met in order to complete things according to schedule. During construction there will be everyday opportunities to influence the process. The electrical and mechanical drawings are schematic at best; there will need to be field interpretation of the documents in many instances. A good PM can find answers to problems before anybody else even becomes aware of them. The PM must repeatedly evaluate the needs of the end users, in order to assure the best results. All the way through the project, the PM will evaluate materials on-site and work in place for payments to the contractors. Cost and schedule savings are likely through good management.

5.1.4   Hierarchy of Decisions

Designed by many, managed by few!

One of the most frequently overlooked aspects of decision making as related to factory design is that of relative importance to the big picture, or *hierarchy*. There needs to be a fixed set of criteria which will help the designers through the process of design. It is usually up to the manufacturing firm to establish, and later to enforce, this critical list. With a clear understanding of the hierarchy of factors, the design team will keep in mind the values of the user group.

By assembling the proper representatives from each group within the manufacturing firm to serve on a building committee, management can achieve many things. The first is representation of all those impacted by design decisions while they have a chance to contribute, and at the right forum. One might think of applying the "golden mean" as ascribed to Aristotle: *Everything in the right proportion creates the best results.* We will take that two steps further by adding that it must be at the right time as well as at the right cost. Construction, like manufacturing, is a five-dimensional animal. There are the regular three dimensions (height, width, and length) that must be considered—along with time and money. A project that comes in on time and on budget is considered more successful than one that comes in exactly the right size. Therefore it is the right proportion of all components which will lead to the best results. Whether we apply that to building the best facility or the best product, it still applies.

Chart of Issues

A typical project might have the following chart of issues, as arranged in decreasing level of relative importance (hierarchy):

1. Completed at a cost not to exceed $125 per square foot
2. Completed within 18 months
3. Utilizes local labor and 25% minority businesses
4. Highest possible reliability of operation
5. Lowest 10-year planning horizon for equipment costs

6. Cash flow during construction not to exceed interest from investments
7. Lowest operating and maintenance costs on equipment
8. Good relationship with local politicians and influential VIPs
9. All byproducts treated and tracked—no dumping

You can see how this would result in a different end result than the following hierarchy chart:

1. Finish as fast as possible; 10 months is best
2. Construction budget not an issue
3. Lowest 5-year planning horizon for equipment and operating costs
4. No permits required; no building inspections required
5. Local labor may be utilized
6. Waste byproducts can be dumped into river

It is very important to establish the hierarchy prior to beginning design, even though it may change during the process. This places a lot of importance on the building committee meetings. It is helpful to distribute copies of all correspondence with the design and build professionals to the building committee. No single person can be expected to make 10,000 right decisions under stress. This simple precaution can prevent a lot of problems later. Often the PM cannot know the impact of all his or her decisions on the ultimate specialized user. Using e-mail to a preprogrammed distribution group can save a lot of time, and much wasted paper.

5.1.5   Importance of Time and Money

This subsection is intended to provide information related to evaluating the importance of time and money for each specific project. Time and money tend to vary more widely than any other construction variable, and certainly they have the greatest impact on the factory design and construction process. The old adage "Time is money" does not provide the total story. There is always a trade-off between time and money. Seldom do we have a finite date for the completion and operation of a factory when we begin the design process. It should be understood that there is a point at which it is no longer cost-effective to try to shorten the delivery process—the point of diminishing return. A good goal would be to design and build the most cost-effective facility that meets the performance criteria set by the committee as spelled out in the building program. The best result is to finish ahead of time, under budget, and still meet or exceed performance expectations.

The Influence of Time

Factors that may have an impact on time as a determining factor might include the market for the end product, its sensitivity to time, the penalties and benefits for delayed delivery, and also the impact on the cost of construction. Construction costs over the long term have increased along with the inflation rate, often influencing the cost of living more than other factors. We have seen the cost of construction components fluctuate more than 100% in less than 90 days. These swings are most often driven by economic changes such as are caused by wars, legislation, or naturally destructive forces such as hurricanes. Labor costs must also be considered. Adding more people to perform tasks does not necessarily improve their performance. There is a maximum number of people who can work on welding one section of pipe, for example, so we must schedule work tasks during construction in such a manner as to maximize their performance. Overtime and additional shifts can be used to shorten construction time more successfully than overstaffing on one shift, but long hours tend to reduce worker yield as well. An experienced construction manager or project manager should know the benefit-cost ratio that is optimal. Major time components in a typical 3-year project schedule for a large, complex facility include:

	Percentage	Weeks
Contract negotiation	6	9
Design development	11	14
Review by owners	4	6
Completion of contract documents	14	22
Permitting	7	11
Bidding and award	5	8
Construction	52	82
Total	100	156

Often projects can be delayed more than 14 weeks due to lengthy review periods or lack of communications during these planning stages. Delays from lengthy owner review periods, or from negotiating insignificant details on a construction contract, can cost time and money. Less obvious but equally important are decisions as to building type, materials, systems, connections, and finishes. Equipment can have more than 10 months lead time for delivery, then require extensive fit-up. The labor element of construction is one that can also affect project delivery time. Brick exterior walls, for example, take more labor hours per square foot than metal panel walls, bolted connections can be quicker than welded ones, and skid mounted equipment is quicker to connect than loose components. Figure 5.2 shows a project planning timetable for a minor project and a major project, with average and expedited times shown.

It can be imprudent to try to expedite the *schematic* design phase very much. We must make efficient use of the available key personnel in the architectural and engineering (A&E) teams' offices. There can really be only a few designers producing effectively during the preliminary/schematic design stage. This is usually a short period in the project schedule anyway, but it has significant implications for the end result. Trying to cut time here is usually counterproductive. Owner review time can be trimmed, especially if there is good interface during the design development. Weekly, or at most biweekly, team meetings combined with continuous design reviews can prevent problems.

The best opportunities for time savings are during the design development and contract documents phases. A large firm can assign more than 30 production staffers full time, and if there are consulting firms under a primary one, there could be more

DESCRIPTION	MINOR PROJECTS		MAJOR PROJECTS	
	Min.	Max.	Min.	Max.
Preliminary				
Preprogram	2	5	4	8
Hire design team	3	5	4	14
Schematic design	2	6	4	10
Review by owner	2	4	4	6
Budget development	1	4	2	6
Site analysis	2	6	2	12
Master plan	3	5	4	8
Analysis & review	1	2	2	4
Financing	2	6	2	10
Design Development				
50% design development	3	5	4	12
Review by owner & CM	1	3	2	4
100% design development	6	8	6	18
Cost estimating	1	4	2	5
Review by owner & CM	1	2	1	4
Pre-permitting review	0	8	1	16
Construction Documents				
50% CDs	4	10	4	14
Review by owner &CM	1	2	2	6
95% CDs	3	8	5	10
Final cost estimate check	2	5	2	6
Final owner & CM review	1	2	2	6
Plans and specs complete	2	4	2	8
Permitting	2	6	4	14
Bidding				
Advertising for bids	2	4	2	6
Selection of firm	1	4	2	8
Bid preparation	3	6	3	8
Evaluation and review of bids	1	2	2	6
Award contracts	0	4	1	8
Shop drawings prep	2	6	4	16
Review, resubmit, approve	0	2	1	8
Building				
Site prep	1	8	4	14
Utilities	1	6	3	12
Foundations	2	5	2	8
Construction	18	32	48	76
Start-up, test & balance	1	4	3	7
Punch list	1	2	1	4
Final completion	2	3	2	10
Project closeout	1	2	2	8
TOTAL IN WEEKS	81	200	145	398

	MINOR PROJECTS		MAJOR PROJECTS	
	Min.	Max.	Min.	Max.
Total number of years	1.6	3.8	2.8	7.7
Number of weeks for reviews	8	27	17	60

FIGURE 5.2 Project planning timetable showing the minimum and maximum times from the preprogram phase through project completion.

than 50 or 60 employees on the project at one time. The cost of this workforce may not be immediately visible, but it should be clear once proposals are evaluated. Providing two alternative schedules for document completion will provide a means for evaluating actual costs for any savings in time. The potential is for cutting more than 2 months from the design process.

The permitting period may be time-consuming or it may not even be an issue, depending on the project. Consultants can often submit early plan sets for review by code officials to prevent lengthy delays. Sometimes nothing can be done toward this end. Local A&E or design/build firms should have the best estimate for the time involved here.

Permitting may also be a good period for final owner review and approval of the design, and for contractors to secure bids from their subcontractors and suppliers. Ideally there will be no major redesign after approval of plans. Usually one addendum can be issued to the plans and specs after bids are in, prior to construction award. An addendum allows changes in the documents to be clarified to all bidders prior to commitment to one contractor. This is really the last chance for a competitive bid situation on known changes. Occasionally a short list of the top proposals will have an opportunity to rebid after clarifications of issues encountered during bid negotiations.

The cost of time is difficult to calculate. There are inflation rates, interest rates, and lost opportunity costs to consider. Usually the fiscal officers for the firm can closely predict short-term trends and long-term levels. Marketing personnel may provide input as to present and future demands for products.

The Cost of Money Changes with Time

The cost of money continues to change, in terms of both interest rates and initiation costs to acquire loans. Similar factors must be weighed if the money is coming from internal sources. What do we have to give up in order to use this money for building? It costs nothing to expedite contract negotiations, unless you make a mistake in your haste. Therefore the issues of time and money are *interdependent* and are extremely important to a successful project. During the design phase, most decisions must be made with complete knowledge of their impact on time and money. The biggest thing to look out for here is the omission of clauses that would cover contract scope under *basic services.* Always take the time to think ahead for *services* such as field representation and inspections, interiors design, landscape and civil design, testing, multiple bid packages, and phasing. Include as many services in the original agreement as possible. Additional services contracts generally cost more than would the same work under a basic service agreement. To prevent contractors from using inflated estimates for loosely defined scopes of work, allowance figures can be developed with not-to-exceed amounts. In addition, agree to terms and conditions for potential future billing rates for out-of-scope work in advance of signing a design contract.

The cost of money is much easier to calculate than the cost of time. This cost can be developed by taking into account loan origination costs, interest rates, duration, and amounts. Another factor might be what other uses could be made with that money. If it is self-financed, the options would include reducing other debt, lost income from the money if invested or working, purchasing revenue-producing equipment, or many other things. For these reasons and more, it is important to have a qualified representative acting as PM to make these decisions. He or she should at least understand them well enough to present the major items to the proper management representatives for decision. Whether you choose to create a simple or complex equation will depend on your firm's specific situation.

Another issue is whether or not to include early equipment purchases in the general contractor (GC) or construction manager (CM) contract scope. If equipment is purchased by the owner, there will be one less markup on the equipment cost and orders can be placed prior to awarding a contract for construction. There are advantages to the subs purchasing the equipment as well, such as sole source responsibility and coordination of installation, which may make up for the markup. A good rule of thumb is to combine all long-lead items that would interfere with project critical path into early bid packages. Whether to buy them direct or through the CM or GC should be a simple decision based on the financial package. Unless your company buys a lot of large equipment direct, the CM can usually get more competitive pricing from major suppliers. It can certainly exceed the 5 to 10% you would pay the CM, and the CM would then be responsible for scheduling and expediting arrivals. This gives the CM additional leverage in negotiations with their subcontractors as well. One good way to assure performance by designers and constructors is to pay them performance bonuses for accelerated delivery, and to penalize them for late delivery. Most professionals will respond to this enticement better than other management techniques such as assessing liquidated damages for late delivery.

The last thing to define is the project budget. Early in the planning stages, it is important to develop accurate project cost estimates. Gross area-based system cost projections can be made from schematic drawings and current prices. It is important that all significant potential cost elements be identified. This should be done early and should be checked all along the way. There will be two or three times when the documents are developed with sufficient detail for the next level of cost estimating. The first budget needs to be general, with projected expenses and revenues. The example in Figure 5.3 shows a one-page summary budget.

The five major issues we have addressed so far are: (1) Appoint a senior manager to head up a building committee. (2) Create a project manager position—hire a good one from outside or assign someone qualified from within. This person can work as head of facilities or plant engineering before, during, or after construction. Expect both managers to work more than 50 hours a week for as long as it takes. (3) Form a building committee of representatives from diverse areas and work together to forge objectives and goals. (4) Establish the hierarchy. (5) Determine preliminary schedule and cost projections.

5.2 SITE SELECTION

We will proceed as though it is the responsibility of the factory facilities staff and plant management to determine the site for the new facility. An alternative procedure

Description	Units	Cost/Unit	Price
Site costs			
Land purchase, acres	20	$21,650	$433,000
Utilities, incl. el. water, sewer, storm, and FP	20	$3,500	$70,000
Clearing, leave trees on 4 acres	18	$650	$10,000
Development, includes fill and compaction	16	$1,512	$24,192
Taxes, permits, fees	20	$6,062	$121,240
	Subtotal	$33,374	$658,832
Other Costs	(sq. ft.)		
Professional soft costs (below)	Percentage	of constr. cost	$965,750
Construction manager fees	Percentage	of constr. cost	$756,480
Buildings, type 1, typical conditioned	170,000	$84	$14,280,000
Buildings, type 2, storage, unconditioned	63,000	$21	$1,338,750
Buildings, type 3, special construction	17,000	$145	$2,465,000
Early equipment purchase	Bid package		$715,475
Parking, paving and planting	6,570	$6	$37,121
Insurance	Percentage	of constr. cost	$126,000
Contingency at 7%	Percentage	of constr. cost	$1,050,000
	Estimated project cost		$22,393,408

Construction Soft Costs
Professional fees, breakdown

A&E		$285,752.50
Mech, elect, structural eng's		$460,087.50
Inspection services		$38,393.75
Surveys and tests		$38,630.00
Systems consultant		$48,287.50
Vibration consultant		$28,972.50
Civil eng. & irrig. design		$51,146.25
Materials handling, removals		$14,480.00
	Sum	$965,750.00

FIGURE 5.3   Preliminary budget.

would be to select the A&E team prior to site determination, and to pay them for their assistance in site evaluation.

Firms with master planning expertise can be helpful in evaluating potential sites. They can often provide the personnel to perform labor-intensive comparisons when the in-house staff does not have the time. If you are adding on to an existing facility site, the fee will be minimal or perhaps the work can be done at no extra cost. Do not assume that it is cheaper to modify the existing facility than to build new, nor assume the opposite. In every case, perform an evaluation of actual cost in time and money prior to determining a course of action. You may be surprised at the outcome.

5.2.1   Expansion of the Existing Site

The first task is to determine the feasibility of adding a new building to an existing site. One must consider circulation, structure, utility needs, and support spaces, as well as the impact on the existing operation. Frequently there is an exterior wall common to the new space that would make no obvious impact on plant operations. If the expansion is not too large, this may be the most cost-effective approach.

In some locations, site coverage may be limited because of impervious areas for percolation of rainfall, or one may be prevented from paving or roofing over green areas. Other regulations must be reviewed in relation to parking spaces, green spaces, numbers and sizes of existing trees, etc. There may be agreements in place that limit the allowable square footage for any building site, such as a "Development for Regional Impact" or similar document. Always take the time to research all applicable rules and regulations before deciding to expand on the existing site. Always look at alternatives, and research available rental space.

Attached or Detached Construction

Once it has been determined that the addition will be made on-site, consider the two obvious options: attached or detached. If previous designs had planned for future attached expansions, there may be planned logical locations and utilities capped for connection to the new spaces. There may be vibration concerns or safety reasons for physical separation. All these issues and more should be considered prior to starting site design.

For tomorrow's manufacturing facility there will be new issues driving the decisions. It is possible that space that was once utilized for storage will be used for staging in the production process, and that overhead conveyors will take the place of pallets and shelving. We must be prepared to look for new means of solving old problems, challenge the traditional obvious solutions, and be ready to address the questions of cost, quality, or other factors.

Existing site utilities must be one of the first considerations. A rough estimate of existing and planned new electrical and mechanical capacities will provide good information. Total area requirements may also dictate the addition of a new building or buildings. It may be that the site cannot accommodate the addition of sufficient built area to be considered. One must look for a better utilization of existing space before thinking of a new building.

5.2.2　New Site Evaluation

The remainder of this subchapter on site selection will proceed as though the decision has been made to choose a new site, and that it will be accomplished without hiring outside consultants. One of the first things to resolve in site selection relates to geographic location. There are a myriad of issues, each of which may be factored into the equation.

First we must look at the materials that will be used in the manufacturing process to see whether there is a cost advantage to locating in a specific area. If a majority of materials originate in one geographic area, or the majority of the market is located in a certain area, these should receive strong consideration. If one city, state, or county is offering tax incentives, that may be another. It is important to look at the big picture when making such a decision. Issues such as being close to the home office may improve the buying power of both plants through combined volume. The answer will depend on the product and the process.

Look at the total cost, net profit, and net-to-gross ratios for mathematical evaluation for a projected planning horizon. Look at the biggest contributors to overall costs over a given period of time. For some factories these will be labor, for others, taxes. For others it could be the cost of operations, water, electrical power, or waste removal.

5.2.3 Analysis of Potential Sites

After looking at the general considerations, it is good to look in-depth at some important factors one at a time. In this manner it is possible to evaluate the pros and cons with more depth. The order of evaluation should start with regional and contextual evaluations, then work toward the site-specific.

Regional Stability

Regional stability can be measured by a number of factors. Some cities and counties have shown their commitment to growth by offering tax advantages, municipal bonds, and other benefits to large potential employers. Donations of land, interest-free loans, and cut-rate utilities are among some of the enticements offered. The most welcome factories of the future will be those offering a stable employment base for a wide variety of skills with a process yielding little or no hazardous waste.

The recent national trend has been for municipalities to charge more each year for every traceable waste product, including condensation water from air-conditioning systems which discharge into the sanitary sewer system. These costs can run upwards of $4000 per month for pure water being processed as waste. Solid and liquid waste costs are escalating across the country, in some regions much faster than others. Among the things to be considered, include these:

1. What is the economic base of the region? Is the economy based solely on tourism or defense industries? Look back 10 to 100 years for the whole picture.
2. What are the demographics? Can we hire 80% local workers, or will we have to recruit 100% from outside the area? Are work ethics, education, service, and support industries in place?
3. What local restrictions apply? Are governmental entities currently considering property tax changes or growth management plans that will significantly change doing business in the future?
4. How are the other businesses in the region performing?

Political Stability

Before committing to a particular site, it is beneficial to find out the political stability of the locale. If today's governor loses a reelection bid, could the entire enticement package be put in jeopardy? Is there a track record of cleaning house every election year, or does the secretary of state serve for decades and usually control such issues? These questions and more can be answered through relatively brief communications with knowledgeable sources. Is new legislation about to be enacted that would affect the site operation? Are there old laws on the books that may?

If you are thinking of hiring a local design consultant, or even interviewing a number of potential ones, there is reason to find out as much as you can from each of them. They may have intimate knowledge that few people have access to. Sound them out for their overall savvy and business acumen. If the project faces extensive permitting and review processes, find out which firms have a reputation for submitting packages that make it through approval with the fewest delays. They are often the best-managed firms, with experienced personnel who are able to produce accurate, complete documentation.

Infrastructure

Now we come to an often overlooked aspect, infrastructure. Tomorrow's factories will require energy, but it is important to survey the major elements of infrastructure that affect any significant project. They include electricity, water, sanitary sewer, storm sewer, steam, gas, roads, and recreation facilities. Can the local utility company provide enough power with current capacities? Is the distribution system large enough? Who pays for the new roads and lighting? The cost of putting up a new electrical substation can run in the millions. See Figure 5.4 for examples of site electrical components. Roadways can cost more than $4 million a mile. If the local municipality will pay the initial cost of providing new services, you can expect them to attempt to recover that

FIGURE 5.4   Examples of the site electrical power components that may have a major impact on the total construction cost.

cost plus some extra through their billings. There are distribution costs, engineering costs, and possibly even land costs to be factored in. Forty megawatts delivered to your site from just a mile away could cost more than you expected. It is better to know these things going into the decision. Work up a projected cost sheet based on current and future estimated utility costs—be certain to include all utilities.

Water has traditionally been one of the determining factors for site selection and site design. At the start of the twentieth century water was important from a transportation point of view. Today the quality and quantity of water delivered to the machines can be more important than port access. Machine, process, or system water is usually highly treated, to the point where the quality at the source is not a factor. Quantities are not often the limiting factor in site design, with storage tanks and wells capable of augmenting that which is available from the water lines. Therefore, the regional water supply and the distribution system in place may be a secondary concern.

Sewer capacities are often more limited than water. Some cities will impose a moratorium on sewer taps for several years. They may require large payments up front just to put your project on the list for future consideration. There are alternatives to every problem encountered, of course, but often the alternative is worse than the original option. Sewer treatment plants may be so regulated as to be cost-prohibitive. Package treatment plants may be the lowest initial-cost solution, yet cost more in the long term or even be prohibited by local law. Most areas have impact fees for every possible system that may be affected by the new facility.

New road systems could be required of your plant, extending beyond the property boundaries. Some entities may mandate miles of roads, all the way to the closest major artery. There may be an acceptable compromise that can benefit both the region and the employer and that can be reached without deceit or dishonesty. Some kind of phasing plan can be developed that plans for growth and pays for it as the real need materializes.

Another type of agreement would be to allow use of existing two-lane roads with the stipulation that the starting and ending times of the workforce be staggered to lessen the impact on existing intersections. Whatever the results of such negotiations in verbal form, always try to get them in writing before you act. The best-meaning committee of city council representatives can drag their feet for years or change their minds overnight upon public scrutiny of a pending agreement.

Recreational Facilities

The issue of recreational facilities should not be overlooked in today's factory site selection. If there are not enough in place to satisfy the need, the new site should contain sufficient area with the right topography to provide adequate space. Statistics have proven that individuals and their families benefit from such activities, and projections indicate that recreational opportunities will become even more important in the future. Hardworking employees need recreation, and even more sedentary people enjoy sitting around a nicely wooded park. A parcourse or walking path will improve the workers' stamina, strength, and health. This benefits everyone. One factor many families consider when relocating is the quality and quantity of recreational facilities. Golf courses, tennis courts, or baseball and softball fields are but a few of tomorrow's recreation activities. Fishing, sailing, and skiing are enjoying growth. Bicycling, climbing, and hiking have always been popular activities. Some of the bigger cities have professional sports teams to add to the entertainment side of recreation. Museums, plays, and more passive activities attract another segment of the population. Some people just like to get away and be alone. The best site allows for many types of activities for years to come.

5.2.4   Site Selection

Example of Geographical Site Selection

As a simplified example, an imaginary company chooses to build a new factory. There is a market for 20,000 widgets per day, which will sell in California only for $28 apiece. The expected material cost per piece is $12.23, and half of all materials come from Pennsylvania. The rest are common materials found everywhere in the continental United States. The factory will consist of 250,000 gross square feet of floor area. With truck docks, parking, and roads, this will require approximately 40 acres. Transportation costs per 1,000 widgets are 35 cents per mile, and for raw materials the cost is 20 cents per mile for 1,000 widgets.

Let us assume that three sites are being considered after preliminary evaluation of more than ten potential locations across the nation. We can create a spreadsheet comparing the potential for net profits as determined by overall long-term expenses. If sales forecasts are reliable, actual projections can be used. Selecting the right planning horizon for cost accounting is critical and will depend on changing laws. For this example we will use 10-year, straight-line depreciation on equipment and buildings, and focus more on the direct-cost side. We will also assume that the list of potential sites has been narrowed to the best three. The paragraphs to follow give brief descriptions of the features of the three. They are in Arizona, California, and West Virginia. The information attributed to each is provided for the sake of illustrating the methodology for site selection. Actual data are available only through research at the time. The information is summarized in spreadsheet form in Figure 5.5, where the three sites are compared for initial and long-term costs. The bottom line is net profits after 10 years.

Phoenix, Arizona, where land is reasonable, climate is great, and utilities costs are minimal. It is 200 miles to the distribution network in California and 850 from the material suppliers. Forty acres would cost $4,000 per acre and annual taxes would be constant at $125,000 per year. Labor costs would be $2 per widget, and the workforce is well trained and has good work ethics. These workers can each produce 18 widgets per day. Electrical supply and waste disposal would cost 65 cents per widget. Construction costs are running $68 per square foot.

Sacramento, California, where land is expensive and utilities costs are higher than average. The site is 5 miles from the sales network and 1200 miles from the Pennsylvania materials source. Forty acres would cost $320,000, and taxes would be

40 acres
250,000 s. f. building
$0.20 per mile raw transport/1000
$0.35 per mile fin transport/1000
500,000 widgets per yr.

10 planning horizon, years

Description	Site A Arizona	Site B California	Site C West Virginia
Cost per acre	$4,000	$25,000	$800
Total cost	$400,000	$2,500,000	$80,000
Annual taxes	$125,000	$0	$80,000
Total taxes	$1,250,000	$0	$800,000
El. & disposal per widget	$0,65	$0.85	$0.50
El. & disp. per year	$325,000	$425,000	$250,000
El. & disp. total	$3,250,000	$4,250,000	$2,500,000
Building initial cost/sf	$68	$74	$50
Building cost	$17,000,000	$18,500,000	$12,500,000
Operating cost/year	$312,500	$686,000	$454,860
Total oper. costs/10 yrs	$3,125,000	$6,860,000	$4,548,600
Transport, raw miles	850	1200	60
Trans for 500,000 widgets, raw/yr	$85,000	$120,000	$6,000
Total raw trans cost/10 yrs	$850,000	$1,200,000	$60,000
Transport finished in miles	200	5	1150
Trans for 500,000 widgets, fin/yr	$35,000	$875	$201,250
Total transp fin/10yrs, flat	$350,000	$8,750	$2,012,500
Labor cost/widget	$2.00	$2.40	$1.50
Total labor/yr	$1,000,000	$1,200,000	$750,000
Labor for 10 yrs	$10,000,000	$12,000,000	$7,500,000
Mat'l cost/widget	$14.40	$14.40	$14.40
Total met'l for 10 yrs	$72,000,000	$72,000,000	$72,000,000
Widgets per day/person	18	16	14
Total employees	111	125	143
Insurance rate/person/yr	$1,200	$1,440	$890
Ins. cost per yr.	$133,333	$180,000	$127,143
Ins cost/10 yrs. flat rate	$1,333,333	$1,800,000	$1,271,429
TOTAL EXPENSES PROJECTED	$109,558,333	$119,118,750	$103,272,529
TOTAL INCOME PROJECTED	$140,000,000	$140,000,000	$140,000,000
EXPECTED 10 YEAR NET PROFITS	$30,441,667	$20,881,250	$36,727,471

FIGURE 5.5   Site comparison spreadsheet used in the site selection process.

waived in order to acquire such a prominent employer. Labor costs would be $2.80 per widget, and worker productivity will produce 15 widgets per day. Electrical supply and waste disposal would cost 85 cents per widget. Construction costs would run $74 per square foot.

Wheeling, West Virginia, which has very affordable real estate. Land costs there are $800 per acre, and taxes will be $80,000 per year. Labor is not highly trained but works cheap. Labor costs would be $1.50 per widget, and each worker can produce 12 widgets per day, average productivity for local industries. Electrical and waste costs would be 50 cents per widget. Construction costs would be $50 per square foot. This site is located 60 miles from the major materials source and 1150 miles from the sales network.

Next it helps to look at the sites on a local level. The key aspects here are the available workforce, economic stability, and available infrastructure. What has been the leading employment source in the past will reveal something about the potential employee base available. If food stores are the largest employer, you may not expect too many technically experienced workers.

It will also be important to look at the educational system, the numbers of technical and vocational schools, community colleges, or universities. Furthermore, you can learn about the courses offered, numbers of students, and degrees earned. This will help provide an indication of the population's level of readiness or commitment to learning. Look at other factors also, especially the demographics. What is the age and sex of the workforce, by percentage? By preparing a customized version of the site selection formula one can look at the most important determinants of all three sites on one sheet of paper. The difficulty is in determining which factors should be most significant. The program can provide such information. Once the site is chosen, final plans can be solidified.

5.3   THE BUILDING PROGRAM

Each new facility will require a unique building program. While the processes and products will be different, the major elements will be similar. A brief excerpt listing the spaces from one such prototype building program is provided in Figure 5.6. A complete building program may exceed 200 pages in length and therefore cannot be included in this book. In the example are 25 areas of the facility that would require in-depth, comprehensive coverage. That information would include gross square footage needs along with performance expectations, lighting levels, utilities, and pertinent information on preferred relationships to other parts of the factory. These 25 areas become the major elements in building design. Space planning information sheets similar to Figure 5.7 can be utilized to provide this information to the design agency. Detailed information relative to each machine in the manufacturing facility is perhaps the most important information in the building program. Accurate plans and all four elevations must be included in the program if the information is known. If only block diagrams are available, it can lead to wasted space, time, and money. Using "basis of design" information can tie down some dimensions, but may result in changes later. (See Chapter 3 for further discussion of requirements planning.)

5.3.1   Elements

From the information in the building program, one can develop an understanding of the elements that are most crucial to the manufacturing process. All future design decisions should be made relative to their ability to support these major functions. Rather than

#	DESCRIPTION	AREA (GROSS S.F.)
1.	ACCOUNTING	1200
2.	ADMINISTRATION	1600
3.	ASSEMBLY AREAS	24000
4.	BREAK ROOMS	600
5.	CASTING	2000
6.	CENTRAL PLANT	4500
7.	CUSTODIAL	800
8.	ENGINEERING	2000
9.	FABRICATION	8000
10.	FIBERGLASS WORK	1450
11.	HAZARDOUS MATERIALS	2400
12.	JANITOR/MECH EQUIP	1200
13.	WORK IN PROCESS STORAGE	1000
14.	MACHINING AREAS	2800
15.	MISCELLANEOUS, CIRC. ETC.	1000
16.	PAINT SHOP. AND OVENS	2000
17.	PAINT PREP	3500
18.	QUALITY ASSURANCE	600
19.	RECEIVING	800
20.	SECURITY	350
21.	SHIPPING	400
22.	SHORT TERM STORAGE	2150
23.	TESTING, MECH. & ELECT.	1400
24.	TOOL CRIB	600
25.	WELDING	1000

FIGURE 5.6   Elements of a program for a twenty-first-century manufacturing facility.

attempt many simultaneous equations with too many variables, it is good to group similar elements. Based on product and process flows, there will be dependent elements that require a certain relationship. First the designer tries to arrange all elements into fewer than five groups. At this stage of design it is better to use abstract geometric shapes to represent the groups of elements. Simple circles and squares are most commonly used, with square elements encircled to form the groups.

Materials flow analysis will often determine adjacencies of these groups. The sample bubble diagram in Figure 5.8 shows how the 25 boxes can be divided into

SPACE PLANNING INFORMATION SHEET

DATE / INITIALS WLW 4/3/92	GROUP NAME OPERATIONS	ROOM NUMBER OP 101	DESCRIPTION
REVIEWED, APPROVED, DATE/ INITIALS 5/1/92 JRG	SECURITY LEVEL BLUE	ROOM NAME CASTING ROOM

HAZARDS: PRESSURE PIPES, HIGH HEAT CURE, MOLTEN METAL
CHEMICAL: NONE
GASEOUS: NONE
OTHER, DESCRIBE: LEAKS, BURST FITTINGS
FUELS, SOLVENTS: 5 GAL. ACETONE
POTENTIAL EXPLOSION: NOT REALLY
WASTE DISPOSAL TECHNIQUE: RECYCLE TO SMELTER

VIBRATION:
SOURCE, WAVELENGTH, FREQUENCY: NOT AN ISSUE
SENSITIVITY TO: N. A. I.
SOUND LEVELS DESIRED: 55 DbA
ACOUSTIC TREATMENT RECOMMENDED: NONE

ARCHITECTURAL:
FLOOR: SEALED CONCRETE — **SPAN REQUIREMENTS:** MIN. 25 FOOT BAYS CLEAR
WALLS: EPOXY PAINT — **FUTURE CONNECTIONS, MODIFICATIONS:** NONE
CEILING: EXPOSED STRUCTURE ABOVE — **WIND LOADS:** EXTERIOR WALLS ONLY
WINDOWS: FIXED GLASS, ALUM. FRAMES — **COLUMN ENCLOSURES:** PIPE BOLLARDS
DOORS: HOLLOW METAL, S. S. KICK PLATES 1/2 — PIPE IN TRAPEZE HANGERS
SPECIAL CONSIDERATIONS:

STRUCTURAL:
UNIFORM LOADS: 100 PSF — **VIBRATION CRITERIA:** NOT AN ISSUE
CONCENTRATED LOADS: 800 PSI — **ISOLATION:** N. A. I.
CRANES, ETC: OVERHEAD — **TYPE CRANE & CAPACITY:** GANTRY, 10 TON
ALLOWABLE DEFLECTION: MINIMAL, LESS THAN 3/4" IN 10' — **NATURAL FREQUENCY:** LOW APPROX. 2 HZ.
EXPLOSION TREATMENT: N.A.I. — **OTHER CONSIDERATIONS:**
REMARKS: LOCATE ON GROUND FLOOR

HEATING, VENTILATION & AIR CONDITIONING:
DESIGN ROOM TEMP RANGE: 72 TO 78 DEGREE F. — **HUMIDITY RANGE:** 35 TO 40% RH
ACCEPTABLE FLUCTUATIONS PER HR.: 3 DEG. — **SPECIAL CLIMATE:** NONE
CLASS FOR CLEAN ROOMS: 1000 — **CFM ALL HOODS:** 1500 — **SHADE USE FACTOR, HOODS:** .25
EST. EQUIP. HEAT GAIN: 9800 WATTS, NIC CURE — **EQUIPMENT USE FACTOR:** .40
SPECIAL FILTRATION: FINAL FILTERS IN MOLD PREP AREA — **AIR CHANGES PER HR.:** 8
OTHER CONSIDERATIONS: CURING AREA NOT CLIMATE CONTROLLED

PIPING, PLUMBING:
FULLY SPRINKLED YES — WET PIPE YES — DRY PIPE CURING NO — DELUGE — STANDPIPE YES — CHEM. EXTINGUISHMENT NONE — FIRE EXTING. CLASS ABC
WATER REQUIREMENTS, DOMESTIC C. W. 6GPM MAX. 2HRS./DAY — HOT WATER SAME — CW SUPPLY H.R. YES 2GPD — RE-IONIZED NO — CHILLED 40 DEG NO — EQUIP. COOLING YES, 12 GPM
GASES COMPRESSED AIR 120 PSI NO — STEAM NO — VACUUM NO — HYDROGEN NO — NITROGEN NO — PHOSGENE NO — NATURAL GAS YES — OTHER NO
WASTE 4" — VENT YES — FLOOR DRAINS 2 — ACID WASTE / VENT NO — INDIRECT DRAIN YES 1-1/2" — REMARKS — CLASSIFICATION OF WASTES ORDINARY

ELECTRICAL SYSTEMS:
LIGHTING LEVEL 65 FC — FIXTURE TYPE PREFERRED H. P.SODIUM — VOLTAGE 277/480 — BACK-UP POWER EXIT — DIVERSITY .8
POINTS 2000 — WATTS 1500 — KVA — DEM. KW 200 — TOTAL SINGLE PH. W — LOAD FACTOR .35
SYSTEMS VOICE YES — DATA YES — FIBER YES — CHIMNEY NO — CATV YES — SPECIAL GROUND NO — BACKBOARDS YES — U. P. S. NO

MISCELLANEOUS COMMENTS: PROXIMITY/ RELATIONSHIPS, MATERIALS FLOW, POWER, WATER, NETWORKS, ETC
NEAR TO MACHINE SHOP, GRINDING AREA AND CONNECTED TO CRANES

FIGURE 5.7 Space planning information sheet.

three groups by function. Within each of the groups, the designer will later transform the squares into more meaningful shapes and determine which of them are on exterior walls, which are closest to the utility plant, etc. At this point, the boxes need only indicate the number of the program element that they represent. Next, more information will be added in order to make other decisions. Each element can be considered

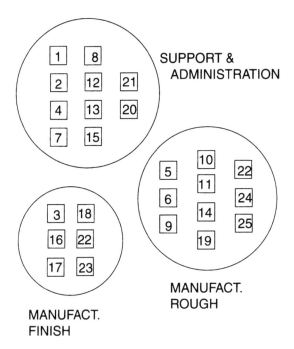

FIGURE 5.8 Bubble diagram showing the 25 major elements divided into three functional groups.

a stand-alone component, even if it is made up of a number of spaces. For example, the administrative area may contain as many private offices as there are staff, yet for elemental design it should be treated as one entity since the general performance criteria are the same. On the other hand, the manufacturing shop may have many areas that have grossly differing needs, and therefore we treat them as different space elements.

These abstract representations of the manufacturing components may be initially combined into two or three groups with similar activities. This will not necessarily dictate the final location for the elements; it provides only a means for representing a complex group of activities. The designers will then prepare as many alternative designs as it takes to exhaust potential solutions. Then all these alternative plans will be evaluated for the best fit for the program performance requirements, considering the advantages and disadvantages of different designs. For example, long linear arrangements can be more costly, while centralized ones may cause material flow problems.

Adjacencies

Another useful study in the programmatic stage is that of preferred adjacencies. One should arrange elements according to what is next to them. Such studies answer questions, such as: Which connections are required, and are they related to product flow or simply driven by stops along a conveyor? Are there safety reasons to be considered? Will we also consider human safety factors? Perhaps there is one element such as a pyrotechnics station where one would construct blast-proof walls and blowout panels. This should be shown as removed from the location of the other building elements. A bubble diagram as shown in Figure 5.9 is an example of the effect of considering potential adjacencies.

Next, consider the mechanical and electrical adjacencies that would best support the manufacturing process. What would be the best arrangement to minimize expensive piping or cable runs? Which of the functions require "clean" conditioned air? Is it better to have them flanking a common heating, ventilating, and air-conditioning (HVAC) unit somewhere, or to have noncentralized clusters for improved redundancy? The designer must consider all of these during the design phase, and try to determine the best options.

BUBBLE DIAGRAMS

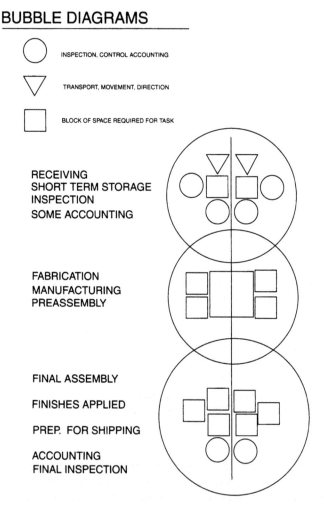

INSPECTION, CONTROL ACCOUNTING

TRANSPORT, MOVEMENT, DIRECTION

BLOCK OF SPACE REQUIRED FOR TASK

RECEIVING
SHORT TERM STORAGE
INSPECTION
SOME ACCOUNTING

FABRICATION
MANUFACTURING
PREASSEMBLY

FINAL ASSEMBLY

FINISHES APPLIED

PREP. FOR SHIPPING

ACCOUNTING
FINAL INSPECTION

FIGURE 5.9   Bubble diagram revised to consider the effects of adjacencies.

Next, one must evaluate choices for locating these systems in the completed building. Will they all be hanging from the roof structure, or under the floor? Is it more cost-effective to build three levels with laminar flow from top to bottom? The addition of pumps would permit us to run all waste and vent piping above the floor. This would enable future modifications to be made without cutting the floor. Traditionally, waste utilities are located under the slabs, while vents and piped gases are run above the equipment. That relationship must be evaluated as well as plan (two-dimensional, planar) adjacencies. It may prove more cost-effective if all piping systems are located on a second level with branches above and below, as in some laboratories. In some process piping systems the cost of special plastic, glass, or stainless steel piping can escalate the building costs to more than $300 per square foot. The use of utility floors, pipe splines, or chases should be a conscious one. Factors such as access for future maintenance and reconfiguration for new process systems should be weighed in the decision along with initial cost. A photograph of an example of piping and cable trays mounted overhead is seen in Figure 5.10.

5.3.2   Relationships

Previous sections have alluded to the relationships required for optimum functional arrangements. There are other issues to be considered as well, such as minimizing expensive piping runs or maintaining adequate displacement for health and safety reasons. With more factory operations being automated, computer centers are controlling machines from centralized locations rather than having personnel in

FIGURE 5.10 Example of piping and cable trays mounted overhead.

potentially hazardous situations. The designers must consider these factors and still provide for flexible operations in the future. Look first at the most critical operations in the process and their preferred adjacencies. In many facilities these will be fabrication and assembly areas. Next look at all the required connections for supply-side materials. Working backward from the final assembly process will often provide the most accurate and complete picture of the important relationships. A matrix that lists major spaces and their preferred adjacencies will provide a graphic illustration (see Chapter 3).

The next level of design includes actual requirements such as area (in gross square feet or meters), orientation, utilization, and other information found on the detailed space plan information sheets. From this second level of information, relationships can be evaluated for their ability to support the primary functional roles in the finished facility. The issues of safety must be factored into the arrangement of elements in the design. This is where the third dimension begins to play a more important role in the equation. Elevation of elements and their support systems must be considered during design development. Section and elevation views must be developed concurrent with the plan.

5.3.3   Order

Once the best three-dimensional relationships are determined, the important issue of *order* should be evaluated for potential impact on the design. Similar to site plan design, the building element order can affect the initial price and long-term operating cost of the facility. Order can be defined for our use here as the arrangement of the elements in a structured system. There are two primary considerations in determining the order for a manufacturing facility. The first is how it will support the necessary functions of the manufacturing process, and the second is cost. Working from the information in Chapter 3, the sequence of manufacturing processes and machine operations is usually quite clear. That sequence will dictate to some degree the general order of the building elements that relate to the process. But what about the other building components? What determines the final layout of the building footprint?

A good place to start is by developing a module. In small areas, less than 100 ft.^{2} (9 m^2), the module usually ranges from 3 to 4 ft., or around 1 m. In contrast, for large areas it may be 100 ft. by 100 ft. In many cases it is approximately 4 ft. for residential-scale buildings, and a minimum of 5 ft. for factories. Some designers will use the width of their conveyance devices, such as automated guided vehicles (AGVs) or fork trucks, to assure that there is sufficient space for circulation. In any case, the module will be used as a design tool in the transition from bubble diagram to schematic plan drawing. This allows the designer to focus more on the big picture without being burdened by pragmatics.

If the manufacturing facility will contain mainly punch presses, each approximately 10 ft. by 20 ft. (including material work room), then that could lead to a 20-ft. by 40-ft. grid. If another area in the shop is mostly drill presses requiring about 5 ft. by 7.5 ft., then a group of four could lead to a 20-ft. by 30-ft. module. Some designers create designs on quadrille pads, then assign some scaling to the grids to achieve the overall area targets. These tend to lead to grids of, say, 25 ft. square. The economical spans of concrete and steel will usually range between 25 and 40 ft. These dimensions would be evenly divisible into 5-ft. modules. Once that is determined, the process of factory ordering becomes easier. The block elements can be arranged so as to function best in sequential operation and still fit within a gridwork of 5-ft. modules.

The configuration of these parts is the issue that affects the building footprint. As mentioned earlier, a square building has less exterior wall surface area than an elongated rectangle of the same total area. Therefore a relatively square building has a lower initial and recurring cost for maintenance. A ratio of about 1.6:1 has been proposed as having the best aesthetic appeal, and therefore the best subliminal acceptance by workers. Long, narrow spaces are perceived as boring, overwhelming, and monotonous. Large, square ones are impersonal and scaleless. The proportions between 1.3:1 and 1.6:1 often yield the best compromise of low cost and high yield. Once these decisions are made, the overall building form is taking shape. Rather than jump ahead to the finished plan and elevation views now (and then try to make the rest work out), it is best to consider the next issues. Studying all aspects first can prevent inefficient use of space and potential problems.

When the designer combines those criteria with site-specific forces and employee needs, there is usually one ordering system that fits better than the rest. Most common in manufacturing facilities are centralized, clustered, radial, and nodal systems. Axial and linear systems are least common because of the increased surface areas and cost associated with their geometries. A successful designer will have the education and experience to make good decisions in this regard. (See Figure 5.11.)

Once the ordering system has been determined, the program yields other significant information such as shared needs for utilities. This leads to groupings of functions. Those groups are arranged by preferred adjacencies, or by methods covered in Chapter 3 on product and process flows. Other support functions, such as storage and administration, are then located for best circulation. See Figure 5.12, which is the natural result of applying adjacencies to the bubble diagram shown in Figure 5.9.

Figure 5.13 shows the natural bridge between this step and the beginning of the actual plant footprint.

Materials Flow

Starting from truck, rail, or waterway access routes, materials receiving and transportation becomes one of the most critical factors to be determined. Quantities of delivered goods will determine the total number of docks required. With just-in-time delivery becoming more prevalent, the number of docks has risen while the area for storage has decreased. Mechanized material transport devices such as AGVs and conveyors help keep both operating and labor costs low. The circulation area required for fork trucks and AGVs must be planned for during design. This can be done using a minimum-width module for the different types of vehicular passageways. The main aisles would permit the biggest fork trucks, of, say, 5 tons and 10 ft. of maximum load width and a 12-ft. height. Secondary paths would be designed for an 8-ft. or 5-ft. module. Each process will have its own conveyance requirements.

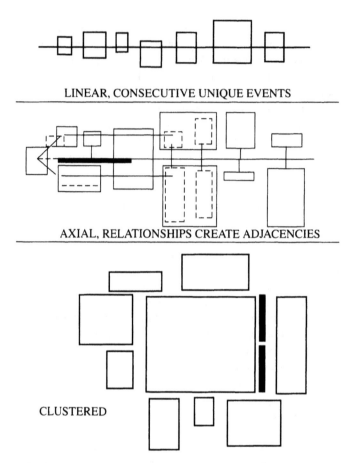

LINEAR, CONSECUTIVE UNIQUE EVENTS

AXIAL, RELATIONSHIPS CREATE ADJACENCIES

CLUSTERED

FIGURE 5.11 Linear, axial, and clustered ordering systems for factory design.

5.3.4 Circulation

A study focusing on circulation can yield some interesting results. It must look at pedestrian, vehicular, and product movement across the entire site and especially as related to support of the product. Simple diagrams can again be used, with symbols to represent the different groups, their trip frequencies, weights, directions, times, etc. (see Chapter 3). This information can then be added to the general block layout already developed. It may show bottlenecks in the flow, areas that need to be widened, or even reveal flaws in earlier design schemes. Areas where pedestrian paths cross vehicular paths must be treated in such a way as to avoid conflict or personal injury while not impeding product flows. Often pedestrian paths are elevated above machine routes to maintain safety. That also implies human's dominance over machines, which can help worker morale. Issues that seem unimportant to some may prove critical to others.

ADJACENCIES DIAGRAM

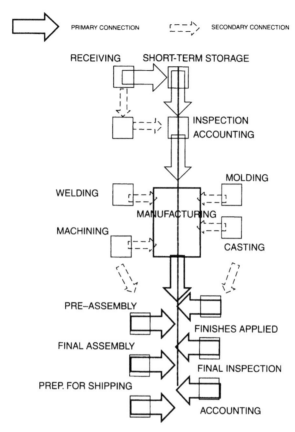

FIGURE 5.12   Results of applying adjacencies to the preliminary design evolution using the bubble diagram in Figure 5.9.

Elevated walkways can also be utilized to provide observation decks for tours without endangering personnel or reducing plant productivity. Employee break areas, cafeterias, restrooms, etc. can share the elevated areas with overhead conveyors and mechanical systems. This is another reason why three-dimensional studies are important in the design process. Coordination of the design disciplines is critical to successful project completion.

5.3.5   Employee Evaluations

Next is the very important aspect of listing information relative to the people who will be working in the facility. It should include the age, sex, educational, religious, philosophical, mental, and physical makeup of the potential workforce. This list can then be combined into summary workforce groups. This can simplify the design

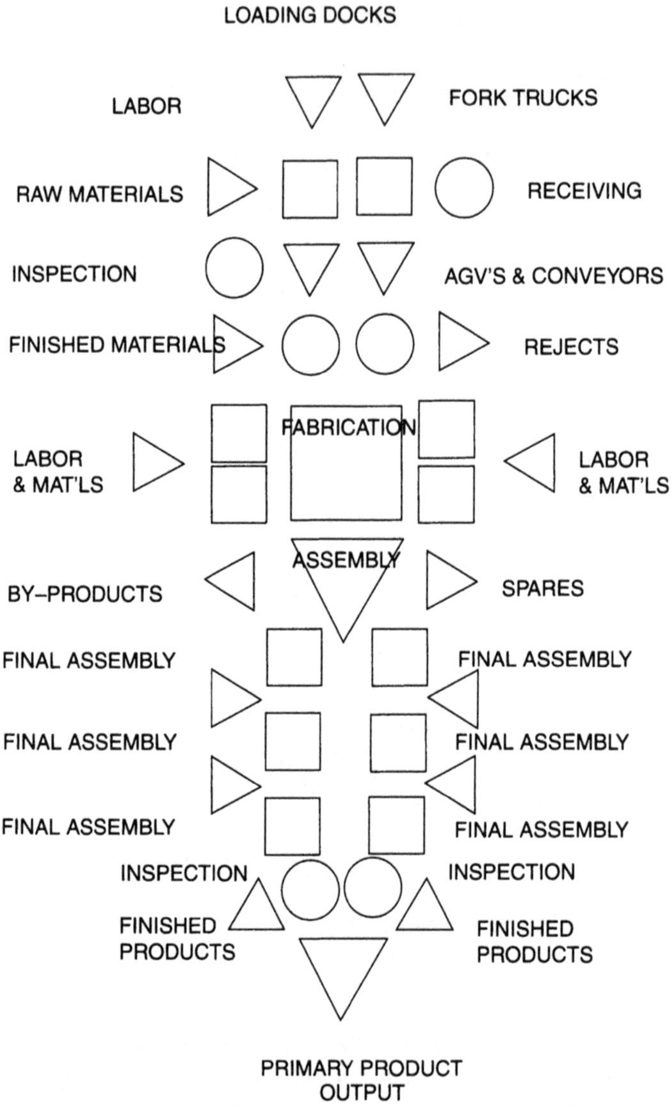

FIGURE 5.13 Example showing the natural bridge between Figure 5.12 and the actual factory layout.

response to the various workforce needs and preferences. The makeup of the workforce can be important in determining the space requirements beyond the production line. A workforce that is predominantly highly religious and well educated will have different needs than a workforce made up of uneducated agnostics. If there are mostly 6-ft.-tall, 250-lb. American males, the aisles will need to be wide. They could be narrower for a predominantly female Asian workforce, for whom the average is about 5 ft.

3 in. tall and slight build. The section in this book on human factors (Subchapter 6.1) will provide useful data on dimensions, strengths, and spatial needs of human bodies in action and at rest.

Physically Challenged

Physically challenged individuals must also be accommodated in the design of new facilities. Ramps can be used for access rather than stairs. Winding, gradual ramps are preferred rather than short, steep ones. A good rule of thumb is that ramps should not exceed 1 ft. of elevation change for every 13 ft. of horizontal distance. Long ramps should not be used without intermediate flat landings of at least 80 in. Passages require a minimum width of 44 in. for one and 60 in. for two persons.

Fire exits must be designed to provide special considerations for those individuals who might need assistance in exiting a burning or smoking building. All elevators are no longer required to proceed immediately to the ground level for fire department use. Some emergency egress is permitted using elevators for assisted exit by challenged individuals. Many rated stairwells are designed utilizing safe zones for assisted egress in cases of emergency. In this way, mobility-challenged persons can wait safely in predetermined areas that do not obstruct others from using the stairwells. These spaces can define where emergency personnel can report to assist them. Be sure to review all safety plans with appropriate fire and police agencies.

5.3.6 Code Review

There are many rules and regulations that will apply to master plan decisions. Prior to initiating design development, the owner and some of the owners' chosen professionals will be required to work together to determine all codes that apply. Some of the codes may change during design and construction, which can affect the operation of the completed facility. It should be the responsibility of all concerned to make continuous efforts to see that all codes are met. We will not attempt to list all of them at this time. Generally speaking, however, they will include:

Uniform, Standard, or Southern building codes
National Electrical and Plumbing codes
American Society of Heating, Refrigerating, and Air Conditioning Engineers (ASHRAE) codes
National Fire Protection Association codes
Life Safety codes
District, county, and municipal building, fire, and zoning codes
Handicap access codes

Other agencies that will also have potential jurisdiction might include the local fire marshals, building inspectors, traffic engineers, the Occupational Safety and Health Administration, water management districts, departments of environmental regulation, departments of natural resources, and many others too numerous to list. Failure to comply with any of them can have significant results. It should be the shared responsibility of the owner and the design team to research applicable codes. (Subchapters 5.2 and 5.3 may offer additional insight here.)

5.4   PROJECT DESIGN

There are many different ways to design and build a facility. The owner can use in-house engineering staff to design and the facilities department to construct. Proper administration of even modest construction projects can require more than five experienced full-time employees. Design of small projects can be accomplished by one registered professional in the firm with the help of specialized staff. Most manufacturing firms will not have an appropriate staff to cope with the time constraints and will therefore hire design professionals. Experienced A&E firms can provide the expertise and personnel to complete hundreds of design documents in a relatively short time. One aspect of their contract requirements should be for them to provide accurate documentation of the existing facility (if not already available), and for them to turn over to the owner a complete set of hard copies and software copies of the design documents when completed. It is common for firms to generate schedules and estimates on Lotus or Excel software, word processing copies of the specifications and key correspondence on Word Perfect or Word software, and drawing files as CAD drawing files (AutoCAD, DataCAD, VersaCAD, etc.). The owner is entitled to copies of all of this at the end of the project. The form and content of the documents may affect future operation and maintenance of the plant and therefore may become very important to the client.

The fees for A&E firms range from 4 to 20% depending on the range and scope of design services. The contracts can vary from minimal schematic design or master planning to complete design and contract administration. For complex projects or tight schedules, A&E firms can provide on-site field representatives to guide construction. In most areas, licensed design professionals are required to seal the construction documents prior to permitting. The choices are made easier by looking at project schedule and cost information developed in the preprogramming phase. Each unique project will create new challenges to the designer, to improve upon what has been done before. With the complexities inherent in preparing hundreds of pages of contract documents, it is often safest to rely on professionals who are in the mainstream of the specialty you need.

5.4.1   Design Firms

Many of today's A&E teams are comprised of specialists who are expert in a very narrow field. A list of specialized consultants for a factory project might include:

Architect
Civil engineer
Landscape architect
Irrigation designer
Structural engineer

Mechanical engineer
Systems specialist, communications, voice, fiber optics, etc.
Process controls engineer
Conveyance systems specialist
Environmental and permitting specialist
Interior design
Soils testing, materials testing, and reports

It will usually be the architect's responsibility to guide the project through design, interfacing with all the consultants and often making recommendations for contract award. The architectural firm will also function as the main point of distribution for questions and answers between the owner and the constructors through the client's PM. The PM for the architect's office will be singularly responsible for implementing the wishes of the building committee. This person must be confident enough to exert authority and experienced enough to understand the implications of his or her decisions.

The project architect and PM in the design firm will be the most important people during document preparation. For these reasons it is important to choose a firm with whom you can communicate completely. The importance of each role in communications can be seen in the earlier diagram (see Figure 5.1). The three PMs are the real points of contact, so choose them with that in mind.

Some manufacturing firms have established strong ties with design teams over time through successful projects. If there is no such relationship, it will be necessary to select the best A&E team or firm for your specific project. This is usually achieved through advertising a fixed scope of work at a fixed fee. The advertisement must include enough pertinent information for the interested firms to understand the expectations of the owner, the scope of the project, and any special experience that would be helpful in designing this project. If the time constraints are atypical, they must be expressed as well. From all applicants, a short list can be developed that contains only the highest-ranked firms based on certain evaluation criteria. A typical list of evaluation criteria would include:

1. Are they technically competent? Can they understand the project requirements?
2. Are their designs cost-effective? Can they be built for the target budget?
3. Can they meet the demands for production? How many employees are available?
4. How many years has the firm been practicing?
5. What is the experience and motivation of their key personnel?
6. List of consultants, past affiliations, and completed projects, with names and addresses.

Once the firms have submitted their preliminary proposals, one is usually able to create a short list of the highest-rated responding firms. Firms on that list will be invited to present their pitch in a scheduled interview format. Intangibles such as fit and attitude can be evaluated from these person-to-person exchanges. Next one should follow up referrals personally, by visiting completed projects if possible. By discussing performance with other clients of the design firm over the past few years, one can get a better picture of the performance of the design firm.

5.4.2   Final Site Analysis

After determining a preferred site, there are a number of site-specific issues that may contribute to the final design decisions. One should start with total area, orientation to transportation, and boundary information (length of each property line and bearings). Building setbacks and easements are critical to development of a site master plan. From these one can determine the net and gross areas of the site and these will lead to the developable area. Topographic information such as elevation above sea level and that of the surrounding areas are important to consider. Flood plain information may apply. There are zoning requirements, utility surveys—a whole host of issues that should be evaluated in depth. The best approach breaks them into subaspects and then studies them as if they were independent entities. Upon completion, the essential important kernels from these analyses are combined, weighting them according to their importance (hierarchy). There are some low-cost tests that may determine the needed foundation system and therefore construction cost. They provide a good starting point for analysis. Following are a few of the most common subaspects for site analysis.

Soil Tests

Soil borings located at 100-ft. intervals in the footprint of the building can provide sufficient data for general subsoil evaluation. The depth should usually exceed two times the maximum projected building height. The testing agency will include a report as to conditions encountered, including groundwater and bearing capacities of the different soils encountered. A minimum bearing capacity of 3000 lb./ft.^{2} at the surface is required for modest floor loading. Significantly higher values will be preferred. It is better to see a consistent profile, rather than levels of high and low bearing capacities in a stratified or mixed manner. If there are voids or very low-density layers in the subsoil, there could be potential problems. If the preliminary tests are inconclusive, testing at 20-ft. intervals will provide better information. If the manufacturing process will be extremely sensitive to vibration, the top 2 or 3 ft. of soil should contain nothing but dry, coarse sand. Clay, rock, or wet soils can increase vibration transmission beneath the slabs. The consultant's report should also contain recommendations as to the type of foundation systems best suited for your specific performance criteria.

Water Quantity and Quality

A number of other site forces need to be taken into consideration. Water quality and quantity in both groundwater and piped supplies should be quantified. Look to see if there is gray water piping, or separate irrigation water systems. Availability of fire protection (FP) water to the site can affect fire-pump requirements. If sufficient water is available, a project can save perhaps $100,000 in pumps and related costs. When sufficient pressure exists, pumps may not be required. Generally, water pressure in excess of 120 psi will be sufficient. Ponds or lakes may function as fire-water reserves in some remote areas. Low-cost elevated tanks can provide pressurized delivery systems without expensive high-capacity FP supply pumps, but they can add to the structural cost. Good designers will be able to determine what effect anticipated growth in the area will have on the water pressure. A factor of 0.75 is generally applied to the test results from fire department flow tests, in order to anticipate future conditions.

Depending on the particular factory processes, water may be a critical requirement. A stream on site may provide a heat exchange source, or a means of transporting treated waste to the properly classified river at its end. Many industrial facilities can operate "closed cycle," with minimal make-up water required. Cooling towers can require more than 50,000 gal. per month to offset evaporation at the basin, while cooling coils in air handlers throughout the project can distill a similar quantity from the inside air of the factory. Presses, ovens, transformers, milling machines, and other machinery can require equipment cooling water. While the quality is not as critical here, some processes require very specific water. Industrial lasers, cleaning machines, and chemical processes have very specific requirements. A great water treatment system can cost in the millions. For some factory processes, the elements present in the water supply can affect initial and operating cost of the treatment system. Commonly we find calcium, iron, and sulfur, but some trace elements can cause the biggest problems.

Other Factors

Some of the other regional criteria to be considered should include storm water, topography, vegetation, and wildlife. If there are endangered or protected wildlife habitats present, there will be more issues to consider. Similarly, historical artifacts found during excavation have been known to delay projects for years. Indian burial grounds, tar pits, or other animal remains on sites further complicate proceedings. The easiest way to protect against these difficult-to-anticipate occurrences might be to include provisions in the land purchase offer. These escape clauses require return of deposits and expenses when such delays take more than 30 or 90 days. In summary, there are mathematical formulas which can provide decision makers with quantitative data. Ultimately, however, no picture is complete, and management must make an educated compromise.

5.4.3   Site Master Plan Development

A *site master plan* is the design document that locates all major elements of the project and deals with gross area calculations. It addresses parking, paving, planting, and all building footprints. It can be thought of as the framework on which all the components are arranged. A good master plan will work for decades, regardless of changes in time or factory processes. It will be a flexible ordering scheme that combines the best natural features of the site with the necessary physical components of the manufacturing facility. It should be a skillful manipulation of the built masses that creates an effective sense of place, where the workers want to be and wherein they all work together toward common goals.

If this seems like too much to ask of the simple task of placing buildings on a site, then one has not grasped the power of design. Design is not simple, and it has a powerful effect on our subconscious minds. Having said this, let us see if we can conjure up one place in our memories that was a joyful place to be. It might be a childhood playground or a religious setting such as Saint Peter's square in Rome. Whatever the place, the arrangement of the components is what stirs the senses. The same can be done in arranging the building elements on a 10-acre site or in a 250-acre industrial park. Both require skillful master planning. Other good examples are Walt Disney World and Golden Gate Park. All were designed to create a *sense of place* where sensory experiences are pleasurable. The stimuli are controlled to heighten the senses of sight, smell, sound, touch, and imagination. Memories of such places are vivid because of the sheer number and intensity of sensory stimuli coupled with the cumulative effect of repetitive elements.

The challenge to create a similar atmosphere in a factory environment has been accomplished successfully in the past. The most important aspect is to consider the senses involved. Sight, smell, and sound are the three that we can easily control and must consider in design (see Subchapter 6.1, on human factors).

Through skillful manipulation we can create a master plan that will keep worker morale and performance high, while keeping illness and turnover low. It takes a careful, thorough examination of the workers and the work to be performed. Only after all this can we expect to make good decisions. We must define the needs of the workers and make conscious decisions to achieve fulfillment of their needs. If you believe that it is not important to consider these things, then get ready for the twenty-first century! Humans and machines work better and longer in the proper environment. Most successful manufacturing firms have been trying to improve their employee performance through improvements in the workplace. The master plan is a great vehicle to start to achieve that improvement.

Physical Studies of the Selected Site

The first study is to evaluate the physical attributes of the area to be built upon. Some of these attributes are the same as we looked at on a regional level, but with more attention to specifics and details. Others were considered less important then, but take on more meaning now. Among those specific site studies are:

Soil types
Topography
Vegetation
Circulation
Prevailing winds
Views
Utilities

Each component should be studied independently and evaluated in terms of its impact for determining the building zones. Building zones are located where there is best subsoil for bearing the anticipated building loads, and where the other factors support cost-effective construction. Excavation and replacement of 3 ft. of topsoil or clay can cost hundreds of thousands of dollars for an average factory footprint. See the example in Figure 5.14 for an introduction to topography and other site forces.

Soil Types

Depending on the type of manufacturing facility, the soil types may be the determining factor for selecting a building zone or zones within the site. Bearing capacity of

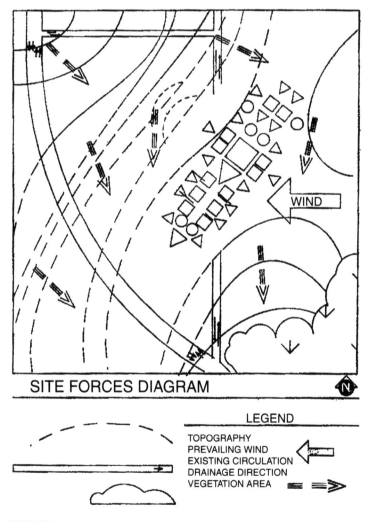

FIGURE 5.14 Preliminary factory siting, considering existing vegetation.

the subsoil should far exceed the future loading potential. For slab-on-grade foundations it can be as low as 2500 lb./ft.^{2}, but for point-loading structural systems you may need more than 5000 lb./ft.^{2}. Other considerations include the permeability of the soil, the amount of clay or rock encountered in the soil tests, and groundwater levels. Each can affect the cost of construction, and should be considered prior to final siting of the buildings. Copies of soil information maps that illustrate the classification and capacities of soils on and around the projected site can usually be obtained from city or county records. More detailed or specific information can be provided through geotechnical consultants.

Topography

Positive natural drainage can usually be achieved through proper topographic considerations. A good master plan design often results in minimal cut and fill operations. The material cut can be used to backfill around the construction. Foundation fill may be obtained from other required on-site operations such as retention pond excavation. On some sites, there is no suitable fill available, and all must be trucked in from off-site at a cost of more than $5 per cubic yard. For a large project this may add up to more than $140,000 for 2 ft. of fill under 250,000 ft.^{2} of floor area. Under-drain systems can be designed to remove surface water from building perimeters also. These systems often cost more than fill, and will require maintenance or replacement over time. The best site plan will avoid costly improvements. Parking lots can be terraced around the building perimeter to retain soil and maintain existing elevations and vegetation.

Vegetation

Vegetation studies usually reveal where the soils are richest and where bearing capacities are lowest. They also indicate where animal life is sustained, which may be a significant issue for people's comfort. There is a subliminal reassuring quality about seeing an animal habitat, especially near a manufacturing environment. It eases people's concerns about their personal safety, and indirectly informs them of the company's values toward life outside the factory line. The once common tradition of simply clearing the entire site prior to design or construction is being replaced by thoughtful and selective clearing concurrent with design. A tree survey of all trees over 4 in. in diameter provides valuable information for use in the process of master plan design.

Twenty-first-century manufacturing facilities should be designed so as to cost-effectively work in harmony with the natural topography. Through careful site planning, amenities such as recreation facilities, fitness trails, parks, and more can be a natural extension of the site. Today's and tomorrow's best employees will weigh such factors in determining future employers. The addition of new plants after site construction is completed can be kept to a minimum, thereby reducing landscape purchase and maintenance costs. See Figure 5.14 for an example of vegetation zone identification.

Circulation

Circulation studies can include vehicular circulation on and around the site, as well as human and animal paths. Even if they are to be consciously altered later, they can provide useful information for the designer. Roads and walks can be designed to allow easy access to loading docks. Waste receptacles, Dumpsters, and parking lots can be sited for best access, or for desired low visibility. Through good design, bottlenecks and dangerous intersections can be avoided, comfortable turning radius corners for tractor trailers can be accommodated, and all users will be happier. A complete study will show more information needed: minimum road width, existing road capacities, parking, sidewalks, etc. For preliminary parking designs, use 10 ft. by 20 ft. for automobile spaces and 25-ft. roadways for two lanes. This does not include a factor for aprons, turning lanes, or circulation. A rule of thumb which is acceptable for preliminary estimates is to use 350 ft.^{2} per vehicle. Double-loaded 90° car parking lots with some walks and retention areas can be designed on 100-ft. centers. Then the preliminary design studies can derive more accurate information, complete with minimum turning radii for different vehicle types, which should take into account trailer parking areas for just-in-time delivery.

Winds

Prevailing winds need to be determined so that fresh air intakes are not located downwind from noxious exhausts. A good rule of thumb is to try to locate all fresh air intakes upwind and on the side of buildings. All exhaust fans should be on downwind rooftops, extended 10 ft. above the roof surface in order to assure mixing. All exterior employee functions should also be located away from the prevailing direction of exhaust discharge stacks.

Views

Views should be diagrammed to determine which areas should be visible from the inside. No longer do we avoid any windows (for theorized production enhancement), as we tried to do in the 1960s and 1970s. We learned that a windowless building *does not* create the best performance environment. Limited access to selected views is a more successful approach for the future. Allowing people to acquire their own information about their outside world is calming and allows them to focus on their work. Humans need to sense if it is dark or light, and if they are facing north or west. These needs are thousands of years old and cannot be ignored without negative results. Visual connection with the outside world is reassuring.

Utilities

Early site studies will have revealed where the existing utility connections must be made. These drawings are often crudely schematic and should always be field-verified prior to design or construction. The quality and quantity of the utilities will affect the eventual project initial and operating costs. At this point in the process of design, detailed information will permit the most cost-effective design solutions. Site planners should always consider the cost of utility extensions along with all the other factors in design. Water, sewer, and electric combined could cost more than $200 per linear foot. Asphalt roadways, by comparison, can cost around $50 per linear foot. Valving, metering, and directions of flow should be shown for water systems, along with size and specification of the conveying materials. Finally all carrying capacities must be determined. Then the designers can make the most economical design decisions. Figure 5.15 shows a possible schematic design that could result from the individual evaluations in the previous sections.

Summary

There is no good reason to avoid any of the simple aspects of site analysis listed above. Cost-effective initial and long-term results depend on it. If you won't do it, hire

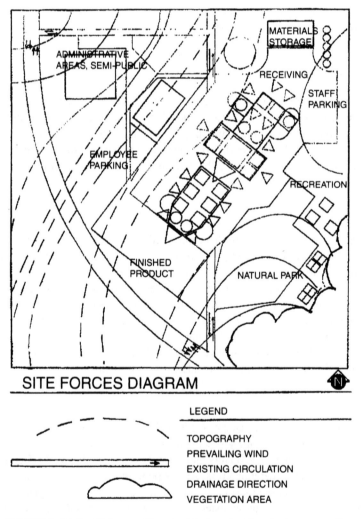

SITE FORCES DIAGRAM

LEGEND

TOPOGRAPHY
PREVAILING WIND
EXISTING CIRCULATION
DRAINAGE DIRECTION
VEGETATION AREA

FIGURE 5.15 Possible overall project design showing the factory outline taking form, and the related support functions.

someone who will. A good master plan starts with a listing of the components of present needs, and should look 10 years or more into the future at potential developments. This will allow for the best arrangement today without erecting permanent structures or infrastructure in areas to be used later for other purposes. It will develop guidelines for an overall land use plan, general arrangement of buildings and spaces, and their connections to vehicular and pedestrian circulation. Net and gross square footage figures are calculated for the site, for total parking, and for each building type. From some of these early numbers, budgets can be accurately developed for economic forecasting. They are also useful in obtaining early review comments from governing regulatory agencies.

Total area of the site will be the first and most limiting factor in a master plan. Zoning, height, setbacks, and minimum parking area will be of secondary importance. Next, one must develop the gross area of each type of building space and specify the general activities to take place within. Then the issues of adjacency, sound levels, safety, physical limitations of equipment, and operational requirements of the facility must be factored in. For some facilities, such as those dealing with explosives or potentially dangerous materials, adjacencies will be the most significant determinants of the master plan. This will generate a widely clustered group of small assembly buildings located hundreds of feet apart.

Facilities that share expensive conveyance devices or that utilize cryogenic systems will normally be rather centralized in order to minimize waste and cost. Each site will generate a unique set of forces that should effect the master plan. Figure 5.16 illustrates the possible impact of topography on a master plan for a manufacturing center that includes employee housing and recreation facilities to complement the factory.

5.4.4 Facility Design

Architecture

The most visible of all decisions are made during the actual facility design. Massing, form, and finish all work together to create an image of the completed facility, but how it fits with the needs of the manufacturing process is what counts most in factory design. This is when the designer takes the programmatic information provided (from Chapter 3, as modified during the programming phase by the A&E firm and the client) and turns it into plans and specifications that yield the final buildings. Decisions made during design affect not only initial cost but how the buildings perform in the manufacturing mode over time. A thorough building program will provide such detailed information relative to the performance requirements that there should be only a few viable organizational schemes that support the optimum manufacturing process.

The traditional manufacturing building resembles a warehouse. The architect is challenged to create a cost-effective method of expressing the corporate image. Obviously, the building must support the specific manufacturing process as described previously. So the length and shape of the product line may be the primary factor in determining the ultimate building form. Steel column and beam construction has been used for hundreds of years for a reason—because of its simplicity. This leads

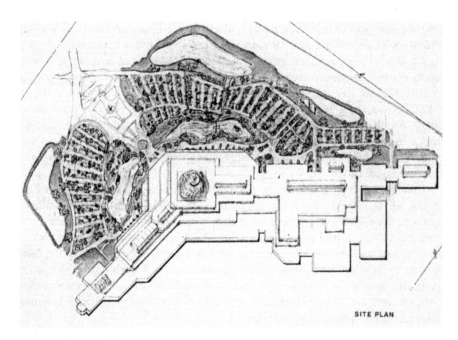

SITE PLAN

FIGURE 5.16 Example of the master plan for a manufacturing plant that includes employee housing and recreation facilities.

to rectangular buildings that are limited in height by money; the higher they are, the more they cost to build and operate. Simple geometry tells us that the closer to square we shape our buildings, the less surface area will be required of exterior walls. Other factors include:

Cost-effective span for joists
Areas that require exterior walls
Height of machines, products, work in process
Material flow
Utilities

The first tool the designer has is the site. First impressions of a place are lasting. Are there visual or physical forces that will affect our perception? If not natural, are there site development requirements that could lead to new features such as retainage ponds, berms, or tanks that will create visual interest? The manner in which we deal with parked cars will have a significant visual impact. The next significant visual aspect of design is the massing of the buildings. Landscaping becomes the finishing touch, softening hard angles. Finally, finishes, color, texture, and details make the last impressions.

There are a few opportunities for the designer to infuse visual interest into a simple box without spending more for it. Through careful evaluation of the necessities of design one can keep the "beast" from growing either too large or too expensive.

Generally, a building using metal wall panels and steel post-and-beam construction will utilize spans of between 25 and 50 ft., with wall height between 8 and 16 ft. Typically the columns are within the outside skin and are not expressed on the exterior. Girts, purlins, and intermediate framing support the skin system. For conditioned buildings there is usually a liner panel with insulation between the two skins to reduce heat loss or gain. Open-web steel joists are most commonly used for secondary structural supports because of their low cost. Mechanical and electrical components are then supported from the bottom chords of the joists, so there is no natural beauty in the materials. It will be up to the skilled professional to create a complex that is more dynamic than the sum of the parts. One of the often overlooked decisions relates to the need for conditioning many of the factory spaces. The range of allowed tolerances for temperature and humidity will dramatically affect the mechanical and electrical systems' initial and long-term costs.

Depending on the particular site, building zones may influence the shape of the building. If land costs are high, it may be more economical to build vertically than horizontally. The shape of the roof is often a powerful image maker in the designer's toolbox. The span will determine the most cost-effective roof structure, and the climate will determine the material. Low-slope metal roofing is cheaper by far than flat single ply, so if the least dimension of the building is not too great, it will be cheapest to design a low-slope roof. If the span exceeds 100 ft., it is common to use internal roof drains and to segment the roof area with sloping insulation around a series of roof drains. The intermediate method is for a low slope roof with a center ridge, and roof gutters and leaders just inside the exterior wall parapet. This has the aesthetic of a higher-cost building without the cost of interior roof drains and leaders. One of the few other elements that has been used historically for decoration is the overflow scupper. Others are exterior lights and lamps, stair rails, and bolted connections.

Visual Impact

One of the low-cost choices with greatest visual impact is the column. Columns and beams that are internal of the skin are not revealed or expressed on the outside, resulting in smoother, planar massing. Columns expressed on the outside of a building can create rhythm, visual interest, and will break the large planes into smaller components. Primary and secondary steel can be painted in such a way as to reinforce the hierarchy of the parts. This will also provide greater visual interest than having all exposed materials the same color.

The choice of skin materials will have to be made with climate and programmatic considerations first. Pragmatics usually cause factory exterior walls to have low initial and maintenance costs. This leaves about four choices, each of which has a different aesthetic. Metal panels can be flat or can have ribs for lateral load resistance. For a planar facade, designers choose flat panels, usually with factory-applied finishes. Metal panels with raised ribs can provide additional resistance to lateral or wind loads. Second choice is a precast concrete panel system or site-cast "tilt-slab" concrete construction. A third choice is a relatively new system known in the industry as EIFS (for exterior insulation and finish system). It is composed of rigid foam panels with a mesh reinforcing layer and synthetic plaster finish. If the wall height is not too great, concrete block exterior walls are a common solution to low-maintenance construction. They can be specified with textures and color from the supplier. "Slump blocks" or split-faced blocks are rough-textured, with the appearance of jagged concrete. These are also available in a variety of colors. Other concrete block choices include smooth, stucco, and scored block. Each will yield a different texture and result in a different contrast in light, shade, or shadow. The skilled designer can combine these basic choices into an effective image-achieving solution. Massing, contrast, and color affect the way we perceive the composition of buildings, while axes and siting affect the power of the image.

One issue that controls how much information we use to form our image of a project is the speed at which we view it. There is not much time for observation at 65 mph. We first see the form and try to compare it with those we have seen before— what is it like? Is it like a barn, or a church, or does it resemble a "blah" warehouse? The three things that give us those lasting visual impressions are the manner in which the building meets the sky (roof shape, slope, scale), the way it breaks up the massing, and the way it meets the ground. Does it appear to be "plopped" on the site, or is it integral with the site? As human society changes with time, so does our perception. While once it was sufficient to spend a few bucks at the entrance in order to give a favorable impression, people today look more at the total picture than before. Few factories have the luxury of being able to spend much of the budget on applied decoration. This places more importance on the few areas where there is little or no additional cost for architectural treatment. It is best to manipulate the necessary items of the building and not rely on applied decoration. Rather than try to hide the systems components behind bulkheads, it is more cost-effective to treat them as elements to be exposed and to make them into the decorations. Spiral duct and cable trays, for example, reinforce the high-tech aesthetic. Leaving them accessible will also reduce maintenance costs and future remodeling costs.

Mechanical Design

Once the programmatic information is understood, it is time to determine the best way to accommodate it. It is also necessary to ask questions. Are the requirements real or perceived? The designer must decide how best to achieve the real needs, while still exceeding all applicable code minimums. The best designers will make use of low-tech, passive systems where possible. They will be able to provide for flexibility of future installations during initial design, without adding substantially to the initial cost. They should also look at many optional solutions for each problem before deciding upon one. This type of brainstorming often results in as many breakthroughs as it does rejects.

For example, rather than conditioning a space 14 ft. high, which is 20 ft. by 40 ft. of floor space around a machine, it may be more cost-effective to construct an enclosure around the machine space, which can be conditioned for 1/10th the initial and 1/100th the 10-year operating cost. In many instances it is possible to cap pipes for possible future installations, and to size them for future anticipated loads. This may result in slightly larger branch piping for about 5% of the distribution network, which will have minimal impact on initial cost. The use of zone isolation valves is another often overlooked aspect of mechanical design. Their initial cost is low compared to the future cost of maintenance, and to draining the entire systems to install a new branch for a new piece of equipment.

The number of air changes per hour, for example, is usually determined by the classification of the use of the space (e.g., office versus industrial use). This affects the FP system design requirements and the type of construction materials allowed. As a rule of thumb, between three and five air changes per hour is a basis for rough estimating. Areas with chemical processes may need more than ten air changes per hour. Another method is to figure 1 cfm supply for every 10 ft.^{2} of building area. Of course, further design studies would take into account calculated heat losses, heat gains, insulation, humidity, and allowable fluctuations over time. The psychometric chart is a good way of understanding the relationships among creature comfort, humidity, and air velocity. Unrealistic design criteria can result in thousands of dollars in mechanical equipment initial and operating cost. This is a critical area in design, when the decision to use one system over another may affect air quality and cost more than at any other time in design or construction. These decisions must be evaluated by the ownership team in the schematic phase, prior to proceeding to construction contract documents.

An example is a requirement to maintain $40\pm2\%$ relative humidity. This could entail the use of expensive humidifiers in the winter and heating coils during the summer months to dry the conditioned air. If the real criteria is that the humidity must stay below 60% year round, that would be much easier and cheaper to achieve. In some instances it is better to increase the velocity of air rather then try to keep the temperature of the air introduced into the room lower. The resultant comfort level for the worker will be the same, and the initial and operating costs will be lower. Similarly, for a positive-pressured building, it may be cheaper to design less exhaust ducting and fans, and to allow perimeter losses through dampered grilles. A comprehensive evaluation of many alternatives should be included in the schematic design phase. Insulation of walls and roof surfaces should then be designed to minimize operational costs. The biggest challenge in mechanical systems design is to get to the essential information. Only through understanding of the intended goal can one design the best solution.

In general terms, the mechanical system should be designed to provide productivity, reliability, and safety. Profit is the goal, and predictable cost forecasting is a good way to get there. For a factory, the project cost of using superior mechanical products may be less than 3% higher than the low-cost competitor. Variable-speed pumps may cost more to purchase, but they can support a wider range of operating conditions. If the demand is relatively constant, fixed-speed pumps should be selected that most closely match the design criteria. There are many decisions that will affect the performance of the plant. Selection of the right mechanical components, and cost-effective layout of the utilities, are critical. See Figure 5.17 for an example of some mechanical installations.

Mechanical system design is far too complex to cover in a few pages of a book on manufacturing engineering. Materials selection alone would require a chapter; piping alone could barely be covered in 50 pages. Specifications and details affect

FIGURE 5.17 Example of typical mechanical installations in a complex project.

construction cost nearly as much as area and scope. Many excellent references provide more comprehensive information on the subject. An overview of both mechanical and electrical equipment systems is presented in McGuiness, Stein, and Reynolds (1980).

Electrical Systems

Electrical systems design is more objective than mechanical design, but there is the same challenge for cost-effective decision making. Starting from the big picture, one should design to utilize natural lighting where practical to augment artificial lighting. Power and lighting distribution should be done at the highest practical voltage. Since electrical systems are the most frequently changed, it is good to use cable trays where possible. This reduces initial and remodeling costs. Some facilities use a lot of 480-V bus duct in lieu of cable in conduit. With recent changes in some manufacturers' marketing schemes, bus is available in standard lengths in a matter of weeks. It is very flexible compared to modifying cable in conduit runs. Bus plugs can be added wherever needed to power step-down transformers at the point of use. This also provides a means of disconnect in lieu of mounting a distribution panel and breaker or main disconnects.

Usually, all large motors will be specified to operate at the highest available voltage. Interruptible or curtailable power can be placed on separate circuits, in order to negotiate a lower annual cost for power. All noncritical systems should be powered on the interruptible circuits. Power factor correction is another important factor in electrical systems design. Capacitor banks can be utilized to offset fluctuations in primary current from the local utility. After distribution and step-down transformers, line conditioners or surge suppressors can be added.

Lighting control systems are available to optimize the efficiency of lighting systems. Experienced designers will choose the most cost-effective fixtures for each space, looking first at the amount of direct and indirect light from the shape of the fixture. Next one would look at the bulbs, striving to achieve the best initial and operating cost for the owner's needs. Lighting consultants can also recommend special designed task lighting in areas for which it would be costly to provide general illumination fixtures.

With the continuing development of ultra-low-energy light sources such as we have seen in the last few years, lighting design is about to be revolutionized. Imagine telephone wire providing enough current to light a small office at 40 foot-candles. The material cost of the installation is going to be low due to the amount of copper in cross section, and there will be no need for conduit since it is all low-voltage. Furthermore, the operating cost will be low due to the current draw, and switches should be cheaper and should last longer. It has been reported that bulbs will last more than 20 times as long as traditional bulbs. Inventory and replacement costs should be negligible. A valuable reference on this subject is Thumann (1977).

Communications Systems

The reliance of business today on information systems will change the role of the manufacturing engineer dramatically. As part of the new team approach to product development, the manufacturing engineer will need skills in computers and information flow. Networks consist of the computers to be connected, the communication media for the physical connection between the computers, and the software for the network communications and management (see Subchapter 7.4).

As the name implies, the local area network (LAN) is geographically the smallest of the network topologies. In general, the LAN is limited to 1 to 2 miles in displacement. The wide area network (WAN) has the capability to allow connecting a LAN to other LANs within the company or with LANs of vendors and customers. The speed of the WAN is determined by the communication media used for the long-distance transmission. At the low end of the speed regime is the modem using regular phone lines. In most cases, however, if you have more than one computer, a modem will not suffice. Most of the data transmitted over long distances is sent over T1/E1 lines.

One of the most important factors to consider in the design of manufacturing support systems is the amount of data to be transmitted over the network and the response time of the data transfer. The speed of the network will depend on the type of connection media (copper wire, coax, or fiber optics), the distance to be covered, and the network software. The actual cabling that connects the various components on the network comes in many varieties. The three major ones are:

Coaxial cable, where one or more concentric conductors are used
Twin twisted-pair cable with two pairs of wires, each pair twisted together
Fiber-optic cable, used for FDDI and can be used with Ethernet

The coax and twisted-pair cables have been used for some time. The fiber-optic cable is fairly new, but this technology is expanding at a very rapid rate. With all of the phone lines being changed over, the price of this transmission medium is dropping to the point where the material is very competitive with wire. The cost of installation of a fiber-optic system is still higher than for the other types, due in large part to the unit cost of converting the signals. The expanded capabilities and the immunity of a fiber-optic system to electrical interference makes it a good choice in many applications.

Once all the elements (i.e. mainframe, mini, PC, and programmable controllers) and all the necessary hardware and software parts for the network are determined, the computer-integrated enterprise (CIE) becomes possible. The CIE consists of a combination of various functional units to perform the desired tasks to operate the business. These include finance, purchasing, material control, manufacturing process control, and reporting. To perform these tasks, the CIE must have certain capabilities. Processing (computer processing unit, CPU), input and output terminals, printers, and an interconnecting network among the units make up the minimum capabilities. A typical architecture for a CIE system is shown in Chapter 7, Figure 7.12. This example assumes that your organization has used computer systems for some time, and you will have a mixture of types of hardware and legacy software programs in a multisite system. Medium to small, or new start-up, manufacturers would have the luxury of designing a system from the ground up. See Figure 7.9 for an example of a new start-up system.

With respect to the multisite system, Figure 7.10, the mainframe is at one location while other elements are at remote locations. The data are transmitted from location to location via hard-line (T1 lines) or in some cases by satellite. For the system to be a true CIE, and to be as productive as possible, the user should input data into the system once. It is then up to the program logic to route and install the data into the correct tables and databases for processing and record keeping. Some of the processing will be handled at the local terminals and servers, while other data will be transmitted to the mainframe for processing. One thing to remember is that in a well-designed system, all applications should be available to the users at local terminals. Timely information at the lowest level of the organization will allow workers to make decisions and improve productivity. Figure 5.18 shows some overhead cable trays containing signal and data cables.

To be first in class, your CIE must provide data to and from your customers, vendors, and across your total factory. For you to accomplish your goals of meeting and beating your competition, you must look at all the factors that will affect your price (and profit). Inventory and work in process must be reduced to an absolute minimum. This can be accomplished by judicious design of your computer interfaces (electronic data interchange, EDI) with your vendors. With a material requirements planning (MRP) or manufacturing resources planning (MRP-II) system, you can effect a just-in-time delivery system to reduce your inventory and work in process.

The definition of the communication systems for your factory must be transmitted to the A&E or other design agency as part of developing the factory requirements. For example, if you plan an Ethernet backbone using fiber optics down the center of

FIGURE 5.18   Typical overhead cable trays.

the plant, you must describe it. If you plan on distribution closets from the backbone at strategic locations throughout the plant, with twisted-pair hardwire into all offices, receiving, stores, shipping, and most of the workstations on the manufacturing floor, you should specify the criteria the building designer will need. Telephones, fax machines, and other systems should also be defined. If there are distributed numerical control (DNC) machines on the floor, you need to define the linkage to the programmers, minicomputers, etc.

5.4.5   Costs

Now that we have discussed the programmatic issues and started the real factory design, it is possible to make an accurate cost estimate based on the actual materials, equipment, and other items specified. The earlier estimates were partly parametric calculations, based on experience with other building elements and projects. The cost-estimating process has been an ongoing activity, as we have progressed through the process leading up to a preliminary design.

Once design documents are past the schematic approval stage, it is time to develop them to the point where specific materials selections and systems components are selected. Cost estimates need to be developed from the entire set of drawings, including scale floor plans and elevations. We can now use actual area information along with lists of major equipment in order to determine how the decisions made during design have influenced costs, and to check current construction cost information relative to our earlier projections. For example, if the costs appear to be 7% over budget at the 50% design development point, reductions can be designed into the final documents. With computer stretch and auto-dimensioning functions, one can easily make area changes. Each subcomponent shown on the plans and specs should be accounted for in this estimate. Figure 5.19 is an example of a warehouse-type factory building estimate, which is representative of the low end of technology and cost. Figure 5.20 is a detailed, multipage example of an estimate for a prototype plant, which is representative of a cost breakdown for a more major high-tech factory building.

5.5   CONSTRUCTION

The two most common methods of building construction are to hire a general contractor (GC) or a construction management (CM) firm. General contracting is the traditional building system, where a firm, fixed-price contract is awarded for a fixed scope of work. Plans and specifications are usually completed prior to the beginning of the bid process. GCs estimate their cost and markup before submitting fixed-price bids. The lowest acceptable bidder gets the award. Many of these firms have full-time employees, carpenters, masons, painters, or electricians. They have an office structure to support continuing business. Most will also have a pool of familiar subcontractors with whom they have established a time-tested relationship. Some of the larger firms have in-house design capabilities. If the project is phased over years, or if there is planned expansion, it favors choosing a local GC firm or CM firm over a large national firm with no local ties.

CM firms are made up of managers and support staff and few actual workers. They often specialize in one building type or project size. Manufacturing facilities have become more sophisticated, with more piping systems and control devices than hospitals. Due to time constraints, many recent projects have utilized CM processes. CMs are usually more experienced in multiple-bid packages and project expediting. Large CM firms have purchasing agents and contract managers who focus on their specialty. Their estimating divisions also have their fingers on the pulse of component costs. CM fees vary from 5 to 20% depending on the project, with an average of about 8% above cost. General conditions, utilities, and overhead for the job-site trailers are in addition to their fees.

The actual selection is made more difficult by the fact that the fees are usually fixed, meaning there is no direct cost competition. Of course, there are hybrids of these pure examples, GCs are always interested in doing cost-plus jobs, and CM firms often have a contracting branch for smaller projects. The choice is seldom clear without careful evaluation.

5.5.1   Selection Process

The first part in CM selection is to determine whether to advertise for competitive bids or to offer it to a limited number of preapproved firms at a fixed fee rate. Understanding the complexity and time constraints will help in making the decision.

TYPICAL WAREHOUSE BUILDINGS				
CONSTRUCTION COST BREAKDOWN				
21,000 SF OF FLOOR SPACE, UNCONDITIONED				
800 SF OF CONDITIONED OFFICE				
Description	Quantity	Units	Unit Price	Price
Building Personnel				
Superintendent	24	wks	$650.00	$15,600.00
Travel	24	wks	$100.00	$2,400.00
Project Engineer partial	24	wks	$200.00	$4,800.00
Clerical support	24	wks	$25.00	$600.00
Permits & Fees				
Electrical Tap fee	1	ump sum (ls)	$1,200.00	$1,200.00
Environmental Permit	1	s	$240.00	$240.00
Water Tap fee	1	s	$900.00	$900.00
Sewer Tap fee	1	s	$2,450.00	$2,450.00
General Conditions				
General Labor	21600	sf	$0.06	$1,296.00
Cleanup	6	mos	$250.00	$1,500.00
Rubbish Removal	6	mos	$350.00	$2,100.00
Dump charges	6	sa	$150.00	$900.00
Temp. toilet	6	mos	$65.00	$390.00
Water & Ice	6	mos	$25.00	$150.00
Pager, mobile ph. partial	6	mos	$65.00	$390.00
Temp. Telephone	6	mos	$120.00	$720.00
Temp. Electric	6	mos	$125.00	$750.00
Temp. water, sewer	6	mos	$45.00	$270.00
Layout	1	s	$250.00	$250.00
Reproduction, etc.	1	s	$450.00	$450.00
Testing	1	s	$550.00	$550.00
Equipment rental	1	s	$1,000.00	$1,000.00
Signage	1	s	$450.00	$450.00
Office Trailer	6	mos	$275.00	$1,650.00
Supplies	6	mos	$25.00	$150.00
Security	3	mos	$500.00	$1,500.00
Site Construction				
Site Clearing	1	s	$13,500.00	$13,500.00
Rough Grading	1.5	acres	$5,000.00	$7,500.00
Finish Grading	22000	sf	$0.25	$5,500.00
Fill & Compaction, 18 inches	2000	cy	$4.50	$9,000.00
Paving	12000	sf	$1.15	$13,800.00
Erosion controls	1	s	$650.00	$650.00
Drainage	1	s	$4,500.00	$4,500.00
Landscaping	1	s	$6,500.00	$6,500.00
Seed & Mulch	2	acres	$1,000.00	$2,000.00
Utilities Improvements	1	s	$6,600.00	$6,600.00
Fencing	1	s	$3,400.00	$3,400.00
Building Construction				
Slab edge forming	1100	f	$0.35	$985.00
Concrete	680	cy	$55.00	$37,400.00
Steel	23000	sf	$0.80	$18,400.00
Concrete finishing	21500	sf	$0.65	$13,975.00
Masonry, block, w/plasters	1400	sa	$1.35	$1,890.00
Ornamental Iron	1	s	$1,200.00	$1,200.00
Metal doors & frames	6	sa	$400.00	$2,400.00
Overhead doors	3	sa	$2,875.00	$8,825.00
Windows	6	sa	$250.00	$1,500.00
Hardware	1	s	$4,250.00	$4,250.00
Glass & Glazing	2	sa	$120.00	$240.00
Metal studs & Drywall	2000	sfsf	$4.35	$8,700.00
Flooring office	320	sf	$1.15	$368.00
Painting	2000	sf	$1.25	$2,500.00
Louvers, vents	1	s	$4,250.00	$4,250.00
Accessories	1	s	$450.00	$450.00
Loading dock, ret. walls	1	sa	$3,125.00	$3,125.00
Pre-engineered Buildings	21500	sf	$5.25	$112,876.00
Plumbing	1	s	$2,500.00	$2,500.00
HVAC, window unit for office	1	s	$650.00	$650.00
Operable shutters & Fans	1	s	$6,500.00	$6,500.00
Electrical	21500	sf	$1.56	$33,540.00
Labor	12	manweeks	$320.00	$3,840.00
Misc'l Material	1	allowance	$6,500.00	$6,500.00
Insurance	1	s	$27,650.00	$27,650.00
Bonds	1	fs	$2,520.00	$2,520.00
Overhead & Profit @ 15% of hard	1	sa	$55,960.00	$55,960.00
TOTAL COST				$470,000.00

FIGURE 5.19 Typical warehouse-type factory building construction cost breakdown.

PROJECT NAME PROTOWIDGET				
Prepared by : W. L. Widget				
Date prepared: 10/12/94				
Based upon Plans and Specs		dated		
DESCRIPTION				
General Conditions	Quantity	Units	Cost per Unit	Cost
Office, trailer etc. cost	10	mos	$650.00	$6,500
Utilities cost, Temp. & connect	10	mos	$14,000.00	$140,000
Submittals, copying, blueprinting	10	mos	$250.00	$2,500
Special Conditions (vary greatly)				
Bid Alternates,deductive alternates , etc.			(be specific)	
Protection. erosion. sound. dust	4	acres	$1,500.00	$6,000
Parking, storage, security	10	mos	$800.00	$8,000
Lay-down space. staging			$0.00	$0
Staff; direct cost plus Insurance	10	mos	$15,000.00	$150,000
				Initial
Site Work	Quantity	Units	Cost per Unit	Cost
Clear and grub	1.8	acres	$2,200.00	$3,960
Misc'l demolition	14,500	sf	$0.05	$725
Cut	21,300	cy	$2.00	$42,600
Fill	5,000	cy	$6.00	$30,000
Haul	16,500	cy	$2.00	$33,000
Rough grade	196,000	sf	$0.05	$9,800
Compact	8,000	cy	$1.50	$12,000
Finish grade	125,000	sf	$0.02	$2,500
Soil treatment	100,000	sf	$0.02	$2,000
Vapor barrier	100,000	sf	$0.05	$5,000
Excavation for footings, piers, etc.	1,200	cy	$4.85	$5,820
Pilings, driven	1,220	lf	$13.50	$16,470
Pilings, sheet	250	lf	$7.50	$1,875
Dewatering	100,000	sf	$0.05	$5,000
Gravel under slab	3,500	cy	$8.00	$28,000
Expansion joints	1,800	lf	$0.28	$504
Waterproofing, sheet	1,200	sf	$4.50	$5,400
Waterproofing, coatings	3,000	sf	$2.90	$8,700
Hand excavation, miscl	45	cy	$45.00	$2,025
Road base, 6" crushed rock, stone	1,600	sf	$0.39	$624
Labor, edge work	2,500	lf	$0.25	$625
1 1-2"Asphaltic Topping	15,000	sy	$3.70	$55,500
Concrete curbs	4,750	lf	$8.24	$39,140
12" reinforced concrete drive, 4000 psi	432	sf	$9.50	$4,104
6" reinforced concrete drive, 4000	40000	sf	$5.80	$232,000
4" reinforced concrete drive, 4000	10000	sf	$3.20	$32,000
4" sidewalks, 3500 w/mesh	3800	sf	$2.35	$8,930
Pavement markings	350	ea	$4.00	$1,400
Signage	6	ea	$125.00	$750
Utilities				
Fiber optics system, backbone	1250	lf	$0.78	$975
Telephone cable. vault and conn.	350	lf	$0.25	$88
Data cable, 16 UTP, from comp's	10250	lf	$2.80	$28,700
Power from property line	125	lf	$3.50	$438
Water line, domestic, 4"	1550	lf	$6.00	$9,300
Water line, domestic, 2"	225	lf	$5.40	$1,215
Fire protection line 6"	500	lf	$15.00	$7,500
Fire prot. 8"	1400	lf	$18.00	$25,200
Post indicator valve	6	ea	$1,325.00	$7,950
Siamese connections	2	ea	$1,500.00	$3,000
Hydrants	5	ea	$1,400.00	$7,000

FIGURE 5.20 Detailed, multipage example of an estimate for a prototype plant. This is representative of a major high-tech factory building.

Thrust blocks	24	ea	$45.00	$1,080
Metering	3	ea	$3,200.00	$9,600
Backflow prevention device	1	ea	$2,100.00	$2,100
Gas piping, 2" steel	125	lf	$9.50	$1,188
Sanitary sewer, 8" PVC	1280	lf	$12.00	$15,360
Sanitary sewer, 6" PVC	200	lf	$8.00	$1,600
Sanitary sewer, 4" PVC	124	lf	$5.00	$620
Storm sewer, culvert	2	ea	$1,500.00	$3,000
Storm sewer, swale	4	ea	$400.00	$1,600
Storm sewer, 15" RCP	1300	lf	$17.50	$22,750
Storm sewer, 12" PVC	250	lf	$15.00	$3,750
Storm sewer, 10" PVC	100	lf	$12.50	$1,250
Storm swere, 8" PET	300	lf	$10.50	$3,150
Storm sewer, 6" PET	1300		$9.50	$12,350
Trench drain	150	lf	$112.00	$16,800
Interceptor	2	ea	$1,250.00	$2,500
Neutralizing tank	1	ea	$2,500.00	$2,500
Treatment tank	3	ea	$2,130.00	$6,390
Liquid sto. 400,000 gal. tank	1	ea	$232,000.00	$232,000
Gas storage tanks	3	ea	$27,500.00	$82,500
Fences, gates, vinyl coated				
	3000	lf	$10.28	$30,840
Site Lighting, poles $ wire				
Well pointing	20	ea	$1,050.00	$21,000
Rock excavation	0	ea	$0.00	$0
	0	tons	$200.00	$0
Landscaping, allowance				
	10000	sf	$12.00	$120,000
Rebar				
Mesh	74	tons	$789.00	$58,386
	9	tons	$800.00	$7,200
Concrete, column footers, caps				
Continuous footings	53	cy	$58.00	$3,074
Pilasters, piers	165	cy	$53.00	$8,745
Isolation mats, slabs	11	cy	$61.00	$671
Pits, elevator etc.	105	cy	$64.00	$6,720
Cooling tower sump	50	cy	$69.00	$3,450
Special reinforced pads	125	cy	$72.00	$9,000
Concrete floor slabs, 4000 PSI	23	cy	$112.00	$2,576
Concrete floor slabs, 3500 PSI	1928	cy	$46.50	$89,652
	35	cy	$45.50	$1,593
Structure	Quantity	Units	Cost per Unit	Price
Structural steel, primary	618	tons, m	$1,173.00	$724,914
Steel, secondary bracing	56	tons	$1,210.00	$67,760
Steel, small members	37	tons	$1,379.00	$51,023
Special frarring				
Roof openings	120	sf	$4.90	$588
Bar joists	12	ea	$1,000.00	$12,000
Floor decking, 2" 20 ga.	0	tons	$1,000.00	$0
Roof decking, 1 1/2" 20 ga.	14000	sf	$1.50	$21,000
Acoustical metal decking	13000	sf	$1.05	$13,650
Misc'l metal supports, etc.	800	sf	$1.85	$1,480
	100000	sf	$0.33	$33,000
Catwalks, etc.				
	4	ea	$750.00	$3,000
Metal wall panels, steel, painted				
Metal wall panels, alum. painted	33700	sf	14.5	$488,650
	4000	sf	$16.00	$64,000
Explosion relief wall system	200	sf		
			$18.00	$3,600

FIGURE 5.20 (Continued)

Masonry wall, 8" CMU	7000	sf	$8.25	$57,750
Masonry 2 hr. wall	2641	sf	$5.50	$14,526
Metal studs & 1/2" gyp., not rated	84500	sf	$2.00	$169,000
Metal studs & 1/2" gyp, 2 hr. rated	6600	sf	$3.00	$19,800
Exterior insul. & finish system	5000	sf	$10.75	$53,750
Glass block	240	sf	$15.00	$3,600
Brick veneer	0	sf	$0.00	$0
Cut stone veneer	500	sf	$6.00	$3,000
Glazed facing tile on block	500	sf	$12.80	$6,400
Miscellaneous				
Roof blocking	3500	lf	$14.00	$49,000
Cant and transition strips	800	lf	$20.00	$16,000
Roof penetrations	200	lf	$10.00	$2,000
Frames at roof drains	64	ea	$125.00	$8,000
Wall frame blocking, deadwood	54	ea	$100.00	$5,400
Telephone backboards	12	ea	$125.00	$1,500
Window sills	2800	lf	$8.00	$22,400
Rough carpentry	100000	sf	$0.50	$50,000
Roof insulation boards, 2" thk	112000	sf	$0.45	$50,400
Roof insul. it w/ concrete, sloped	4800	sf	$2.85	$13,680
Sealants, caulking	100000	sf	$0.45	$45,000
Finish carpentry	100000	sf	$0.80	$80,000
Casework	15000	sf	$12.00	$180,000
Furnishings, desks, chairs, cab's	25000	sf	$1.75	$43,750
Shelving, metal, heavy	450	lf	$18.00	$8,100
Shelving, metal, light	220	lf	$9.00	$1,980
Fume, flow hoods, thru roof	8	ea	$6,750.00	$54,000
Special hardware for connections	8	ea	$1,000.00	$8,000
Dock levelors, w/bumpers	6	ea	$5,250.00	$31,500
Lockers	60	ea	$180.00	$10,800
Benches	12	ea	$140.00	$1,680
Marker boards	14	ea	$240.00	$3,360
Fire extinguishers & cab's	36	ea	$125.00	$4,500
Wall corner guards	2000	lf	$15.00	$30,000
Access flooring, raised	1200	sf	$10.50	$12,600
Ramps, steps	325	sf	$20.00	$6,500
Crane, 20 ton X-Y, remote, 500 sf	2	ea	$110,000.00	$220,000
Crane, 10 ton X-Y per 500 sf	2	ea	$80,000.00	$1,60,000
Telephone enclosures	0	ea	$1,000.00	$0
Pallet storage system	2000	sf	$45.00	$90,000
Safe storage cabinets, containm't	600	sf	$115.00	$69,000
Suspended ceiling systems	22000	sf	$1.45	$31,900
Acoustical treatment	1200	sf	$8.00	$9,600
Curtain wall glazing system, ext	2000	sf	$31.00	$62,000
Glass panels, interior	1000	sf	$12,00	$12,000
Epoxy floor coatings, fgl reinf	800	sf	$3.50	$2,800
Epoxy floor paint, 2 coat	60000	sf	$2.50	$150,000
Concrete stain, seal	15000	sf	$1.75	$26,250
Ceramic tile, incl. cove & base	2400	sf	$10.00	$24,000
Carpet	18000	sf	$1.50	$27,000
Acid resist surfaces, w/backer bd.	4000	sf	$14.00	$56,000
Painting, ext. masonry	1200	sf	$0.88	$1,056
Paint, int. masonry	120000	sf	$0.78	$93,600
Block filler	120000	sf	$0.54	$64,800
Epoxy on walls	50000	sf	$0.75	$37,500
Latex walls	20000	sf	$0.50	$10,000
Ceilings	8000	sf	$0.75	$6,000

FIGURE 5.20 (Continued)

Description	Quantity	Units	Cost per Unit	Price
Accent paint colors	4000	sf	$0.85	$3,400
Color coded piping, safety striping	75000	sf	&0.33	$24,750
Miscl iron & stairs, rails	100000	sf	$0.17	$17,000
Elevator	0	ea	$50,000	$0
Mechanical Equipment Allowances				
Wells & pumps, 1000 GPM cap.	1	system	$30,000	$30,000
Cooling towers, per 30000 gpm	2	ea	$125,000	$250,000
Chillers, per 2000 ton unit	2	ea	$925,000	$1,850,000
Primary water pumps, 500 gpm	2	ea	$27,000	$54,000
Secondary pumps	6	ea	$21,000	$126,000
Strainers & filtration, varies greatly	1	allowance	$50,000	$50,000
CUP condenser water piping	1	allowance	$200,000	$200,000
Water treatment system	1	allowance	$50,000	$50,000
DI system, 5 megohm, polish, demin.	1	allowance	$350,000	$350,000
Dual air compress sys., complete	1	allowance	$50,000	$50,000
Heat exchanger, dual @ 1000gpm	1	allowance	$200,000	$200,000
Process fluid pumping, meter'g etc.	1	allowance	$60,000	$60,000
Process gases, bottles, tanks etc.	1	allowance	$50,000	$50,000
Reclamation system, scrubbers, etc.	1	allowance	$75,000	$75,000
Boiler, 6,000 MBH, gas, steam gen.	1	allowance	$60,000	$60,000
Boiler, gas, hot water, 1,500 MBH	1	allowance	$25,000	$25,000
Furnace, elect. 30 MBH, w/ controls	1	allowance	$750	$750
Oven, 5' horiz. gas fired, @ 450 C	1	allowance	$12,500	$12,500
Vibration isolation for process piping system	1	allowance	$35,000	$35,000
Fire pump system, 2 500 GPM	1	allowance	$55,000	$55,000
Circ'g pumps, 1/12 to 1 hp, total 10	1	allowance	$5,000	$5,000
Vertical turbines, 250 psi, 50' head	1	allowance	$25,000	$25,000
Fume hood, 100 cfm, thru roof	1	ea	$5,000	$5,000
Fan, exhaust, roof curb mt, 500 cfm	1	ea	$2,500	$2,500
Fan, exhaust, roof curb mt, 5000 cfm	1	ea	$5,000	$5,000
Fan, wall exhaust, 2500 cfm	1	ea	$1,200	$1,200
Fan, recirc. fan coil, 10,000 cfm	1	ea	$10,000	$10,000
Smoke evac. fan & controls sys	1	ea	$7,500	$7,500
Fan, floor or base mt, 6 by 6, 120v	1	ea	$150	$150
Piping, black steel, screw, hung, 2"	1000	lf	$15	$15,000
Piping, black steel, welded, 12", insul	160	lf	$120	$19,200
Piping, black steel, welded, 8"	5000	lf	$60	$300,000
DI water piping, stainless, 4" w/ insul	2000	lf	$150	$300,000
Water piping, PVC, solvent weld, 4"	2000	lf	$20	$40,000
DI water piping, polypropylene, 2"	1000	lf	$15	$15,000
Copper piping, 2", soldered with valves	1250	lf	$20	$25,000
Borosilicate pipe, 2" acid waste	800	lf	$40	$32,000
PVC sch40, 6" w/ hangers	600	lf	$25	$15,000
PVC sch40, 2"	1500	lf	$14	$20,250
PVC sch40, 1/2"	600	lf	$9	$5,400
Grooved steel, 6" 150 psi victaulic	1000	lf	$40	$40,000
PVC, sch 10, 48" duct from hood	400	lf	$36	$14,400
FP water piping, 4", 150 psi, plastic	4000	lf	$10	$40,000
Galv. metal ductwork,	2000	lf	$9	$18,000
PVC lined metal duct, 4" by 14', slip jts	200	lf	$12	$2,400
Stainless steel duct, 4" by 14", welded	400	lf	$18	$7,200
Spiral duct, 12", flex, strap hangers	1000	lf	$8	$7,500
Plumbing				
Emerg. eye wash	2	ea	$75	$150
Eye wash with shower, plastic or metal	4	ea	$350	$1,400
Shower, ceramic tile surround, 3' by 4'	4	ea	$750	$3,000
Lavatories, sinks, etc.	14	ea	$175	$2,450
Water closets, uri's, white vc, flush v.	10	ea	$500	$5,000
Water drinking coolers, 5 gpm, inst.	3	ea	$450	$1,350
Drinking water fountains, installed	4	ea	$45	$180

FIGURE 5.20 (Continued)

Description	Quantity	Units	Cost per Unit	Price
ADA approved stall, shr, lav, wc, mirror	2	ea	$1,000	$2,000
Loop heat exch's 10 gpm, 1200w/hr	2	ea	$3,500	$7,000
El. water heater, 200 Gal, 180 deg	2	ea	$15,000	$30,000
Point of use water heaters, 1500 w	6	ea	$150	$900
Accessories	25000	sf	$0.20	$5,000
Electrical, summary of takeoff				
Description	Quantity	Units	Cost per Unit	Price
Temporary service, 200 Amps, 2 loc's	2	ea	$750	$1,500
Primary svc. 13,200 volts, in conduit	1	lump sum	$450,000	$450,000
Power factor correction, 95%	1	lump sum	$140,000	$140,000
Main switchg'r, double ended, w/enclo.	1	lump sum	$250,000	$250,000
Motor control centers	1	lump sum	$3,000	$3,000
Distribution system, 480v, power	1	lump sum	$650,000	$650,000
Distribution at 208v, power	1	lump sum	$845,500	$845,500
Distribution at 110v, power	1	lump sum	$115,750	$115,750
El. generator, diesel w/tank, 150Kw	1	system	$40,000	$40,000
Lighting distribution panels	1	lump sum	$15,000	$15,000
Lighting circuits, conduit & wire, inst.	1	lump sum	$215,000	$215,000
Uninterruptible power system, 150A	1	lump sum	$165,000	$165,000
Controls, energy mgm't, security & HVAC	1	lump sum	$125,000	$125,000
Distributed controls system, dual, digital	1	lump sum	$1,600,500	$1,600,500
Systems programming, start up	1	lump sum	$45,000	$45,000
Fire alarm controls, smoke evac	1	lump sum	$25,000	$25,000
Line conditioning, point of use	10	ea	$1,500	$15,000
Surge protector	100	ea	$150	$15,000
Power towers, to 12', unistrut, cable&cond.	20	ea	$250	$5,000
Transformers, 480 to 208, 10 KVA	12	ea	$1,000	$12,000
Lightning protection system	100000	sf	$0.35	$35,000
Equipment ground system	100000	sf	$0.15	$15,000
480 V busway, w/plugs, 2000A	2700	lf	$460,00	$1,242,000
Solid copper bus bars, 6000A,	400	lf	$600,00	$240,000
Cable in conduit, 500 MCM 15KV in tray	1000	lf	$54.00	$54,000
Cable in conduit conductors, 250 MCM	2000	lf	$25.00	$50,000
Cable in conduit, 5 KV w/PVC jacket, #2	5000	lf	$13.00	$65,000
Cable in tray conductor, #12, 3wire	15000	lf	$2.25	$33,750
Telephone wire, 4UTP, from closet to set	4000	lf	$1.20	$4,800
Data cable, plenum 16pr. UTP w/fiber	8000	lf	$3.00	$24,000
Data cable, coax only in cable tray	1000	lf	$2.00	$2,000
Security camera & cable, to closet	100000	sf	$0.50	$50,000
Multiplexer, rack mtd. w/monitor	1	ea	$2,000.00	$2,000
Magnetic door locks, interlocks	4	sets	$1,000.00	$4,000
Fixtures, 2 by 4 lay-in flou, high effic., 40w	1000	ea	$125.00	$125,000
Fixtures, incandescent task lights, 100w	40	ea	$75.00	$3,000
Fixtures, sodium vapor, 400w	60	ea	$585.00	$35,100
Fixtures, exterior, mercury vapor, 500w	10	ea	$500.00	$5,000
Fixtures, exterior, parking on poles	12	ea	$1,250.00	$15,000
Fixtures, exterior ground, 35 w	10	ea	$300.00	$3,000
Fixtures, conference area, spots	14	ea	$250.00	$3,500
Fixtures, explosion proof	4	ea	$450.00	$1,800
Fixtures, emergency exit lights	25	ea	$125.00	$3,125
Fixtures, special	2	ea	$1,200.00	$2,400
				$0
Cable tray, ladder type, alum, heavy	5000	lf	$20.00	$100,000
Cable tray, ladder type, alum, light	2000	lf	$13.00	$26,000
TOTAL BUILDING HARD COST				$17,028,213

FIGURE 5.20 (Continued)

If the performance criteria are very high, such as for a clean-room-environment batch plant, it may be better to have a fixed fee and have the firms compete based on their ability to perform. If the budget is tight and time is not as critical, it may be better to develop a short list of preapproved CMs and let them submit sealed bids in a competitive situation, with fully completed plans and specifications, then award to the low bidder. This logic is very general, however, and for each project careful consideration must be given to the selection method.

Either way, the next step is to write a notice of intent (call for bidders). You must first define the scope of the project. It should include a brief description of the facility to be built, along with the criteria to be used in selection. It must spell out the type of contractual agreement being sought, fixed price or guaranteed maximum price (GMP). On a GMP contract, the owner recovers all savings derived though the process of value engineering during construction. A fixed-price contract implies that total project costs are subject to change only through changes in the drawings or specs. A formal notice to CMs would contain the following information:

Notice to Construction Managers

1. Name and address of the owner, complete with the branches and main offices involved
2. Name, number, and location for the project
3. Gross area description of major facility components, complete with anticipated timetable and special requirements
4. Brief description of systems to be included
5. Phasing or bid alternatives desired
6. Specific expectations as to the role of the CM
   a. Cost feedback during completion of design documents
   b. Value engineering expectations
   c. Integral member of the design team
   d. GMP
7. Selection criteria

(Defining the selection criteria is most important. Usually the fee is fixed, and it can be difficult to determine how much of a presentation is smoke and how much is real.)

8. General information

It can be helpful to provide an explanation as to the method of notification for interviews and selection. If there are special requirements or circumstances that would be unique to this project, or are crucial to its ultimate success, they should be mentioned. The wording of the advertisement sets the tone for negotiation and selection of the CM. It must be specific. A project fact sheet should accompany all requests for bids, containing information as shown in the following example.

Project Fact Sheet

Project Name and Number

Ownership, Company Name, and Address

Project Description:

This project consists of a complex of facilities for the manufacturing of composite electronics devices for use in converting photovoltaic DC energy into 120 volts AC. Phase
One is currently planned for 50,000 gross square feet of factory space and 25,000 gross
square feet of support spaces. It is projected to start in August 1998 and be complete
within ten months. The main area will have a height of 40 feet with the support spaces
at 25 feet. Primary structural system is anticipated to be post-and-beam steel framing
with a ribbed metal panel exterior wall system.

The Second Phase is the remodeling of 250,000 GSF of existing manufacturing and
office spaces. Phase 2 will start in July 1999, and will modify a brick and block facility
built in 1950.

The design of these facilities will require knowledge and experience with the following:

- Special glass & plastic process piping installations
- Clean environment construction
- High-pressure stainless steel cooling loops
- Large cooling towers, chillers. & pumps
- The effects of high-voltage harmonics, power correction
- Conveyance systems design and installation
- Distributed controls, integrated systems, fiber optics
- Fast-track projects

The first phase of the project will require the firm to work closely with the design team
in order to provide input during the completion of plans and specifications. Long lead
time items such as chillers, electrical switchgear, and water treatment equipment will
be expedited through the use of early bid packages. After testing and startup of Phase 1,
the existing factory will be upgraded to include connection to the Phase 1 Utility Plant
system. The selected construction manager will provide value engineering during the
design and construction processes. Upon negotiation of a Guaranteed Maximum Price
(GMP) the CM will become a single point of responsibility for the performance of the
construction contract for the project.

Firms will be evaluated and selected for interviews on the basis of their experience
and ability—experience in a proven track record for the firm and also for the individuals
on the team that will be responsible for this particular project. The chosen firm will have
illustrated record keeping ability. estimating, cost control, scheduling, quality control,
and the ability to close out jobs.

The most important factors for comparison will be:

1. Knowledge and experience in the construction of large-capacity process piping
   systems
2. Knowledge and experience in the construction of facilities for the manufacture
   of composites
3. Experience in accelerated design & construction

4. Experience in GMP projects, not-to-exceed features
5. Experience or knowledge of local environmental, growth management, legislative or permitting processes

General Information:

All applicants will be notified of the results of the short list selections in writing. Finalists will be informed of the interview times and dates along with any additional project information that may be pertinent. The selection committee will make a final recommendation to the chairman of the board. The board will inform all finalists after successfully negotiating a contract with their highest rated firm.

Responses

Determination should be based on both the tangible and intangible aspects of those expressing interest. In order to obtain enough information to make such a selection, one must require that each firm complete a standard response form. The notice to contractors and fact sheets can be submitted through electronic mail to the distribution list rather than mailed. Since signatures are required, returns must be made through the mail. Soon there will be certifications and signatures done through electronic media that will be binding. This will save time and trees. After reviewing the completed packages, one may be able to compile a short list of six to eight firms for interviews. This is where one learns of the intangibles offered. The forms must be typed, and must include a copy of a license for contracting in the project state. A letter of intent from their surety company will acknowledge bonding for the job. The insurer should be licensed in the state and have an "A" rating or better.

Sample Response Form

1. Project name and number.
2. Firm name and address:
   a. Address of office in charge
   b. Telephone and fax numbers, e-mail addresses
   c. Federal ID number
   d. Corporate charter in the state
3. Number of years the firm has been providing CM services.
4. Applicant's personnel to be assigned:
   a. List total number by skill group, department.
   b. Name all key personnel, and describe in detail experience and expertise of each member (*Note:* key personnel are to be assigned for the duration of the project unless excused).
5. Consultants, if any affiliated for this project, and why.
6. Experience, complete with:
   a. Owner name and phone number; if joint venture list all
   b. Project name
   c. Status, complete or in process, phase
   d. Completion date

    e. Project architect

    f. Services provided

    g. Approximate value of GMP or contract

    h. Role of the staff as submitted in number 4 above

7. Services offered; does the firm also do developing, general contracting, or real estate, for example?

8. Describe cost control methodologies: what type, and how often do you update them? Provide examples of how they were used and what degree of accuracy was achieved in past projects.

9. Describe the types of records, reports, monitoring systems, and information management systems used. Provide examples.

10. Describe methods used to avoid and prevent conflict.

11. Explain how you achieve quality control; give examples.

12. Discuss any special considerations, and why you feel your firm is the best qualified for the job.

13. It is hereby acknowledged that all information included herein is of a factual nature, and is certified to be true and accurate. All statements of future action will be honored.

14. It is understood that if the information is found to be, in the opinion of the selection committee, substantially unreliable, this application may be rejected. The selection committee may reject any or all applicants and may stop the selection process.

15. This completed package must be received by the owner at their published address on or before 3:00 PM MST on Wednesday, December 1, 1997. The undersigned certifies that they are a principal or officer for the firm applying for consideration, and that they are authorized to make the above acknowledgments and certifications for the firm.

5.5.2 Construction Process

After selection of the best proposals, interviews should be scheduled for the limited number of firms that stood out. It is best not to exceed six; except for unusual circumstances, four is enough. After interviews, final selection should be made within a few days. The selected firm should then be notified and the work can begin.

Permitting and utilities connection can be the first of the challenges to face the team. Then shop drawings will be submitted and reviewed for compliance with the specifications, substitutions will be presented, and decisions made. Those with the longest lead time may be processed first. Detailed review of shop drawings can prevent subtle changes that could significantly lower the quality of the finished product. Architects and engineers play a key role during this phase, as well as the owner's PM and team. Finally, it is time to order equipment and materials. Opportunities for value-engineering alternatives can lead to significant cost savings. By this time the subcontractors will know actual costs for the anticipated products. Final negotiations often relieve schedule and cost problems. Both the owner and the contractor can benefit from honest, open exchange of ideas. The architect will issue change orders (COs) for any agreed-to deviations in scope or plans and specs. Each CO will have cost and schedule implications clearly spelled out.

Project coordination is the ultimate responsibility of the builder. The designers show line diagrams for three-dimensional systems. Experienced firms will often prepare complete coordination drawings prior to beginning installation of the first component. It costs a lot to go back and remove installed components; it costs very little to plan ahead. Usually the A&E did not have enough time between design and construction to evaluate sloping piping systems passing through areas with a lot of ducts or conduits. Conflicts can be avoided.

During construction the contractor will present payment requests for partial payment for completed portions of the work. They can be submitted monthly or as the project dictates. Fast-track projects could actually cost less if the payments were known in advance to be available every two weeks. The bidders would be able to reduce their exposure to material and labor costs.

After most construction is complete, a punch list will be generated that lists all known deficiencies. Upon completion of that list, the contractor can be considered to be substantially complete with the project. There may be some start-up or balancing of minor components remaining, but the major parts should all be up and running. At this time the contractor can expect to have around 5 to 10% of the total contract amount held in retainage, but the rest of the contract amount is payable. After final completion and project paperwork close-out, the retainage can be released. A complete test and balance report should be prepared to check the operation of the mechanical system. It is best to verify that there are no outstanding liens against the project, and that materials suppliers have been receiving payment throughout construction. If a GMP-type contract is used, all allowance figures will have to be accounted for and adjustments made to the contract sum. From this time forward, the contractor will still be obligated to correct any deficiencies or deviations encountered in the first year of operation without argument. Often the manufacturers of the system components offer longer warranties.

5.6   REFERENCES

Baumeister, T., Avallone, E., and Baumeister, T., III, *Marks Standard Handbook for Mechanical Engineers*, McGraw-Hill, New York, 1978.

BCR for Reinforced Concrete, ACI 138.

Comer and Stephens, *Internetworking with TCP/IP*, Prentice Hall, Englewood Cliffs, N.J., 1993.

CRSI Working Stress Design, ACI Code.

Greene, *Mind and Image*, University Press of Kentucky, Louisville, 1962.

Higgins and Morrow, *Maintenance Engineering Handbook*, McGraw-Hill, New York, 19xx.

Littleton, *Industrial Piping*, 2nd ed., McGraw-Hill, New York, 1962.

Lynch and Hack, *Site Planning*, 3rd ed., MIT Press, Boston, 1984.

McGuiness, Stein, and Reynolds, *Mechanical and Electrical Equipment for Buildings*, 6th ed., John Wiley, New York, 1980.

Moffat, *Plant Engineer's Handbook for Charts and Tables*, Prentice Hall, New York, 1974.

Placing Reinforcing Bars, CRSI.

Ramsey and Sleeper, *Architectural Graphics Standards*, John Wiley, New York, 1978.

Sanders and McCormick, *Human Factors in Engineering and Design*, McGraw-Hill, New York, 1987.

Sanoff, *Methods of Architectural Programming*, DH&R.

Schodek, *Structures*, Prentice Hall, New York, 1980.

Sheth, V., *Facilities Planning and Materials Handling*, Marcel Dekker, New York, 1995.

Steel Construction Manual, American Institute of Steel Construction (AISI), 1973.

Swinburne, *Design Cost Analysis*, McGraw-Hill, New York, 1980.

Thumann, A., *Plant Engineer's and Manager's Guide to Energy Conservation*, Van Nostrand Reinhold, New York, 1977.

Urquhart, *Civil Engineering Handbook*, 4th ed., McGraw-Hill, New York, 1959.

White, E. T., *Introduction to Architectural Programming*, Architectural Media, Tucson, Ariz., 1972.

6 Human Elements in the Factory

Jack M. Walker

with

Timothy L. Murphy

Jeffery W. Vincoli

and

Paul Riedel

6.0 INTRODUCTION TO HUMAN ELEMENTS IN THE FACTORY

To be effective, manufacturing strategy must recognize and address the interests of all three types of stakeholders, customers, investors, and employees. There is an increasing interest in producing goods and services that provide greater customer satisfaction. This shows up in the "new" considerations demanded in product design, product quality, competitive pricing, and continuing customer support. We are also quite aware of the importance of making a profit for our investors and the difficulty of achieving this in our global economy. This is seen in some of the drastic short-term cost-cutting solutions such as reducing the workforce and consolidating plants, as well as new improved machines, tools, and factory operations.

Written material, including this handbook, is available which talks about how best to achieve better product design, a better manufacturing process, better factory design, and better manufacturing operations for the long term. Computer systems that give us the ability to assess and control all of these functions are becoming more integrated, permitting us to solve problems by taking the positive actions needed to achieve total customer and investor satisfaction.

This chapter focuses on the interests of the third element, which in broad terms is the key to achieving satisfied customers and investors, the employee as a stakeholder.

To begin, we will examine a typical factory population, to get an idea of who the people are, and to talk about their individual and collective needs, and the skills they must have:

1. Of all "value-added" and "value-related" employees in the United States, 85% are employed by small companies, as opposed to the giants such as the automobile companies.
2. Most plants, even the very large ones, buy most of their parts and subassemblies from outside suppliers. Ford, Chrysler, etc. usually perform only about 12 to 16% of the direct labor in-house. Other companies, such as McDonnell Douglas, Boeing, etc., typically perform 5 to 10% in-house. All subcontract the remainder to specialty shops that can do a better job for less money.
3. In a typical factory of 1000 employees, fewer than 40% are on the shop floor performing "touch labor"—which adds value directly to the product—by fabricating parts or performing assembly operations. Figure 6.1 shows the labor distribution. As we move toward more automation in the factory, the percentage of "touch" employees will decline even further.

Historically, a product was developed based primarily on function—something that would perform its intended purpose. The same pretty much described the production process, and the tools and machines involved. People were hired and expected to perform the required production tasks. Sometime later, skills training, mostly on the job, was needed in order to allow employees to perform the intended task. Still later, minor changes were made in the process to permit employees to perform assigned tasks somewhat better. This was usually to improve the cost or quality of the product.

Let's look at today's shop worker. Most companies expect an entry-level worker to have a high school education, but this is simply not always the case. Many people

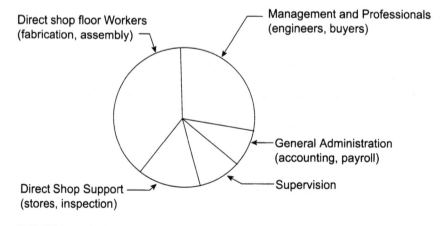

FIGURE 6.1   Labor distribution in a typical factory of 1000 employees.

now employed in small, low-technology shops are not even able to read and write. Some of them may need to learn English. Their opportunity for growth—and even continuing employment—is almost nonexistent. Learning to read and write, followed by a high school equivalency, is their only alternative to unemployment and welfare. These classes may be sponsored in some fashion by the employer, but only after the normal work shift, on the employees' own time. Classes in improving mathematical skills, blueprint reading, etc. may be given after hours by a paid company instructor. An entry-level machinist must have high school equivalency, reasonable math skills, and experience (at least in school) in basic machine operations such as drilling, turning, and milling—although it may be on older, manual machines. Classes on the newer computer-controlled machines may have to be furnished by the employer, and probably during working hours (with pay). We see this worker required to continually improve his or her skills, not only in the entry skill, but in developing new and broader skills. These skills will include the use of computers, knowledge of statistics, and the thought processes required to contribute to "continuing improvement" teams striving to improve quality and drive down costs. It boils down to fewer direct value-added workers in the manufacturing industry, but higher wages for the skilled ones that remain.

The other key people in manufacturing are those workers who support the production shop. They include engineers who design the products, establish the manufacturing process, design tooling, design and equip the factory, etc. There are also employees who purchase materials, maintain stock levels, and move parts around the factory. Add to these the workers in accounting, contract administration, quality control, training, safety, and others. In general, these workers have different training and education needs, but they must improve their individual job skills while broadening their knowledge of all the other operations in the factory. Advanced materials and processes, computer skills, communication skills, government requirements, and others are included.

The promise of lifetime employment security with one company may no longer be possible. Firms realize that they can no longer commit to providing jobs for the rest of an employee's life. They would perhaps like to, but in situations where they have to close down factories, combine divisions, or move them to another location, this is no longer a viable business option. On the other hand, workers are being given more power to expand their jobs and skills to permit them to be competitive, and therefore more useful to a company for a longer period of time. Some companies are sponsoring training programs for employees to expand their skills so that they can stay competitive in the job market—both within the company and outside the company. In return, corporations expect bigger contributions from a higher-skilled staff. Employees feel that they can jump ship for a better job, and still be able to return to the previous employer for advancement should conditions warrant it. Training is concerned primarily with *adapting people to machines or working situations*. We are not *just* teaching entry-level shop workers the elementary job skills, but rather trying to staff an entire plant with a population of employees requiring widely diverse skills.

Human-factors engineering designs machines and working situations *to meet the capacities of people*. While training is essential, often the major gain is in job process improvement. Ergonomics, the "science of work," is a field of technology that considers human capabilities and limitations in the design of machines and objects that people use, the work processes that they must follow, and the environments in which they operate. It is a multidisciplinary system of how people interact with work tasks, workstations, and work area environments.

The productivity of an employee depends on the state of his or her work environment. If work area conditions are hazardous, or even perceived to be hazardous, morale will suffer and people will not perform effectively. Moreover, if conditions are so bad as to cause worker illness or injury, the consequences could affect the company severely. If the work environment provides a healthful and safe means to allow people to produce, then the employer will achieve profitable results.

Just about all of us have heard the statement that "safety is everyone's business," but few realize that this is more than just a cliché or a clever approach to safety management. The requirement for a safe work environment is actually a mandate on industry and its employees from the federal government. Safety and health regulations for general industry can be found in the U.S. Code of Federal Regulations. These codes, which address literally thousands of situations, working conditions, hazard control requirements, and worker safety and health protection standards, are all intended to make the work environment safer.

Although numerous areas are addressed and a minimum level of worker protection is ensured by implementation of these many standards, the major driving factor for worker safety is contained in Section 5 of the Occupational Safety and Health Act of 1970. This factor, better known as the "General Duty Clause," simply states that:

> Each employer shall furnish to each of his employees employment and a place of employment which are free from recognized hazards that are causing or are likely to cause death or serious physical harm to his employees, and [each employer] shall comply with the occupational safety and health standards promulgated under this Act.

"Worker's compensation" is the name commonly applied to statutes that give protection and security to workers and their dependents against injury, disease, or death occurring during employment. The statutes establish the liability for benefits, which are usually financed through insurance bought by the employer.

Today, all 50 states have worker's compensation acts. The fundamental concepts of the compensation acts are to provide disabled employees with medical, economic, and rehabilitation support. The manner in which this support is handled has caused great distress between workers and their companies. Some employees feel that companies should do more for them, and many companies feel that they are providing more benefits than they can afford.

The escalating cost of worker's compensation is staggering. Payments for worker's compensation programs rose to an all-time record of $35.3 billion in 1990. The *average* cost of worker's compensation insurance per wage hour worked ranges from $1 in some states to more than $6 in others. The actual cost runs from less than 1% of wages to about 50% of wages, depending on the job classification and the history of claims payments. This can become a very significant element of payroll cost to a manufacturer, since the insurance company raises the worker's compensation insurance premium to cover the actual costs per year, plus the administrative cost of handling the claims.

The manufacturing engineer, along with his or her coworkers, needs to be aware of some of the problems and responsibilities regarding people in the factory today and the factory of the future. This chapter looks at ergonomics, safety, industrial hygiene, and worker's compensation. In the highly competitive environment of today's business world, success is achieved through the collective talents, skills, and contributions of all the individual employees of the organization. The business enterprise with the right combination of skills and clear, focused objectives will often outperform its competitors in every category. It cannot be disputed that people are a company's greatest resource. Therefore, it should be equally clear that the assurance of a safe and healthy workplace, within which this resource must perform, is as important to a company as any issue related directly to production, sales, or profit.

6.1 ERGONOMICS

6.1.0 Introduction to Ergonomics

The term *ergonomics* can be defined simply as the study of work.[1] Ergonomics helps adapt the job to fit the person, rather than forcing the person to fit the job. Adapting the job to fit the worker can help reduce ergonomic stress and eliminate many potential ergonomic disorders. Ergonomics focuses on the work environment and items such as the design and function of workstations, controls, displays, safety devices, tools, and lighting to fit the employees' physical requirements and to ensure their health and well-being. It may include restructuring or changing workplace conditions to reduce stressors that cause repetitive motion injuries.

Major causes of many current ergonomic problems are technological advances such as more specialized tasks, higher assembly-line speeds, and increased repetition, plus a lack of ergonomically designed technologies. Consequently, workers' hands, wrists, arms, shoulders, backs, and legs may be subjected to thousands of repetitive twisting, forceful, or flexing motions during a typical workday. Some jobs still expose workers to excessive vibration and noise, eye strain, repetitive motion, and heavy lifting. In many instances, machines, tools, and the work environment are poorly designed, placing undue stress on workers' tendons, muscles, and nerves. In addition, workplace temperature extremes may aggravate or increase ergonomic stress. Recognizing ergonomic hazards in the workplace is an essential first step in correcting the hazards and improving worker protection.

Training is concerned primarily with adapting people to machines or working situations. Ergonomics helps design machines and working situations to meet the capacities of people. While training is essential, often the major gain is in job-process improvement.

Historically, a product was developed based primarily on function—something that would perform its intended purpose. The same pretty much described the production

1. The author would like to thank Timothy L. Murphy, group manager, human relations, McDonnell Douglas Corporation, and Alvah O. Conley, mechanical and safety engineer in OSHA's Chicago office, for their assistance with this chapter.

process, and the tools and machines involved. People were hired and expected to perform the required production tasks. Sometime later, skills training, mostly on the job, was needed in order to allow employees to perform the intended tasks. Still later, minor changes were made in the process to permit workers to perform assigned tasks somewhat better. This was usually to improve the cost or quality of the product.

Common sense, safety, the workers' general welfare—unfortunately, these all needed someone carrying a big stick to remind industry that this is good business. Considering only the costs of OSHA citations, willful acts of negligence on the part of employers can cost up to $70,000 per incident, accompanied by jail terms of up to 2 years, according to Raymond E. Chalmers, senior editor of *Forming and Fabricating* magazine. Industrial safety classes need to be a part of any management agenda. Good ergonomics depends on mobility, flexibility, and adaptability—blending together tool design and managerial attitudes, assembly-line specifications, and company policies.

"It's not uncommon to think that by the end of the decade 50 cents of every worker's compensation insurance dollar will go for soft-tissue disorders, unless we come up with some reasonable intervention strategies," according to OSHA's chief ergonomist, Roger Stephens. The good news, however, is that many soft-tissue disorders can be prevented through ergonomic intervention. By simply redesigning the job to fit the employee, not vice versa, workers can be spared a lot of pain and discomfort, and the company can save a lot of money without necessarily making a large investment.

The preferred method for control and prevention of work-related musculoskeletal disorders is to design the job to match the physiological, anatomical, and psychological characteristics and capabilities of the worker. In other words, safe work is achieved as a natural result of the design of the job, the workstation, and the tools; it is independent of specific worker capabilities or work techniques.

Worker selection or hiring based solely on physical capacities is generally illegal, as a result of the U.S. Federal Rehabilitation Act of 1973 and the recent Americans with Disabilities Act of 1990. However, once a worker is offered a job, he or she can be tested to determine his or her capabilities as a prelude to specific job placement within the firm. Good records can defend a company against complaints by OSHA or workers. Bad records mean bad faith. Companies that falsify records constitute the worst violators in the mind of OSHA and the workers.

6.1.1   Ergonomics

"Ergonomics is basically common sense," says Dr. Hal Blatman, who specializes in muscle pain and is an adjunct professor at Miami University's Center for Ergonomic Research.

"Efficiency is the by-product of the work, and the company that manufactures no sore backs, shoulders, wrists, legs and behinds has a competitive edge over the ones that do," according to Blatman. As the science of work, ergonomics is a field of technology that considers human capabilities and limitations in the design of machines and objects that people use, the work processes that they must follow, and the environments in which they operate. It is a multidisciplinary system of how humans interact with work tasks, workstations, and work area environments.

The time and motion studies of Frederick W. Taylor, and of Frank and Lillian Gilbreth, laid the groundwork for the field. Ergonomics—then called human-factors engineering—grew in importance during World War II as engineers sought help from psychologists and physiologists in designing military equipment adaptable to a wide variety of users. Since that time, ergonomics has also played a role not only in the design of better workplaces and tools, but also in the design of consumer goods— automobiles and stoves, for instance.

Industrial and organizational psychology is the field of applied psychology that studies the behavior of persons in organizational work settings. Its areas of concern are personnel selection, the mapping of organizational processes, training, improvement of employee morale and working conditions, and improvement of both individual and group productivity.

Ergonomics is concerned with making purposeful human activities more effective. Most of the activities studied can be called "work," although there are topics such as "ergonomics of sport," "ergonomics in the home," and "passenger ergonomics."

The focus of study is the person interacting with the engineered environment. This person has some limitations which the designer should take account of. Complexity arises from the nature of people and the variety of the designed situations to be considered. These latter can vary from relatively simple ones such as chairs, the handles of tools, and the lighting of a bench, to highly elaborate ones such as aircraft flight decks, control rooms in the process industries, and the artificial life-support systems in space or under the sea.

Ergonomic Essentials

Consultants practicing ergonomics hail from a variety of backgrounds—health sciences, industrial hygiene, safety, industrial engineering, psychology, and others. Regardless of an ergonomic consultant's primary area of expertise, two things are necessary to solve ergonomics problems in the workplace: an understanding of the *demands of the task* and an understanding of the *capabilities of the human operator,* said Jim Forgrascher, director of ergonomics for the Ohio Bureau of Worker's Compensation, Division of Safety and Hygiene.

Traditionally, those who practiced ergonomics specialized in studies of the human as a physical engine or the human as an information processor. The "physical engine" experts, often industrial engineers, studied human biomechanics and the physical elements of a task and workplace. The "humans as information processors," often experts in psychology, examined the reception, processing, and sending of information from human to human, and from human to machine.

These two areas of expertise within ergonomics have blended and multiplied, creating many specialty areas within the field. Ergonomic specialties include biomechanics, environmental design, human-computer interaction, human performance, training, and simulation.

The people studied can vary from fit young men in military systems to middle-aged housewives in the kitchen, from the disabled of many kinds to the very highly selected such as racing drivers. Sometimes the relevant characteristic is essentially biological and unchanging except with age, as in the limits of dark adaptation, for example. Sometimes it varies with sex and race—for example, body dimension. Sometimes it varies with degree of economic and social development—for example, acceptable working hours.

The limits or boundaries of ergonomics itself are not entirely agreed upon. For example, most ergonomists would agree that energy expenditure studies are within ergonomics, but the design of diets to provide the energy is not, and that physical hazards such as heat stress are part of ergonomics, but that chemical hazards such as carcinogenic substances are not.

Ergonomic Tips

The following tips may resolve many ergonomic concerns in the workplace, without having to resort to the more formal anthropometric studies that are presented in the next subchapter.

1. *Avoid extreme reach.* Integrate all workplace components and surfaces so that frequent task motions occur in front of the body and between the hip (waist) and the shoulders. The optimum height is within 1 in. (2.5 cm) from the elbows. Having adjustable work surfaces and seats (if they are used) will make the integration possible and effective.
2. *Avoid excessive force.* Reduce muscular exertions (forces) by minimizing the weight handled, improving motion patterns, eliminating extreme joint motions (range of motion), avoiding awkward and maximum reaches, counterbalancing and suspending tools, eliminating differential heights among product/component transfer points, using mechanical aids (powered conveyors, gravity feed devices) for moving products to and from the workplace, presorting and aligning work pieces (components), relaxing the fit tolerances for the assembly of components, and reducing pace or speed of motion (activity).
3. *Avoid static posture.* Job method should not impair circulation and should offer the muscles sufficient time to recover after exertion. Static loading can be eliminated through supporting body segments that are not involved in task motions. Seats and arm rests are examples of body supports. Use mechanical aids (jigs, special fixtures) for positioning and holding workpieces or components. The workplace should allow for flexible positioning of workpieces (adjustable height, angling, tilting, and rotation). Protective clothing should allow for freedom and ease of body movement and offer no impediment to circulation. Rest periods should be planned and scheduled to overcome any residual effects of static loading.
4. *Avoid stress concentration.* Job activities should be shared by more than one muscle group. In so doing, static loading and excessive force may be eliminated as a by-product. Use a large contact surface for hand-held tools, and use simultaneous arm motions. Change pinching tasks to gripping actions, wrist motions to arm motions, and arm motions to shoulder

motions. Eliminate pressure at points of contact between the body and
the work surface, tools, machines, or any hard surface. For standing jobs,
use cushioned floor support or modify the job method to promote the
use of "muscle pumps," especially in the lower extremities. Tight-fitting
shoes and high heels (good examples of stress concentration) should be
avoided.

5. *Avoid faulty human-machine interface.* Machine controls should be
designed based on the limiting values of human strength, body joint range
of motion, and speed and accuracy of the required response. Identification,
grouping, and the relationship among controls as well as displays should
conform to population norms and stereotypes. Eliminate the potential for
ambiguous information feedback from the machine. Avoid having the
machine operator process complex pieces of data before making an appropriate response.

Display (visual or auditory) designs should take into consideration the
potential user population and their skills and degrees of familiarity with the
equipment. Time, frequency of use, and the environment (noise, vibration,
light under which equipment will be used) should also be considered.

6. *Avoid extremes of the environment.* Extreme environmental conditions are
harmful because of the physiological and stress responses they elicit from
the body, as well as for their potential to cause damage to body tissues.
Regardless of the type of offending environment, from noise to vibration,
from hot to cold, from brightly lit surfaces to high-intensity lights, the
control strategies are the same. Basically, the following control options
can be used separately or collectively.

(a) Apply engineering changes and modifications to bring the level of the
offending source to an acceptable level. Examples include the substitution of different material or energy sources, reducing the intensity (level) of the current source, or bypassing the source all together
through elimination of total enclosures.

(b) Separate in space and time the offending source from the exposed
workers (receivers). Increasing distance from the source and shielding
the receiver are examples of achieving the desired separation.

(c) Eliminate any potential harm through limiting or minimizing the
exposure of those working close to the offending source. The use of
personal protective equipment is an example of this control strategy.

Some common ergonomic problems and commonsense solutions are listed in
Figure 6.2.

6.1.2 Anthropometry

It is important to have information about the people who are working in today's manufacturing industries. Their capabilities and limitations are among the most basic of
our considerations in designing products, processes, and equipment. *Anthropometry*
is the study of human body measurements. Important to the field of ergonomics,

1. Repetitiveness
    a. Use mechanical aids
    b. Enlarge work content by adding more diverse activities
    c. Automate certain tasks
    d. Rotate workers
    e. Increase rest allowances
    f. Spread work uniformly across workshift
    g. Restructure jobs

2. Force/Mechanical Stress
    a. Decrease the weight of tools/containers and parts
    b. Increase the friction between handles and the hand
    c. Optimize size and shape of handles
    d. Improve mechanical advantage
    e. Select gloves to minimize efects on performance
    f. Balance hand-held tools and containers
    g. Use torque control devices
    h. Optimize pace
    i. Enlarge corners and edges
    j. Use pads and cushions

3. Posture
    a. Locate work to reduce awkward postures
    b. Alter position of tool
    c. Move the part closer to the worker
    d. Move the worker to reduce awkward postures
    e. Select tool design for work station

4. Vibration
    a. Select tools with minimum vibration
    b. Select process to minimize surface and edge finishing
    c. Use mechanical assists
    d. Use isolation for tools that operate above resonance point
    e. Provide damping for tools that operate at resonance point
    f. Adjust tool speed to avoid resonance

5. Psychosocial Stresses
    a. Enlarge workers' task duties
    b. Allow more worker control over pattern of work
    c. Provide micro work pause
    d. Minimize paced work
    e. Eliminate blind electronic monitoring

FIGURE 6.2   Examples of ergonomic interventions.

anthropometric measurements aid designers and manufacturers in the production of safe, efficient consumer goods that are comfortable to use.

The design of the workplace must ensure the accommodation of, and compatibility with, at least 90% of the workforce population. Generally, design limits should be based on a range of the 5th percentile to the 95th percentile values for critical body dimensions. For any body dimension, the 5th percentile value indicates that 5% of the population will be equal to or smaller than that value, and 95% will be larger. The 95th percentile value indicates that 95% of the population will be equal to or smaller than that value, and 5% will be larger. Therefore, use of a design range from the 5th to the 95th percentile values will theoretically accommodate 90% of the user population for that dimension. This is considerably different than the "average" worker.

When using the anthropometric data that is presented in this section, the following must be considered:

1. The nature, frequency, and difficulty of the specific worker task
2. The position of the worker's body during performance of this task
3. Mobility or flexibility requirements imposed by this task
4. Increments in the design-critical dimensions imposed by the need to compensate for obstacles, projections, etc.
5. Increments in the design-critical dimensions imposed by protective clothing, packages, lines, padding, etc.

Anthropometric data is available from several sources. One source (MIL-STD-1472C) is a study of 14,428 military men and 1,331 women from all branches of service, by the U.S. Army and Air Force. These data provide one representation of worker population—with a few notes of caution. According to Alvah O. Conley, a mechanical and safety engineer in OSHA's Chicago office, we must remember that the figures are for military personnel, and do not include the "real" worker population, which contains more older workers, obese workers, and an increasing Asian population. This consideration would change some of the sizes of workers, etc., but would probably not affect measurements of range of motion, field of vision, and the like. The best source might be your personnel department, which could at least furnish the age, weight, height, and sex of your employees. This could be used to factor the anthropometric data that is available. Conley indicated that the standard data might be accurate for some regions of the country, while other regions would require more caution and consideration of the local workforce makeup. There may be an influence due to management style in a particular application, which can make a significant difference in motivation and discipline.

Under the U.S. Department of Health and Human Services, the National Institute for Occupational Safety and Health (NIOSH) develops and recommends criteria for preventing disease and hazardous conditions in the workplace. NIOSH also recommends preventive measures including engineering controls, safe work practices, personal protective equipment, and environmental and medical monitoring. NIOSH then transmits these recommendations to the Occupational Safety and Health Administration (OSHA), under the U.S. Department of Labor, for use in promulgating legal standards. Some of their studies and reports are available from NIOSH at no cost, while others can be ordered from the U.S. Department of Commerce, National Technical Information Service (NTIS), or the U.S. Government Printing Office, Superintendent of Documents.

Since some of the data are presented in the English system of measurement and others in the metric system, a conversion table is included as Figure 6.3.

TO CONVERT FROM:	TO:	MULTIPLY BY:
DEGREE (ANGLE) (deg)	RADIAN (rad)	1.745 329 E—02
FOOT (ft)	METER (m)	3.048 000 E—01
FOOT2 (ft²)	METER2 (m²)	9.290 304 E—02
FOOT3 (ft³)	METER3 (m³)	2.831 685 E—02
FOOTCANDLE (ft—C)	LUX (lx)	1.076 391 E+02
FOOTLAMBERT (ft—L)	CANDELA PER METER2 (cd/m2)	3.426 391 E+00
INCH (in. OR ")	METER	2.540 000 E—02
INCH2 (in.²)	METER2 (m²)	6.451 600 E—04
INCH3 (in.³)	METER3 (m³)	1.638 706 E—05
MINUTE (ANGLE) (min)	RADIAN (rad)	2.908 882 E—04
OUNCE—FORCE (ozf)	NEWTON (N)	2.780 139 E—01
OUNCE—INCH (ozf·in.)	NEWTON METER (N·m)	7.061 552 E—03
POUND (lb) AVOIRDUPOIS	KILOGRAM (kg)	4.535 924 E—01
POUND—FORCE (lbf)	NEWTON (N)	4.448 222 E+00
POUND—INCH (lbf·in.)	NEWTON METER (N·m)	1.129 848 E—01
SECOND (ANGLE) (sec)	RADIAN (rad)	4.848 137 E—06

PREFIXES

NANO	n	10^{-9}	CENTI	c	10^{-2}
MICRO	μ	10^{-6}	KILO	k	10^{3}
MILLI	m	10^{-3}	MEGA	M	10^{6}

TEMPERATURE CONVERSION

$$°C = \frac{5}{9} (°F - 32)$$

$$°F = \frac{9}{5} °C + 32$$

NOTE: EACH CONVERSION FACTOR IS PRESENTED AS A NUMBER, BETWEEN ONE AND TEN, TO SIX DECIMAL PLACES, THE LETTER E (FOR EXPONENT), A PLUS OR MINUS SIGN AND TWO DIGITS FOLLWING THE NUMBER, REPRESENT THE POWER OF 10 BY WHICH THE NUMBER IS TO BE MULTIPLIED.

FOR EXAMPLE: $3.048\ 000\ E—01 = 3.048\ 000 \times 10^{-1} = 0.3048000$
OR $1.076\ 391\ E+01 = 1.076\ 391 \times 10^{1} = 10.76391$

EXAMPLES OF USE OF TABLE:
   TO CONVERT $2ft^3$ TO m^3, MULTIPLY 2 BY 2.831 685 E—02
                                    $2 \times 0.028\ 316\ 85 = 0.056\ 634\ m^3$
   (TO CONVERT $2m^3$ TO ft^3, DIVIDE 2 BY 2.831 685 E—02)
                                    $(2/0.028\ 316\ 85 = 70.629\ 325\ ft^3)$

A MORE COMPLETE LISTING AND DISCUSSION MAY BE FOUND IN ASTM E 380—76

FIGURE 6.3   Metric equivalents, abbreviations, and prefixes.

Body Dimensions

Studies of body dimensions are made by several authorities in the field, but some controversy exists as to the validity of any of them. The following data are presented as an example of military personnel body dimensions. Used as a baseline, these data can be modified by examining the population of your individual manufacturing facility and making proper adjustments. An approach would be to refer to the following data, and use the conservative limits that would apply to the most of the factory population. Figure 6.4 defines the standing body dimensions, and gives the values for men and women. Figure 6.5 is a similar chart for sitting body dimensions. Anthropometric data shown in Figure 6.6 shows the 5th and 95th percentile reaches for men and women. This describes the body in several positions.

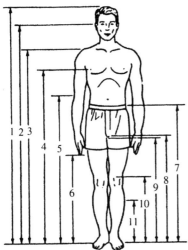

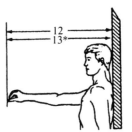

*Same as 12, however, right shoulder is extended as far forward as possible while keeping the back of the left shoulder firmly against the back wall

	Percentile values in centimeters					
	5th percentile			95th percentile		
	Ground troops	Aviators	Women	Ground troops	Aviators	Women
Weight (kg)	55.5	60.4	46.4	91.6	96.6	74.5
Standing body dimensions						
1. Stature	162.8	164.2	152.4	185.6	187.7	174.1
2. Eye height (standing)	151.1	152.1	140.9	173.3	175.2	162.2
3. Shoulder (acromiale) height	133.6	133.3	123.0	154.2	154.8	143.7
4. Chest (nipple) height*	117.9	120.8	109.3	136.5	138.5	127.8
5. Elbow (radiale) height	101.0	104.8	94.9	117.8	120.0	110.7
6. Fingertip (dactylion) height		61.5			73.2	
7. Waist height	96.6	97.6	93.1	115.2	115.1	110.3
8. Crotch height	76.3	74.7	68.1	91.8	92.0	83.9
9. Gluteal furrow height	73.3	74.6	66.4	87.7	88.1	81.0
10. Kneecap height	47.5	46.8	43.8	58.6	57.8	52.5
11. Calf height	31.1	30.9	29.0	40.6	39.3	36.6
12. Functional reach	72.6	73.1	64.0	90.9	87.0	80.4
13. Functional reach, extended	84.2	82.3	73.5	101.2	97.3	92.7

FIGURE 6.4   Standing body dimensions.

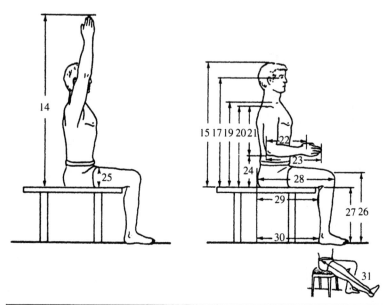

Seated body dimensions	Percentile values in centimeters					
	5th percentile			95th percentile		
	Ground troops	Aviators	Women	Ground troops	Aviators	Women
14. Vertical arm reach, sitting	128.6	134.0	117.4	147.8	153.2	139.4
15. Sitting height, erect	83.5	85.7	79.0	96.9	98.6	90.9
16. Sitting height, relaxed	81.5	83.6	77.5	94.8	96.5	89.7
17. Eye height sitting, erect	72.0	73.6	67.7	84.6	86.1	79.1
18. Eye height sitting, relaxed	70.0	71.6	66.2	82.5	84.0	77.9
19. Mid–shoulder height	56.6	58.3	53.7	67.7	69.2	62.5
20. shoulder height, sitting	54.2	54.6	49.9	65.4	65.9	60.3
21. Shoulder elbow length	33.3	33.2	30.8	40.2	39.7	36.6
22. Elbow–grip length	31.7	32.6	29.6	38.3	37.9	35.4
23. Elbow–fingertip length	43.8	44.7	40.0	52.0	51.7	47.5
24. Elbow rest height	17.5	18.7	16.1	28.0	29.5	26.9
25. Thigh clearance height		12.4	10.4		18.8	17.5
26. Knee height, sitting	49.7	48.9	46.9	60.2	59.9	55.5
27. Popliteal height	39.7	38.4	38.0	50.0	47.7	45.7
28. Buttock– knee length	54.9	55.9	53.1	65.8	65.5	63.2
29. Buttock– popliteal length	45.8	44.9	43.4	54.5	54.6	52.6
30. Buttock–heel length		46.7			56.4	
31. Functional leg length	110.6	103.9	99.6	127.7	120.4	118.6

FIGURE 6.5   Seated body dimensions.

	Percentile values in centimeters			
	5th percentile		95th percentile	
	Men	Women	Men	Women
1. Weight–clothed (kilograms)	58.6	48.8	90.2	74.6
2. Stature–clothed	168.5	156.8	189.0	178.7
3. Functional reach	72.6	64.0	86.4	79.0
4. Functional reach, extended	84.2	73.5	101.2	92.7
5. Overhead reach height	200.4	185.3	230.5	215.1
6. Overhead reach breadth	35.2	31.5	41.9	37.9
7. Bent torso height	125.6	112.7	149.8	138.6
8. Bent torso breadth	40.9	36.8	48.3	43.5
9. Overhead reach, sitting	127.9	117.4	146.9	139.4
10. Functional leg length	110.6	99.6	127.7	118.6
11. Kneeling height	121.9	114.5	136.9	130.3
12. Kneeling leg length	63.9	59.2	75.5	70.5
13. Bent knee height, supine	44.7	41.3	53.5	49.6
14. Horizontal length, knees bent	150.8	140.3	173.0	163.8

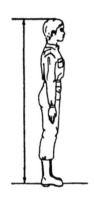

② Stature (clothed)

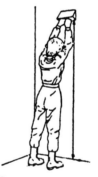

⑤ Overhead reach height

④ Functional reach, extended

⑨ Overhead reach, sitting

FIGURE 6.6   Anthropometric data for common working positions.

Body Strengths

Static muscle strength data are shown in Figure 6.7. This includes both standing and seated. Arm, hand, and thumb-finger strengths are shown in Figure 6.8. Figure 6.9 shows leg strengths at various knee and thigh angles of the 5th percentile male. Finger- or hand-operated push buttons are shown in Figure 6.10. Where data are shown for male workers and comparable data are not available for female workers, the Army Research Institute of Environmental Medicine has performed a study of 1500 women, with the following results:

For upper extremities, women's strength is 56.5% of men's.
For lower extremities, women's strength is 64.2% of men's.
For trunk extremities, women's strength is 66.0% of men's.

These numbers may serve as a preliminary design guideline until more up-to-date information becomes available.

NIOSH Lifting Equation (Excerpts from NIOSH Publications)

NIOSH convened an *ad hoc* committee of experts who reviewed the current literature on lifting, recommended criteria for defining lifting capacity, and in 1991 developed a revised lifting equation. Although this equation has not been fully validated, the recommended weight limits derived from the revised equation are consistent with or lower than those generally reported in the literature. The 1991 equation reflects new findings, provides methods for evaluating asymmetrical lifting tasks and objects with less than optimal hand-container couplings, and offers new procedures for evaluating a larger range of work durations and lifting frequencies than earlier equations. The objective is to prevent or reduce the occurrence of lifting-related low back pain (LBP) among workers. An additional benefit is the potential to reduce other musculoskeletal injuries associated with some lifting tasks, such as shoulder or arm pain.

Three criteria (biomechanical, physiological, and psychophysical) were used to define the components of the lifting equation. Figure 6.11 shows the criteria used to develop the new equation. Although the lifetime prevalence of LBP in the general population is as high as 70%, work-related LBP comprises only a subset of all cases. Some studies indicate that specific lifting or bending episodes were related to only about one third of the work-related cases of LBP.

Since NIOSH and OSHA have released very few actual recommendations for industry to consider (or comply with), the author feels that an introduction to this specific study is worthwhile, not only to set a standard for lifting tasks, but as an indication of what will eventually be a massive set of recommendations and government regulations.

The NIOSH formula for calculation of recommended weight limit (RWL) is a follows:

$$RWL = LC \times HM \times VM \times DM \times AM \times FM \times CM$$

STRENGTH MEASUREMENTS	PERCENTILE VALUES IN POUNDS			
	5th PERCENTILE		95th PERCENTILE	
	MEN	WOMEN	MEN	WOMEN
A STANDING TWO–HANDED PULL: 15 in. LEVEL				
MEAN FORCE	166	74	304	184
PEAK FORCE	190	89	323	200
B STANDING TWO–HANDED PULL: 20 in. LEVEL				
MEAN FORCE	170	73	302	189
PEAK FORCE	187	84	324	203
C STANDING TWO–HANDED PULL: 39 in. LEVEL				
MEAN FORCE	100	42	209	100
PEAK FORCE	113	49	222	111
D STANDING TWO–HANDED PUSH: 59 in. LEVEL				
MEAN FORCE	92	34	229	85
PEAK FORCE	106	42	246	97
E STANDING ONE–HANDED PULL: 39 in. LEVEL				
MEAN FORCE	48	23	141	64
PEAK FORCE	58	30	163	72
F SEATED ONE–HANDED PULL: CENTERLINE, 18 in. LEVEL				
MEAN FORCE	51	24	152	88
PEAK FORCE	61	29	170	101
G SEATED ONE–HANDED PULL: SIDE, 18 in. LEVEL				
MEAN FORCE	54	25	136	76
PEAK FORCE	61	30	148	89
H SEATED TWO–HANDED PULL: CENTERLINE, 15 in. LEVEL				
MEAN FORCE	148	54	274	173
PEAK FORCE	148	64	288	189
I SEATED TWO–HANDED PULL: CENTERLINE, 20 in. LEVEL				
MEAN FORCE	118	46	237	142
PEAK FORCE	134	53	267	157

FIGURE 6.7   Static muscle strength data.

Component	Metric
LC = load constant	23 kg (normally the maximum)
HM = horizontal multiplier	(25/H)
VM = vertical multiplier	$1 - [0.003 (V - 75)]$
DM = distance multiplier	$0.82 + (0.5/D)$
AM = asymmetric multiplier	$1 - (0.0032A)$
FM = frequency multiplier (from Figure 6.13)	
CM = coupling multiplier (from Figure 6.12)	

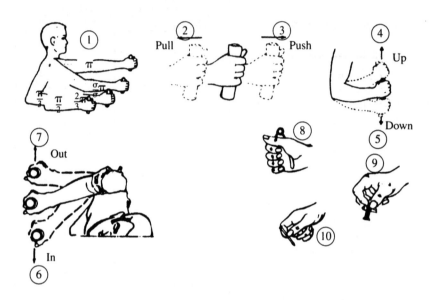

Arm strength (N)													
(1)	(2)		(3)		(4)		(5)		(6)		(7)	
Degree of elbow flexion (rad)	Pull		Push		Up		Down		In		Out	
	L**	R**	L	R	L	R	L	R	L	R	L	R
π	222	231	187	222	40	62	58	76	58	89	36	62
$\frac{\sigma}{\sigma}\pi$	187	249	133	187	67	80	80	89	67	89	36	67
$\frac{2}{3}\pi$	151	187	116	160	76	107	93	116	89	98	45	67
$\frac{1}{2}\pi$	142	185	98	160	76	89	93	116	71	80	45	71
$\frac{1}{3}\pi$	116	107	98	151	67	89	80	89	76	89	53	76
Hand and thumb–finger strength (N)													
	(8)			(9)			(10)						
	Hand grip			Thumb–finger grip (palmer)			Thumb–finger grip (tips)						
	L		R										
Momentary hold	250		260	60			60						
Sustained hold	145		155	35			35						

*Elbow angle shown in radians
**L= Left; R = Right

FIGURE 6.8 Arm, hand, and thumb-finger strength (5th percentile male data).

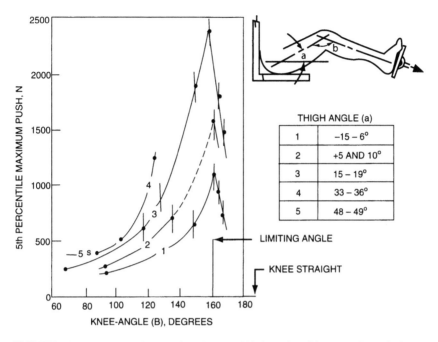

FIGURE 6.9   Leg strengths at various knee and high angles (5th percentile male data).

where

    H = horizontal distance of hands from midpoint between the ankles. Measure at
        the origin and the destination of the lift (cm).
    V = vertical distance of the hands from the floor. Measure at the origin and
        destination of the lift (cm).
    D = vertical travel distance between the origin and the destination of the lift (cm).
    A = angle of asymmetry—angular displacement of the load (left or right) from
        the sagittal plane. Measure at the origin and destination of the lift
        (degrees).
    F = average frequency rate of lifting measured in lifts/min. Duration is defined
        to be:

$\leq$ 1 hr.; $\leq$ 2 hr.; or $\leq$ 8 hr., assuming appropriate recovery allowances.

Coupling modifier: Loads equipped with appropriate couplings or handles facilitate lifting and reduce the possibility of dropping the load.
Psychophysical studies that investigated the effects of handles on maximum
acceptable weight of lift suggested that lifting capacity was decreased in
lifting tasks involving containers without good handles. The coupling modi-
fiers are displayed in Figure 6.12.

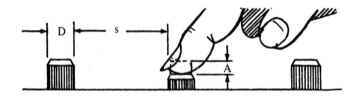

Dimensions			Resistance		
	Diameter D			**Different**	
	Fingertip	Thumb or Palm	Fingertip	Fingers	Thumb or Palm
Minimum	9.5 mm (3/8 in)	19 mm (3/4 in.)	2.8 N (10 oz.)	1.4 N (5 oz.)	2.8 N (10 oz.)
Maximum	25 mm (1 In.)		11 N (40 oz.)	5.6 N (20 oz.)	23 N (80 oz.)

	Displacement	
	Fingertip A	**Thumb or Palm**
Minimum	2 mm (5/64 in.)	3 mm (1/8 in.)
Maximum	6 mm (1/4 In.)	38 mm (1-1/4 In.)

	Separation			
	Single Finger	Single Finger S Sequential	Different Fingers	Thumb or Palm
Minimum	13 mm (1/2 in.)	6 mm (1/4 in.)	8 mm (1/4 in.)	25 mm (1 in.)
Maximum	50 mm (2 in.)	13 mm (1/2 in.)	13 mm (1/2 in.)	150 mm (6 in.)

Note: Above data for barehand application. For gloved hand operation, minima should be suitably adjusted.

FIGURE 6.10 Push buttons, finger or hand-operated.

Frequency modifier: For the 1991 lifting equation, the appropriate frequency multiplier is obtained from Figure 6.13 rather than from a mathematical expression. The committee concluded that the frequency multipliers provide a close approximation to observed and predicted effects of lifting frequency on acceptable workloads.

From the NIOSH perspective, it is possible that obese workers may exceed the energy expenditure criteria for lifts from below the waist. In addition, there are some circumstances in which local muscle fatigue may occur even though whole body fatigue has not occurred. This is most likely in situations involving lifting at high rates for longer than 15 min., or prolonged use of awkward postures, such as constant bending.

Discipline	Design Criterion	Cut–off value
Biomechanical	Maximum disc compression force	3.4 KN (770 lbs)
Physiological	Maximum energy expenditure	2.2–4.7 kcal/min
Psychophysical	Maximum acceptable weight	Acceptable to 75% of female workers and about 99% of male workers

FIGURE 6.11 Criteria used to develop the NIOSH lifting equation.

Couplings	$V < 75$ cm (30 in)	$V \geq 75$ cm (30 in)
	Coupling multipliers	
Good	1.00	1.00
Fair	0.95	1.00
Poor	0.90	0.90

FIGURE 6.12 Coupling modifier (CM) for NIOSH lifting equation.

	Work duration					
	≤ 1 h		≤ 2 h		≤ 8 h	
Frequency lifts/min	$V < 75$	$V ? 75$	$V < 75$	$V ? 75$	$V < 75$	$V ? 75$
0.2	1.00	1.00	0.95	0.95	0.85	0.85
0.5	0.97	0.97	0.92	0.92	0.81	0.81
1	0.94	0.94	0.88	0.88	0.75	0.75
2	0.91	0.91	0.84	0.84	0.65	0.65
3	0.88	0.88	0.79	0.79	0.55	0.55
4	0.84	0.84	0.72	0.72	0.45	0.45
5	0.80	0.80	0.60	0.60	0.35	0.35
6	0.75	0.75	0.50	0.50	0.27	0.27
7	0.70	0.70	0.42	0.42	0.22	0.22
8	0.60	0.60	0.35	0.35	0.18	0.18
9	0.52	0.52	0.30	0.30	0.00	0.00
10	0.45	0.45	0.26	0.26	0.00	0.00
11	0.41	0.41	0.00	0.23	0.00	0.00
12	0.37	0.37	0.00	0.21	0.00	0.00
13	0.00	0.34	0.00	0.00	0.00	0.00
14	0.00	0.30	0.00	0.00	0.00	0.00
15	0.00	0.28	0.00	0.00	0.00	0.00
>15	0.00	0.00	0.00	0.00	0.00	0.00

Notes:
  values of V are in cm; 75cm = 30in.

FIGURE 6.13 Frequency modifier (FM) for NIOSH lifting equation.

Visual Field

The worker's visual field is controlled by eye rotation, head rotation, or a combination of eye and head rotation. Nominal, optimum, and maximum values are given in Figure 6.14.

Range of Human Motion

Figures 6.15 and 6.16 illustrate and give the ranges, in angular degrees, for all voluntary movements the joints of the body can make. The designer should remember that these are maximum values; since they were measured with nude personnel, they do not reflect the restrictions that clothing would impose. The following general instructions apply to the dimensions in Figure 6.16.

1. The lower limit should be used when personnel must operate or maintain a component.
2. The upper limit should be used in designing for freedom of movement.

All operating positions should allow enough space to move the trunk of the body. When large forces (more than 13.6 kg) or large control displacements (more than 380 mm in a fore-aft direction) are required, the operator should have enough space to move his or her entire body. Figure 6.17 is a continuation of the range of human motions. The following definitions may be useful in discussions of body positions and motions.

Flexion: Bending, or decreasing the angle between parts of the body
Extension: Straightening, or increasing the angle between parts of the body
Adduction: Moving toward the midline of the body
Abduction: Moving away from the midline of the body
Medial rotation: Turning toward the midplane of the body
Lateral rotation: Turning away from the midplane of the body
Pronation: Rotation of the palm of the hand downward
Supination: Rotation of the palm of the hand upward

OSHA believes that ergo-hazards are prevented primarily by effective design of the workstation, tools, and job. Seated eye-height measurements may be reduced as much as 65 mm when personnel sit in a relaxed or slumped position. This slump factor should be considered when selecting the range of movement for adjustable seats, as well as in locating displays, optics, and vision ports.[2]

---

2. The examples in this subchapter are taken largely from MIL-STD-1472 and from MIL-HDBK-759. They are considered fairly accurate, and are used in the design of military hardware. The author recommends that the data be used with caution, and supplemented with more recent studies and in-plant measurements wherever possible. While the data presented do give a starting point for good ergonomic workplace design, they should not be considered the final authority should discussions between management and workers become elevated during contract negotiations regarding working conditions—rather, local studies and conditions should take preference.

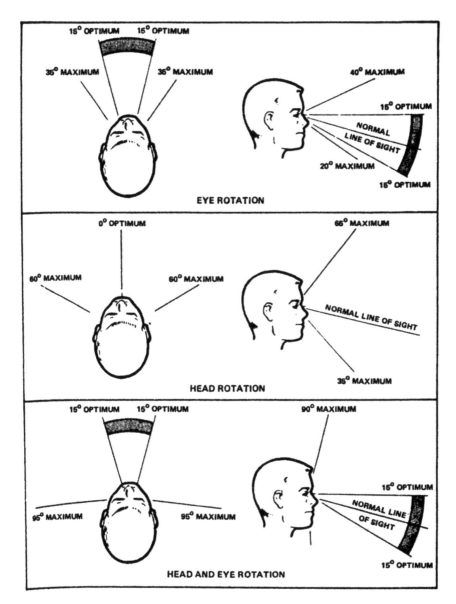

FIGURE 6.14 Vertical and horizontal visual fields.

6.1.3 Factory Applications

The work area environment consists of its total atmosphere. This includes climate, illumination, ventilation, noise, odor, vibration, congestion, and isolation, and may include high elevations or subterranean locations. Each element of the work area environment will consciously or unconsciously affect the worker and

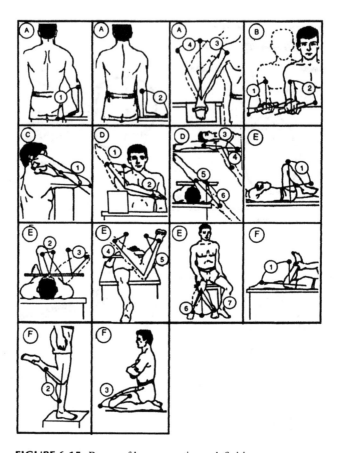

FIGURE 6.15 Range of human motion—definition.

his or her performance. Levels of illumination in the workplace are shown in Figure 6.18.

Ventilation requirements for persons working in enclosed spaces are shown in Figure 6.19. Vibration exposure criteria for both longitudinal and transverse directions with respect to body axis are shown in Figure 6.20.

The worker should not be expected to adjust to all these elements. Without some ergonomic controls, productivity suffers and individual employees may also sustain various work-related injuries and illnesses.

Workers are various sizes and ages, and may have physical or mental limitations. The work tasks that people perform require a variety of skill levels. The job assignment may be repetitive, or may require strength and physical endurance. Good and bad work practices on the part of the worker are sometimes more important than equipment selection by the manufacturing engineer. Figure 6.21 shows a screwdriver used properly and improperly. Wrist posture is determined by the elevation and orientation of the work surface with respect to the worker and the

Body Member	Movement	Lower Limit (degrees)	Average (degrees)	Upper Limit (degrees)
A. Wrist	1. Flexion	78	90	102
	2. Extension	88	99	112
	3. Adduction	18	36	27
	4. Abduction	40	54	47
B. Forearm	2. Supination	92	113	135
	2. Pronation	53	77	101
C. Elbow	1. Flexion	132	142	152
D. Shoulder	1. Lateral Rotation	21	34	47
	2. Medial Rotation	75	97	119
	3. Extension	47	61	75
	4. Flexion	176	188	190
	5. Adduction	39	48	57
	6. Abduction	117	134	151
E. Hip	1. Flexion	100	113	126
	2. Adduction	19	31	43
	3. Abduction	41	53	65
	4. Medial Rotation (Prone)	29	39	49
	5. Lateral Rotation (Prone)	24	34	44
	6. Lateral Rotation (Sitting)	21	30	39
	7. Medial Rotation (Sitting)	22	31	40
F. Knee Flexion	1. Prone	115	125	135
	2. Standing	100	113	126
	3. Kneeing	150	159	168

These values are based on the nude body. The ranges are larger than they would be for clothed figures.

FIGURE 6.16 Range of human motion—values.

shape of the tool. Stress concentrations should be distributed evenly over muscle eminences.

Workstations usually consist of the tools and equipment required to perform specific tasks. The workstation could be a factory with assembly lines and large manufacturing equipment, or an office environment with office equipment. The workstation environment fits the worker properly if the worker feels comfortable and natural while performing the work task. If the work task requires excessive reaching, bending, lifting, twisting, stooping, or working with the arms over shoulder height, this unhealthy work environment will result in poor productivity. Figure 6.22 displays the preferred angle of inclination for ramps, stairs, etc.

Until changes in the design of the workstation or tooling are made, rotation of workers in an unfriendly environmental work area may be an option. Some managers believe that rotation of workers will help prevent an individual from being burdened with an unpleasant task. The real truth of rotation is that everyone shares in the misery. The solution to the problem is to correct the environment—which increases

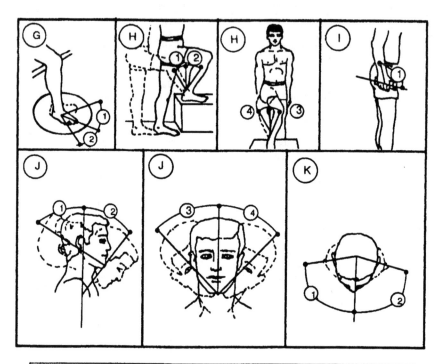

Body Member	Movement	Lower Limit (degrees)	Average (degrees)	Upper Limit (degrees)
G Foot Rotation	1. Medial	23	35	47
	2. Lateral	31	43	55
H. Ankle	1. Extension	26	38	50
	2. Flexion	28	35	42
	3. Adduction	15	33	24
	4. Abduction	18	30	23
I. Grip Angle		95	102	109
J Neck Flexion	1. Dorsal (back)	44	61	88
	2. Ventral (forwared)	48	60	72
	3. Right	34	41	48
	4. left	34	41	48
K. Neck Rotation	1. Right	65	79	93
	2. Left	65	79	93

FIGURE 6.17 Range of human motion—continued.

*LUX (Ft-C)

WORK AREA OR TYPE OF TASK	RECOMMENDED	MINIMUM
Assembly, general		
medium	810 (75)	540 (50)
precise	3220 (200)	2155 (200)
Bench work		
medium	810 (75)	540 (50)
extra fine	3230 (300)	2155 (200)
Electrical equipment testing	540 (50)	325 (30)
Screw fastening	540 (50)	325 (30)
Inspection tasks, general		
medium	1075 (100)	540 (50)
extra fine	3230 (300)	2155 (200)
Machine operation, automatic	540 (50)	325 (30)
Storage		
live, medium	325 (30)	215 (20)
live, fine	540 (50)	215 (20)
Office work, general	755 (70)	540 (50)
Reading		
large print	325 (30)	110 (10)
handwritten reports, prolonged reading	755 (70)	540 (50)
Business machine operation	1075 (100)	540 (50)

NOTE: As a guide in determining illumination requirements, the use of a steel scale with 1/64 in. divisions required 1950 lux (180 ft-c) of light for optimum visibility.
*As measured at the task object or 760 mm (30 in.) above the floor.

FIGURE 6.18 Illumination levels for work areas.

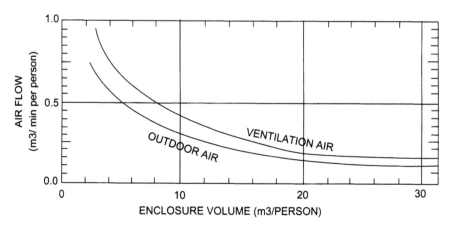

FIGURE 6.19 Ventilation requirements.

productivity for the entire shop. An electromyographic study of five jobs where job rotation had been introduced concluded that job rotation may be useful for reducing stress associated with heavy dynamic tasks rather than for reducing static muscular load in "light" work situations (Jonasson, 1988).

Factory Machines and Workplace Safety

In the past, progress in assembly conditions was necessitated by complaints, errors, accidents, absent workers, or other problems in operation of some operator-machine combinations. Education and training are two obvious means of improving workplace safety. Assembly-line operations need to be regarded as human line operations that create monotony, fatigue, excessive noise, and vibration. Noise levels for various types of work are given in Figure 6.23, plotted as ambient noise level versus distance from the speaker to the listener.

According to Dennis R. Ebens, president of Rockford Systems, Inc., in Illinois, updating power presses to comply with OSHA requirements means that employers must consider all five of the following requirements: safeguarding, lockout/tag-out devices, starters, and covers.

One of the most common elements of work is activation of machinery or equipment by hitting control buttons. It is not uncommon for an operator to activate equipment every 4 or 5 sec. Allowing for rest intervals and other tasks, this may mean that the hands must perform the motion pattern more than 5000 times per day. It becomes imperative, therefore, that consideration be given to both the hand position and the force required to engage the control. Different controls may vary considerably in the amount of force required, and this should be given due attention in the selection of different devices. Touch-sensitive controls now available virtually eliminate forces. Rexcon Controls (Mt. Clemens, Michigan) has developed a button activated by sensors, requiring no touch at all.

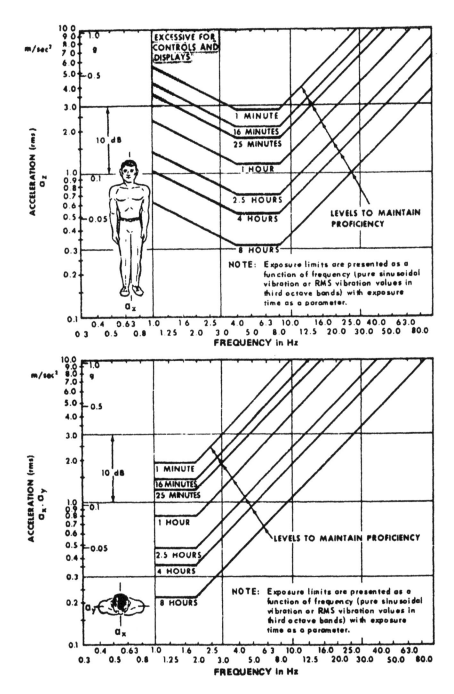

FIGURE 6.20 Vibration exposure criteria for longitudinal (upper curve) and transverse (lower curve) direction with respect to body axis.

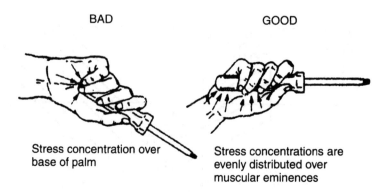

BAD   GOOD

Stress concentration over   Stress concentrations are
base of palm   evenly distributed over
                                         muscular eminences

Select tools that spread stress areas evenly over muscular eminences.

FIGURE 6.21 Wrist posture is determined by the elevation and orientation of the work surface with respect to the worker and the shape of the tool.

Proper design of knobs and handles is becoming more important in the factory as we become aware of the ergonomic (and productivity) benefits of all elements of the work process. Figure 6.24 shows recommended knob dimensions. Palm grasp is preferred over thumb-and-finger encircles, and both are preferable to fingertip actuation. Figure 6.25 shows the recommended dimensions for handles of various designs. Gripping efficiency is best if the fingers can curl around the handle to any angle of 120° or more. The diameter of the handle increases as the weight of the item increases. The strength requirements of a task are affected by the size and shape of the handles, as also shown in Figure 6.25.

Makers of hand tools are trying to make their tools easier on users. Redesign efforts aim particularly at avoiding stress and strain to the hand and wrist. One wrist disorder, carpal tunnel syndrome, strikes 23,000 workers a year and costs employers and insurers an average of $30,000 per injured worker. U.S. Labor Department figures show that the vast majority of hand and wrist disorders occur among tool-using workers.

Engineering Controls

Engineering controls are the preferred method of control, since the primary focus of ergonomic hazard abatement is to make the job fit the person, not force the person to fit the job. This can be accomplished by ergonomically designing workstations and tools or equipment.

Workstations should be ergonomically designed to accommodate the full range of required movements. Moreover, they should be designed so that they accommodate the workers who are actually using them to perform the job—not just the "average" or "typical" worker. Examples of good and bad designs for containers are shown in

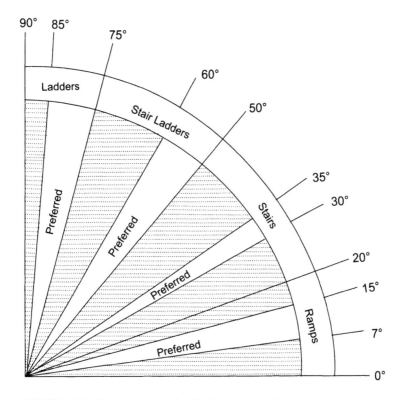

FIGURE 6.22 Type of structure in relation to angle of ascent.

Figure 6.26. All edges that come in contact with the worker should be well rounded. Changes of this type are very low-cost to implement, and are usually cost-effective as well as ergonomically correct. Figure 6.27 shows some of the elements of workbench and jig designs. The jigs should be oriented so that parts can be assembled without flexing the wrist.

The workstation should be designed to permit the worker to adopt several different but equally healthful and safe postures that still permit performance of the job; sufficient space should be provided for the knees and feet. Worktables and chairs should be height-adjustable to provide proper back and leg support. Seat cushions can be used to compensate for height variation when chairs or stools are not adjustable. There should be definite and fixed space for all tools and materials. Machine controls should be reachable and equally accessible by both right- and left-handed operators.

Most hand tools are designed for only occasional use, not for repetitive use over prolonged periods. When acquiring tools for regular use in an industrial setting, an employer should consider the following ergonomic features:

Tools should be light in weight, and handles should be designed to allow a relaxed grip so the wrists can remain straight.

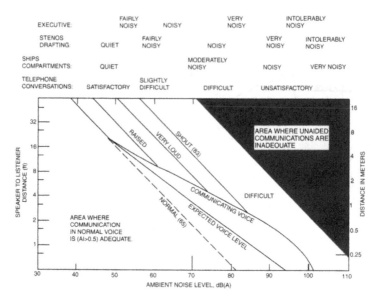

FIGURE 6.23 Permissible distance between a speaker and listeners for specified voice levels measured 1 m from the mouth.

Tools should be designed for use with either hand and be of various sizes so they can be used by both women and men.

Tool handles should be shaped so that they contact the largest possible surface of the inner hand and fingers, and should be fitted to the functional anatomy of the hand. Tool handles with sharp edges and corners should be avoided.

Power tools should be used to reduce the amount of human force and repetition required. Whenever possible, the weight of the tool should be counterbalanced to make it easier to handle and to reduce vibration.

To reduce tool vibration, special absorbent rubber sleeves can be fitted over the tool handle.

Wrist posture is determined by the elevation and orientation of the work surface with respect to the worker and the shape of the tool. Figure 6.28 shows some good and bad designs for powered drivers and drills. An example of the old versus the new shape of hand drill motors is shown in Figure 6.29. Cooper Power Tools is one of the companies that is very conscious of the importance of ergonomic design of their equipment. The new grips are modeled on vibration, torque, noise, and human geometry data. The same type of equipment now comes in different size grips, for the different hand sizes of workers. Cooper's new drill line was designed by Biomechanics Corp. of America, a company that measures how users interact with products, tools, and workplaces. John Lawson, aerospace marketing manager at Cooper's power tool division,

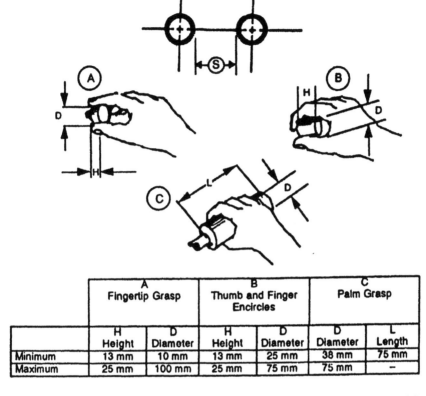

	A Fingertip Grasp		B Thumb and Finger Encircles		C Palm Grasp	
	H Height	D Diameter	H Height	D Diameter	D Diameter	L Length
Minimum	13 mm	10 mm	13 mm	25 mm	38 mm	75 mm
Maximum	25 mm	100 mm	25 mm	75 mm	75 mm	-

	S Separation		Torque	
	One Hand Individually	Two Hands Simultaneously	*	**
Minimum	25 mm	75 mm	-	-
Optimum	50 mm	125 mm	-	-
Maximum	-	-	32 mN·m	42 mN·m

*Up to/including 25mm diameter.
**Greates than 25mm diameter.

FIGURE 6.24 Recommended knob and handle design criteria.

says, "There's more and more concern in the power tool industry for worker safety. Ergonomics is the future." Figure 6.30 shows one of Cooper's new drill motors.

Personal Protective Equipment

Personal protective equipment may also be necessary to help prevent or reduce ergonomic hazards. For example, when vibrating tools are used, vibration may be dampened

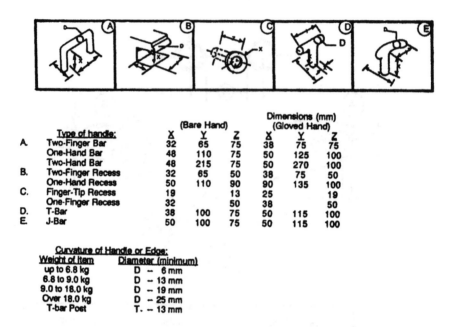

FIGURE 6.25 Recommended handle dimensions using bare and gloved hands.

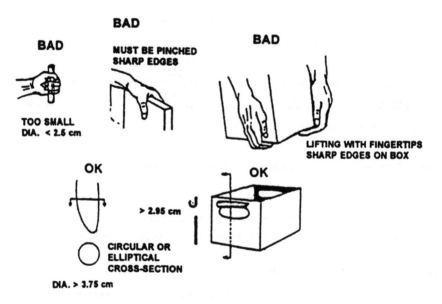

FIGURE 6.26 Strength requirements of a task as affected by the size and shape of the handles.

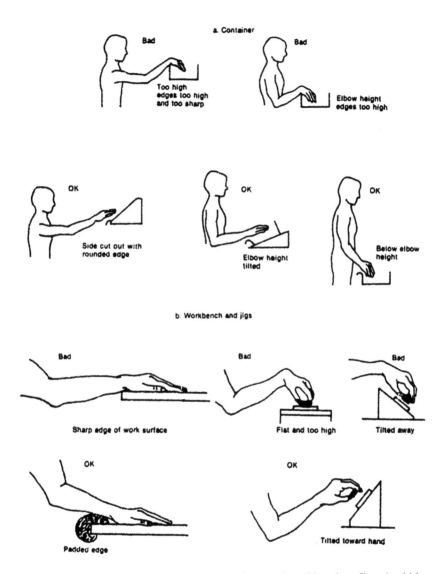

FIGURE 6.27 Good and bad designs for containers and workbenches. Jigs should be located and oriented so that parts can be assembled without flexing the wrist.

by using rubber-backed, low-pile carpet sections on the work surface. Gloves may also be worn to reduce the effects of vibration and force.

A long-term study in Norway found that employee turnover dropped to 10% from 40% after a plant installed new equipment. The company realized an 852% return on its initial capital investment—money saved primarily on training, recruitment, and employee sickness and injuries.

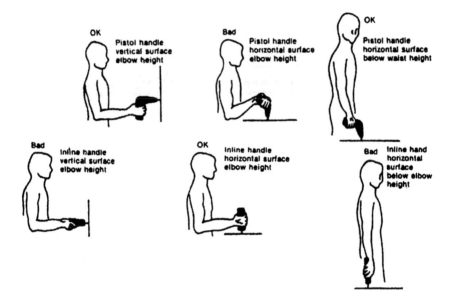

FIGURE 6.28 Good and bad designs for powered drivers.

6.1.4 Office Applications

With the increased usage of computers, there has been a rise in work-related injuries and illnesses in this field. The same attention to a correct workstation should apply to office areas as to the manufacturing shops.

In order for equipment to be utilized most efficiently, it must be designed for the specific user population. We must design for people who, while actually working, will be under conditions of stress and fatigue from many causes. A performance decrement may arise while working not so much because the workers are basically unable to perform, but because the individual is overloaded—both physically and mentally.

Health and safety features are foremost in ergonomic design. Other considerations, such as the information processing and perceptual limitations of humans, must be taken into account in the design of communication devices, computer displays, and numerous other applications. We must design equipment and processes that are as simple as possible to use. The job should not require intellectual tasks, such as transforming data, which might prove distracting to the worker. While training and practice can improve a person's proficiency, this should not be considered a substitute for good design. Figure 6.31 shows a good layout for a computer workstation, with factors to consider.

NIOSH has done considerable research on the use of video display terminals (VDTs). They reviewed the possible effects of radiation, fear of possible cataracts,

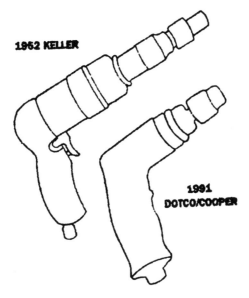

FIGURE 6.29 Comparison of 1952 drill motor modeled after the German Luger, and 1991 grips modeled on vibration, torque, noise, and human geometry data. (Courtesy of Dotco/Cooper, Inc. With permission.)

psychological stress and strain, reproductive concerns, and musculoskeletal strains. As a result of these studies, most problems were found to be nonexistent or minor—with the exception of the ergonomic considerations. Their recommendations include the following.

1. *Workstation design.* Maximum flexibility should be designed into VDT units, supporting tables, and operator chairs. VDTs should have detachable keyboards, worktables should be height-adjustable, and chairs should be height-adjustable and provide proper back support.
2. *Illumination.* Sources of glare should be controlled through VDT placement (i.e., parallel to windows as well as parallel and between lights), proper lighting, and the use of glare-control devices on the VDT screen surface. Illumination levels should be lower for VDT tasks requiring screen-intensive work and increased as the need to use hard copy increases. In some cases, hard-copy material may require local lighting in addition to the normal office lighting.
3. *Work regimens.* Continuous work with VDTs should be interrupted periodically by rest breaks or other work activities that do not produce visual fatigue or muscular tension. As a minimum, a break should be taken after 2 hr. of continuous VDT work, and breaks should be more frequent as visual, mental, and muscular burdens increase.

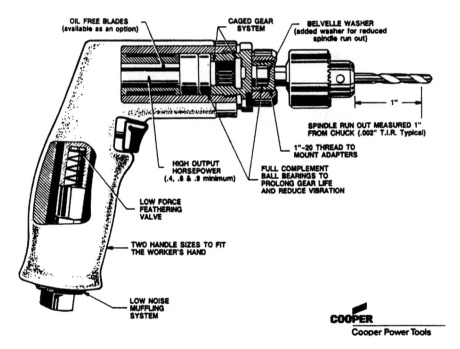

FIGURE 6.30 Details of Dotco® ergonomic drill. New lightweight model comes in two handgrip sizes. (Courtesy of Dotco/Cooper, Inc. With permission.)

4. *Vision testing.* VDT workers should have visual testing before beginning VDT work and periodically thereafter to ensure that they have adequately corrected vision to handle such work.

In a review of the literature, the World Health Organization found that "musculoskeletal discomfort was commonplace during work with VDTs," and that "injury from repeated stress to the musculoskeletal system is possible." It suggests that the primary emphasis for reducing musculoskeletal strain has been on improving the workstation environment by applying well-established ergonomic principles. However, some authorities suggest that ergonomically designed workstations are an incomplete prescription for preventing musculoskeletal discomfort in VDT work because they do not correct for a major contributory factor, namely constrained postures. Some authorities feel that exercise programs designed to reduce this discomfort are valuable.

Some interesting work on "fitting the task to the man" was done by Dr. E. Grandjean, director of the Institute for Industrial Hygiene and Work Physiology, at the Swiss Federal Institute of Technology, Zurich. The frequency of bodily aches among sedentary workers in an office is shown in Figure 6.32, based on 246 persons questioned. The results are as follows.

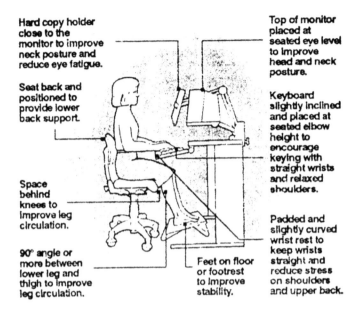

Hard copy holder close to the monitor to improve neck posture and reduce eye fatigue.

Seat back and positioned to provide lower back support.

Space behind knees to improve leg circulation.

90° angle or more between lower leg and thigh to improve leg circulation.

Top of monitor placed at seated eye level to improve head and neck posture.

Keyboard slightly inclined and placed at seated elbow height to encourage keying with straight wrists and relaxed shoulders.

Padded and slightly curved wrist rest to keep wrists straight and reduce stress on shoulders and upper back.

Feet on floor or footrest to improve stability.

FIGURE 6.31 Office specifications. (Courtesy of Ergo Tech, Inc. With permission.)

1. Seat heights between 38 and 54 cm were preferred for a comfortable posture of the upper part of the body.
2. Aches in the thighs were caused mostly by putting the weight on the thighs while working, and only to a lesser extent by the height of the seat.
3. A desk top 74 to 78 cm high gave the employees the most scope to adapt it to suit themselves, provided that a fully adjustable seat and footrests were available.
4. Of the respondents, 57% complained of back troubles while they were sitting.
5. Of the respondents, 24% reported aches in neck and shoulders and 15% in arms and hands, which most of them, especially the typists, blamed on too high a desk top (cramped hunching of the shoulders).
6. Of the respondents, 29% reported aches in knees and feet, mostly among small people who had to sit forward on their chairs because they had no footrests.
7. Regardless of their body length, the great majority of the workers liked the seat to be 27 to 30 cm below the desk top. This permitted a natural posture of the trunk, obviously a point of first priority with these workers.
8. The frequency of back complaints (57%) and the common use of a back-rest (42% of the time) showed the need to relax the back muscles from

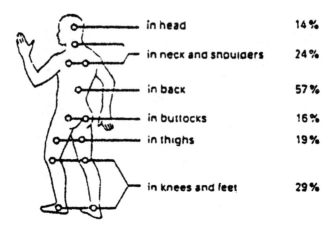

in head	14%
in neck and shoulders	24%
in back	57%
in buttocks	16%
in thighs	19%
in knees and feet	29%

FIGURE 6.32 Frequency of body aches among sedentary workers. (From Grandjean and Burandt. With permission.)

time to time, and may be quoted as evidence of the importance of a well-constructed backrest.

Muscle groups commonly requiring relaxation of activation after continuous VDT work are shown in Figure 6.33. They are:

1. Chronically tensed scapular elevators require stretching and relaxation.
2. Spinal extensors of the lumbar, thoracic, and cervical regions are over-stretched and require activation.
3. Muscles of the anterior thoracic region are shortened and require stretching.
4. Forearm flexors are chronically tensed and shortened, and require stretching and relaxation.

A 1986 study by the Army Corps of Engineers documented a 20.6% improvement in office employee productivity 1 year after ergonomic furniture was installed.

6.1.5 Ergonomic Checklists

Try checking off this list from Atlas Copco Tools (Farmington Hills, Michigan). "Yes" answers to most items means that an ergonomics effort is in order. "Don't know" means that it is time for managers to take a look at their records and their assembly areas.

FIGURE 6.33 Muscle groups commonly requiring relaxation of activation after continuous VDT work. (From Grandjean and Burandt. With permission.)

Do production reports show decreased or already low efficiency?
Do inspection and rejection reports indicate deteriorating product quality?
Is absenteeism increasing in certain areas—or everywhere?
Are accident rates up sharply?
Do medical reports show many back or repetitive stress injuries?
Do workers on certain lines log frequent medical visits?
Is plant turnover high or increasing?
Is turnover in specific jobs high?
Does training new workers take too long?
Do workers make a lot of mistakes?
Is there too much material waste?
Are operators damaging or not maintaining equipment?
Are operators often away from their workstations?
Do workers make subtle changes in their workplaces?
Is the plant moving to two- or three-shift operation?
Are plant engineers ignorant about, or suspicious of, ergonomic principles?
Do you use incentive pay to increase productivity?
Are workers exercising hands, fingers, and arms to relieve stress?

A Sample Ergonomic Checklist to Identify Potential Hazards in the Workplace (Things to Look for as You Tour a Factory)

General Work Environment

Workforce characteristics
Age, sex, anthropometries (body size and proportions)
Strength, endurance, fitness
Disabilities
Diminished senses
Communication/language problems (e.g., non-English-speaking or illiterate workers)

Lighting
Climate
Noise level
Health and safety safeguards
Job and workstation design
Location of controls, displays, equipment, stock

Accessibility
Visibility
Legibility
Efficiency of sequence of movements when operating or using
Use of pedals
Posture of workers
Sitting, standing, combination
Possibility for variation
Stooping, twisting, or bending of the spine
Chair availability, adjustments
Room to move about
Work surface height

Predominately dynamic or static work

Alternation possible
Use of devices such as clamps or jigs to avoid static work
Availability of supports for arms, elbows, hands, back, feet

Muscular Workload/Task Demands

Repetitiveness

Frequency
Force
Availability of rest pauses
Possibility for alternative work
Skill, vigilance, perception demands
Efficiency of organization (supplies, equipment)

Use of hand tools

Hand and wrist posture during use
Work surface height
Size and weight of tool
Necessity, availability of supports
Shape, dimensions, and surface of handgrip
Vibratory or nonvibratory

Physical strength requirements

Strength capabilities of employee
Working pulses/respiratory rate
Loads lifted, carried, pushed, or pulled

Manner in which handled
Weight and dimensions of objects handled

6.2 WORKER'S COMPENSATION

Timothy L. Murphy, McDonnell Douglas Corp., Titusville, Florida

6.2.0 Introduction to Worker's Compensation

Worker's compensation was traditionally called workman's compensation. It is the name commonly applied to statutes that give protection and security to workers and their dependents against injury, disease, or death occurring during employment. The statutes establish the liability of benefits—usually financed by insurance bought by the employer—including hospital, other medical payments, and compensation for loss of income.

Factory safety regulations should enable industry to produce and market products without causing the illness, injury, or death of employees. Workers have the right to return home to their loved ones in a healthy condition after a day's work. However, the proper working conditions were not always provided by all employers.

6.2.1 Compensation Systems

It was not until the latter part of the American Industrial Revolution, in 1908, that the U.S. Congress passed the first Workman's Compensation Act to cover federal employees. Today, all of the 50 states, territories, commonwealths, and the District of Columbia have worker's compensation acts. The Federal Employees' Compensation Act covers federal civilian employees. The Longshoreman and Harbor Workers' Compensation Act covers maritime employees. The fundamental concepts of the compensation acts are to provide disabled employees with medical, economic, and rehabilitation support. The manner in which this support is handled has caused great distress between workers and their companies. Some employees feel that companies should do more for them, and many companies feel that they are providing more benefits than they can afford.

Companies are responsible for providing medical care and financial support to their employees who have work-related injuries or occupational illnesses. Each state identifies the payment scale that should be followed while the injured or ill employee is recuperating. The company is also responsible for the rehabilitation of disabled employees. These employees must be rehabilitated so they can return to their previous job or be retrained for another position. Benefits covered by worker's compensation acts cannot be disallowed because a worker was either partially or fully responsible for the injury. The only exception to this rule is if the worker deliberately injured himself or herself or was under the influence of alcohol or drugs.

Claims

When an employee files a worker's compensation claim, he or she is filling out an application for benefits for an occupational injury or occupational disease/illness. A work-related injury is usually considered to be caused by a single incident. An occupational disease/illness is considered to have occurred from the result of exposure to an industrial hazard. A worker's compensation claim is valid when the injury or illness has occurred "in the course of" and "arising out of" employment. The phrase "in the course of" is defined as an injury or illness incurred during work time and at work. The phrase "arising out of" employment is defined as an injury or illness that was caused by working. Most states do not provide compensation for illnesses that are "ordinary illnesses of life," or illnesses that are not "peculiar to or characteristic of" the employee's occupation.

Four primary benefits are covered under worker's compensation: cash benefits, medical benefits, rehabilitation benefits, and death benefits. The most cost-effective monetary method of handling worker's compensation is to prevent an accident or illness from occurring in the first place. Consequently, companies that have inadequate safety programs usually pay a very high worker's compensation insurance premium, due to large payments to disabled employees.

6.2.2   Compensation

There are four basic disability compensation classifications:

1. *Temporary partial disability* (TPD): A disability caused by a work-related injury or occupational disease for a definite period that does not prevent the employee from gainful employment. A TPD employee is expected to regain most of his or her former capability.
2. *Temporary total disability* (TTD): A disability caused by a work-related injury or occupational disease for a definite period, as a result of which an employee cannot return to gainful or regular employment. A TTD employee will not be able to regain his or her former capability.
3. *Permanent partial disability* (PPD): A disability caused by a work-related injury or occupational disease where the employee is able to return to work but has not attained maximum medical improvement.

4. *Permanent total disability* (PTD): A disability caused by a work-related injury or occupational disease that completely removes the employee from employment, due to loss of use of designated part or parts of the body, or from amputation.

Cash Benefits

Cash (indemnity) benefits include both disability benefits and impairment benefits. Disability benefits are paid when there is a work-related injury or an occupational illness and a wage loss. Impairment benefits are paid only for certain specific physical impairments. Most states provide indemnity payments of 67% of the employee's weekly wage with a maximum weekly benefit allowance. A few states will pay 100% of the employee's weekly wage. Most states require a waiting period of 3 to 7 days before payment of indemnity benefits. The waiting period is intended to eliminate minor medical and first aid cases.

Medical Benefits

Medical benefits include first aid treatment; services of physicians, chiropractors, dentists, outpatient nursing, and physical therapists; surgical and hospital services; medication; medical supplies; and prosthetic devices. Every state law requires the employer to provide medical care to work-related injured or occupationally ill employees. There is usually no dollar limitation or time limitation on medical benefits. The states are evenly divided over the practice of who is allowed to choose the attending physician. Some states supply an approved list of physicians.

Rehabilitation Benefits

Rehabilitation benefits are divided into two classes, medical rehabilitation and occupational rehabilitation.

Medical rehabilitation has two phases:

1. The acute phase, requiring hospitalization and/or surgical management
2. The postoperative phase, when multiple sessions per week are held with doctors, nurses, and/or therapists to restore the employee to his or her fullest physical capacity

Occupational rehabilitation is the transition phase in which evaluation of work capacity, training, counseling, and job placement are performed. Some workers may require retraining to enter completely new career fields. Rehabilitation for a seriously injured worker can be lengthy and extremely expensive. Many states have enacted a second injury fund to help promote the hiring of disabled workers. If a company knowingly hires a disabled worker and that employee reinjures himself or herself, the state will help to reimburse the company for the extra worker's compensation costs. Every state receives funding from the Federal Vocational Rehabilitation Act to sponsor vocational rehabilitation for disabled industrial workers. Employers who choose modified work (light duty) for injured employees show a reduced total recovery period. The employees who work at modified workstations usually return to their former positions more quickly than those who remain out of work for the entire period. This method of reinstating employees in a timely manner rewards both the employee and the employer. If an employee is not working because of a work-related injury, the cost of that injury will be about 15 to 25 times greater than for an employee who goes for medical treatment and immediately returns to work.

Death Benefits

Death benefits are generally paid to a spouse until he or she remarries and to the children until a specified age, usually 18. Many states continue benefits for dependent children while they are attending college as full-time students. The benefits include a burial allowance as well as a portion of the worker's former weekly wages. In most states, the duration of benefits is unlimited. Death benefits are more than 14% of all total indemnity benefits. Although death is the ultimate personal loss, the economic loss associated with death cases is often less than for a permanent total disability.

6.2.3   Abuses in the System

Showing concern for an injured employee usually reduces the time that the employee is on disability leave. Most of the injured employees who seek legal counseling do so because no one from the company has talked with them and shown concern for them. In some states, a new industry has been created from their liberal worker's compensation systems. Many daytime television commercials are advertisements for worker's compensation attorneys.

Recruiters actually seek out potential worker's compensation candidates from the unemployment lines. The unemployed candidates are advised of their potential benefits from the worker's compensation system. If the candidates feel that they may qualify for worker's compensation benefits, the recruiter escorts them to a medical clinic. These clinics then process the people through a series of medical evaluations. When necessary paperwork is completed, a new candidate has joined the ranks of the state worker's compensation system. In the past, back injuries were most common. Recent experience shows a higher incident of other skeletal problems, most notably carpal tunnel syndrome (see Subchapter 6.1). Some states now recognize stress resulting not only from the workplace environment, but from traffic driving to work, as an injury or disease "arising out of employment."

6.2.4   Costs

The escalating cost of worker's compensation is staggering. Payments for worker's compensation programs rose to an all-time record of $35.3 billion in 1990. California leads the nation with $6.05 billion in medical and indemnity payments. Medical benefits account for approximately 30% of all worker's compensation costs.

The cost to a company for worker's compensation insurance premiums is based on payroll cost (wages). Recently, all U.S. wages and salaries are as follows (in billions):

Salaries	$2917
Supplemental	$608
Composite	$3525 per year

This says that 17% of payroll cost is due to the supplemental benefits, including worker's compensation. The gross average wages in the U.S. manufacturing industry are $470 per week. The median income for machine operators and assemblers is $350 to $400 per week.

There is a wide range of cost based on worker classification. Some examples from the state of Florida are as follows:

Clerical:	$0.75/$100 of payroll
Electrical apparatus assembler:	$4.40/$100
Iron and steel fabrication shop:	$20.00/$100
Roofer:	$48.00/$100

The average cost of worker's compensation insurance is $3.50 per $100 of payroll in Florida, and ranges from slightly over $1 in some other states to more than $6 in California. The variation depends on the average wages, types of work classification (industrial versus agricultural, etc.), and the history of claims payments.

Typical U.S. Payroll Costs

A new start-up company would pay rates in the above standard ranges. This is known as a "modification factor" of 1.0. As the insurance carrier has experience with the number and cost of claims, the "standard" rate for each employer is adjusted by changing the modification factor. The rates can be adjusted downward (or upward) based on the average of the previous 3 years' history. There is no cap on the upper limit that can be charged. An example of this can be made for a manufacturing plant with 1000 employees. Approximately half of the workers are employed in the shop. An average for a well-managed factory would be 160 to 170 claims per year. Of these, 120 or so claims are for very minor injuries, perhaps requiring a couple of stitches—and back to work. This cost might total $40,000 per year.

The other 40 claims, when the worker goes to a clinic and stays off work for some time, may cost the company more than $600,000 per year in lost wages, medical bills, rehabilitation, etc. This can become a very significant element of payroll cost to a manufacturer, since the insurance company raises the worker's compensation insurance premium to cover the $650,000 per year, plus the administrative cost of handling the claims. This total of $750,000 per year can be a significant factor in the profitability of a small manufacturing firm, and warrants positive measures to keep it as low as possible.

6.2.5   Cost Reduction Methods

The most important positive actions that a company can take are to (1) reduce the number of claims, and (2) lower the cost of medical expenses on each claim. Employee training and education are important elements contributing to reduced claims. Each job or work element in the manufacturing plant should be analyzed for proper ergonomic considerations (See Subchapter 6.1). An analysis of several medium-sized factories, where the work is relatively light and clean, shows that more than 40% of the "lost-time" cases are due to strain or overexertion; 25% are due to cumulative, repetitive job injuries; 15% are due to falls; and the remaining are due to all other causes. Looking at the cost of these categories, 40% are due to strain; 25% are due to falling; 17% are due to cumulative injuries, and 12% result from being struck by something (or striking something). Proper employee and supervisor training plus careful review of the process design can reduce both the number and severity of these lost-time cases.

Most states offer a 10% premium reduction if the firm has an organized safety program in place (see Subchapter 6.3). Also, a 5% reduction is available if the employer has a "drug-free workplace" program implemented (see Subchapter 6.4).

6.3   INDUSTRIAL SAFETY

Jeffrey W. Vincoli, CSP, REP, Titusville, Florida

6.3.0   Introduction to Industrial Safety

In the highly competitive environment of today's business world, success is achieved through the collective talents, skills, and contributions of all the individual employees of the organization. In fact, the business enterprise with the right combination of skills and clear, focused objectives will often outperform its competitors in every category. It cannot be disputed that people are a company's greatest resource. It is clear that people are literally and figuratively the lifeblood of any organization. Therefore, it should be equally clear that the assurance of a safe and healthy workplace, within which this resource must perform, is as important to a company as any other issue related to production, sales, and profit.

Just about all of us have heard the statement that "safety is everyone's business," but few realize that this is more than just a cliché or a clever approach to safety management. The requirement for a safe work environment is actually a mandate on industry and its employees from the federal government. Safety and health regulations for general industry can be found in the U.S. Code of Federal Regulations (CFR) at Title 29 (Labor) Part 1910, and for the construction industry at Title 29, Part 1926. These codes, which address literally thousands of situations, working conditions, hazard control requirements, and worker safety and health protection standards, are all intended to make the work environment safer.

Although numerous areas are addressed and a minimum level of worker protection is ensured by implementation of these many standards, it is important to note that the major driving factor for worker safety is contained not in the Code of Federal Regulations but in Section 6 of the Occupational Safety and Health Act of 1970. This factor, better known as the "General Duty Clause," Section 6(a)(1), simply states that:

> Each employer shall furnish to each of his employees employment and a place of employment which are free from recognized hazards that are causing or are likely to cause death or serious physical harm to his employees, and [each employer] shall comply with the occupational safety and health standards promulgated under this Act.

This section goes on, in Part (b), to state that *each employee* shall comply with safety and health requirements. It is from this legislation where the statement "safety is everyone's business" becomes a legal as well as a moral obligation.

Having established this fundamental principal of industrial safety, this subchapter will briefly explain the required actions an organization must take to ensure compliance with the intentions of the law regarding employee safety and health. The reader should also note that many individual states have obtained approval from the federal Occupational Safety and Health Administration (OSHA) to regulate their own state-wide safety and health programs, as long as these programs are *at least as stringent* as those mandated on the federal level. However, because so many individual state safety and health laws now exist, it would not be possible to discuss the particulars of each in the space provided here. This chapter will therefore focus on the federal level, with the understanding that these requirements are the minimum standards which, in many cases, must be exceeded in order to ensure the proper protection and preservation of employee safety and health. As Figure 6.34 shows, those proactive companies that anticipate and exceed these minimum regulatory requirements will be the leaders of tomorrow's highly competitive market. Conversely, those who are simply reactive to the requirements will more than likely fail.

By establishing workplace objectives for accident and illness prevention, management can demonstrate both commitment to worker safety and action toward that commitment. Without exception, the realization of these safety objectives should be considered as important as those established for sales, productivity, engineering, quality, and so on. In fact, how could any organization justify placing the safety and health of its workers second to production and sales? The truth is, few employers who have tried this approach have been very successful in avoiding investigation by federal OSHA or state safety inspectors.

6.3.1   The Business of Safety

While the General Duty Clause does require both employers and employees to ensure safety at work, the language of the law makes it quite clear that the employer has a special responsibility to its workers to provide a safe and healthy workplace. Since it is no secret that we in the United States live in one of the most litigious societies on earth, employers who wish to remain in operation today must also be in the business of safety.

Under the law, the management of an organization becomes the responsible element for ensuring that the safety process within the organization is in compliance

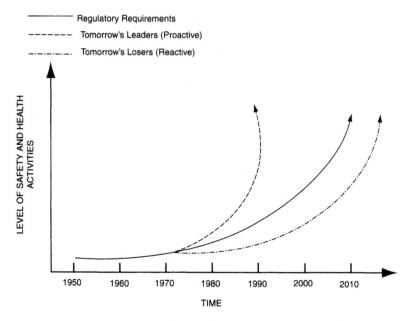

FIGURE 6.34 Staying ahead of the regulatory curve pays off.

with applicable requirements. However, it must be clearly understood that compliance with these various codes and regulations basically equates to compliance with only the minimum acceptable requirements. Therefore, in the practice of modern manufacturing engineering, "design by code" is no substitute for intelligent engineering. Since codes only establish minimum standards, actual design must often exceed these codes to ensure an adequate level of safety performance. The successful management of the safety process means a commitment of company resources. Time, money, and talent are required not only to properly identify and control hazards in new and existing operations and processes; such resources must also be expended to eliminate, limit, or control these hazards to their absolute minimum. But the real cost of safety goes far beyond that which is readily identifiable as *direct* cost. There are also numerous *indirect* expenses which affect the profitability of any given enterprise as a result of poor or ineffective safety policies and programs. This concept is best illustrated using the familiar iceberg graphic, as shown in Figure 6.35.

While the visible (or direct) costs are easily seen on the surface, there are often many more costs (indirect) resulting from inadequate safety performance which are not easily detectable at first glance but are there nevertheless. Only when an organization accepts the cost of safety as an important cost of doing business will the axiom "safety first" have any real meaning for the workforce. In a manufacturing setting, a combination of effort must often be employed to ensure that these established safety

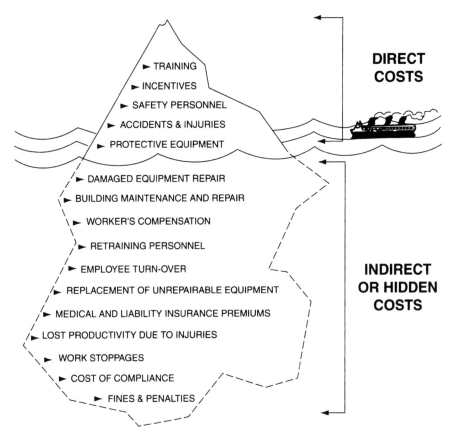

FIGURE 6.35 The true costs of safety.

objectives are achieved. For example, engineering controls may have to be installed, personal protective equipment (PPE) might be required, and/or specialized training may have to be provided by the company.

In short, as with every other aspect of a successful business endeavor, safety in the workplace will never be realized without a firm commitment beginning with top management. Evidence of this commitment will be visible in every decision the company makes and in every action it takes. The management personnel responsible for the proper execution of this commitment throughout the organization must be clearly identified to each employee at every level through established policies. One way in which company safety policy is communicated to employees is through an effective written injury and illness prevention program, more commonly referred to as a safety manual. Historically, when employees come to understand that their company has made a sincere commitment to the safety of the workforce, overall performance is typically improved and the organization works as a cohesive team toward company goals.

6.3.2   Responsibility and Accountability

As stated in the previous paragraph, once management has established objectives designed to ensure the safety of their personnel, products, and services, and after they have committed the necessary and required resources to achieve these goals, they must also clearly define those responsible within their organization who will be held accountable for the success of the safety program.

The manufacturing engineer is often uniquely situated within the organizational structure to act as the primary implementing force behind effective safety practices. Whether the engineer is occupied with the complexities of designing factory layout, establishing operating specifications for specific machinery, retooling existing equipment, or writing process sheets, the responsibility to consider the impact on safety resulting from any of these actions is a key to the success of an overall safety program. For the manufacturing engineer, the ability to recognize obvious as well as insidious conditions that could potentially threaten the safety of personnel and/or property is of paramount importance. However, this is often a difficult thing to do, since most engineers are not specifically trained in the process of hazard recognition, evaluation, and control. This is where the concept of a "working team" becomes critical. Specifically, the manufacturing engineer should form a working partnership with the company safety professional. Together, regardless of the task, the hazards associated with production can be successfully identified, understood, and hopefully eliminated (or at least controlled). In most industrial settings, effective interdepartmental cooperation can mean the difference between the success and failure of the entire business enterprise. Therefore, in order to be truly effective, such working relationships should be a matter of company policy and not simply one of convenience. Practically speaking, this "teaming approach" to ensuring a safe and healthy work environment has proven to be the most effective course of action in the manufacturing setting.

The process of assuring safety at work is really a critical element of the manufacturing engineer's overall job. In fact, because engineering personnel are normally located within the typical organization as a line function, effective safety management becomes a primary task. For example, in a typical line and staff organization (as shown in Figure 6.36), the task of implementing safety requirements is always a line function. This means that, while company safety personnel may provide recommendations, advice, and assistance to the line managers/personnel regarding compliance issues, it is still the line managers and supervisors who have the authority and responsibility to implement those recommendations. The task of safety should therefore be approached with the basic understanding that workplace safety programs are implemented as a *line* responsibility.

In comparison, the company's professional safety representative performs a *staff* function only. The safety department is really a service organization that must be an effective member of any working team in order to ensure the appropriate infusion of safety concepts and considerations into work process. Among other things, the safety department representative will research applicable requirements, develop implementation strategies, work with line personnel to ensure proper implementation, monitor that implementation to ensure compliance, and conduct periodic

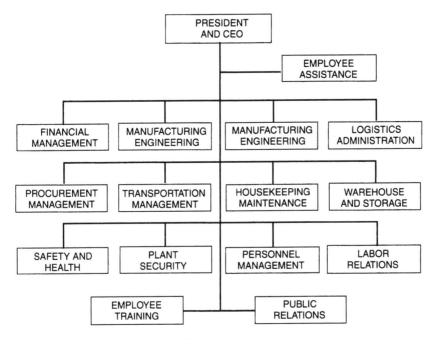

FIGURE 6.36 Typical line and staff functions of an organization.

assessments of the program to measure degree of success. But it is still the line managers, supervisors, and associated personnel, such as the manufacturing engineer,
who must actually implement and enforce the safety program in their designated
work area(s). Therefore, safety as a task must clearly be the function of the line, or
operational safety will not succeed. This approach to safety management is based
on a fundamental concept that stipulates that line management (especially first-line
supervisors, but including all levels from the top down) is absolutely responsible for
all operations that occur within its assigned area(s). There are very few line managers
or supervisors who would dispute this position or who would have it any other way.
It is therefore logical that this responsibility include the safety of those operations
and the employees required to perform them. This is an extremely important concept
that must be clearly understood and accepted through all levels of the organization.
As an example, Figure 6.36 shows how the safety department is located in a typical
organizational chart. Notice how staff functions, such as safety, training, security,
personnel, and labor relations, are distinct from the various line management responsibilities. This structure is generally found throughout industry.

6.3.3   Safety in the Manufacturing Process

Typical manufacturing operations occur in a work environment that contains many
inherent hazards to both personnel and equipment. The very process of manufacturing

requires the performance of hazardous tasks and procedures on a routine basis. These may include welding, painting, electrical work, heavy lifting, crane operations, radiation, robotics, machining, and the use of hazardous chemicals or even explosives. While safety professionals will examine the many work procedures that contain hazards to personnel and equipment and recommend controls to limit exposure to these hazards, it is the responsibility of the manufacturing engineer to implement these controls to ensure safe and successful operations. Simply stated, if the manufacturing engineer does not implement these safety recommendations on a daily basis, a safe workplace will not be achieved. To assist with this objective, safety representatives can perform audits and inspections, which serve to verify the level of compliance in each work area. When deficiencies are identified in the manufacturing process, the task of abating these discrepancies falls to the engineering supervisor responsible for that work area. The supervisor should look to the company safety professional for advice and assistance in compliance issues. However, they should never assume that safety program requirements will be met by the safety department alone. Safety must be a *team effort*. In the manufacturing industry, or any other for that matter, it just does not work any other way.

To ensure safe operations and the safety of personnel in any manufacturing process, adequate planning for workplace safety and health is an important aspect of everyday business operations. To effectively reduce the costs and risks associated with occupational accidents and injuries, workplace safety policies must be well considered, clearly written, comprehensive, and, above all, properly disseminated through every level of the organization. To assist line management with their safety responsibilities and to ensure a consistent approach to safety program implementation, a company safety manual should be developed and distributed to all levels of the workforce. The safety manual should contain general as well as specific safety rules, regulations, policies, and stated objectives for all employees to understand and follow. Also, to ensure its adequacy, the safety manual should not be developed without thorough planning and effective evaluation of company tasks and operations.

Safety and health planning is considered effective when at least the following factors have been considered by management during the development of its safety policy:

1. General safety rules should be written and constructed so as to apply to every member of the organization. Rules to address requirements such as the use of PPE, appropriate clothing for the workplace, expected behavior and conduct while on company property, and special emergency procedures all fall within the general nature of written company safety policy. Employers and employees should perform regular reviews of these rules to ensure an up-to-date and fair approach to safety and health assurance. Also, whenever new processes or products are introduced into the work environment, new safety rules and procedures must be written to clearly define the hazards and controls required to perform any new tasks.

2. The safe and healthful work practices required for specific tasks and procedures must be developed and differentiated from any general rules, if applicable. In most cases, specialized training may be required on these

particular safety considerations before employees are permitted to perform the job. These specific tasks or procedures may include examples such as laser operations, robotics, chemical processing, use of mechanical lifting equipment (cranes, hoists, forklifts), assembly and/or use of explosives and explosive devices, and other special equipment procedures and operations.

3. Procedures that explain the company's enforcement and reward recognition policies will help ensure that all safety and health rules and work requirements are practiced adequately at all levels in the organization. Rewards or other positive reinforcement tools such as bonuses, incentives, or employee recognition programs provide for positive motivation for compliance with company safety rules and procedures. Likewise, when clear violations of established safety rules have occurred, the company must be equally prepared with written methods of disciplinary action. Unbiased and consistent enforcement of both negative (disciplinary) and positive (recognition) policies provide a clear message of the company's firm commitment to the safety process.

4. There should be a written plan to address all types of emergency situations that could possibly occur at the manufacturing enterprise. The plan should include a set of procedures that respond to each possible situation. In fact, some emergency procedures, such as those covering medical emergencies and fire evacuation, are required by OSHA laws.

5. There should also be adequate provisions in the safety plan to ensure employee participation and involvement. This includes a system for effectively communicating safety program plans and directives to employees at all levels in the organization. Safety plans, newsletters, incentive programs, poster contests, and safety committees are a few examples of methods used to obtain employee participation.

6. Management must include a system to assure that the proper record-keeping and safety program documentation requirements of OSHA are fulfilled. Examples of required record-keeping activities include employee medical records, injury and illness statistics, inspections and corrective actions, equipment maintenance logs, and employee safety training.

7. Considerations for accident prevention and investigation must be an integral factor of the overall safety process. Since there is no way an organization can hope to prevent accidents, which only downgrade the efficiency of the business enterprise, without investigating those that do occur, there must be a comprehensive accident investigation and loss control program established as an integral part of company safety policy.

6.3.4 Developing a Specific Safety Program

In most cases, general safety rules focus on concerns that are germane to almost every type of work environment. These rules usually concentrate on housekeeping safety requirements, which form the basis of most generic safety and health programs.

Hence, general rules will provide an excellent foundation on which to build a comprehensive occupational safety and health program. While some rules may seem to be nothing more than common sense, employers usually discover the hard way that their employees often have very different perceptions, understandings, and interpretations of commonsense issues. It is therefore always best to provide the workers with these requirements in a clear, written, and well-documented fashion.

Once the general rules for safe conduct and safe operations have been developed, the organization must then determine those specific rules and regulations that are required to ensure the safety and health of all workers. Before any selection of these specific safe work standards can be made, as much information as possible must be obtained regarding the current conditions of the workplace and the adequacy of existing safe work practices. This information will not only help identify existing problem areas, it will also help determine the requirements necessary to resolve them.

This safety assessment should be a team effort conducted by the engineering manager or supervisor responsible for ensuring workplace safety and health, together with the company safety representative or professional occupational safety and health consultant. As a minimum, the workplace safety assessment should entail a detailed survey of the entire facility with a comprehensive evaluation of all required tasks and procedures. This survey should evaluate workplace conditions regarding:

Applicable safety and health regulations
Generally recognized safe work practices
Physical hazards of any equipment or materials
Employee work habits with regard to safety and health
A discussion with employees regarding safety and health

As shown in Figure 6.37, the safety survey should, as a minimum, also address the following five specific areas of concern (if applicable):

1. *Equipment safety.* The surveyor should develop a list of all the equipment and tools that are in the facility, and include their principal area(s) of use. Special attention should be given to inspection schedules, maintenance activities, and the physical layout of the facility. This information will greatly facilitate the development of specific safety program objectives.

2. *Chemical safety.* If it is not already available, a list of all chemicals (hazardous and nonhazardous) must be developed. The material safety data sheet (MSDS) for each chemical must be obtained and reviewed to ensure adequate precautions are being taken to preclude hazardous exposures. The surveyor should also attempt to identify and list the specific location(s) within the workplace where each chemical is being used, the processes involved with that use, and the possibility of employee exposure. This information will assist in the establishment of specific safety rules to prevent injury and illness resulting from the improper handling of hazardous chemical substances.

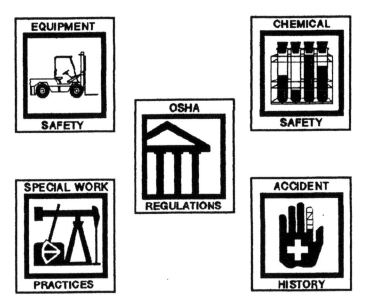

FIGURE 6.37 Specific areas of concern to investigate during a safety survey.

3. *Work practices.* All specific work practices involving the use of any equipment, tools, chemicals, and/or materials not previously identified must be examined in detail to ensure that appropriate and adequate precautions are in place. Special concentration on concerns such as the use of PPE, machine and equipment guarding, ventilation, noise, illumination, emergency procedures, and the use of appropriate tools for the job will help identify additional problem areas in need of attention. Also, any industry-specific safety hazards, such as potential exposure to biological hazardous wastes (medical industry), asbestos (automobile repair), or radiation (medical, nuclear, and aerospace industries) must be evaluated and properly addressed.

4. *OSHA requirements.* The safety professional should review all rules (state and federal) that are applicable to the specific operations, equipment, and processes in work at the subject facility. Since these rules normally establish only the minimum requirements for workplace safety and health, simple compliance may not always equate to absolutely safe operations. Therefore, a detailed evaluation of the these minimum safety requirements, as they apply to the subject facility, will help determine the overall applicability of the rule as well as identify those areas where more attention is required to ensure a safe workplace. Also, all OSHA-required record-keeping activities should be evaluated for compliance with the minimum standards established in the CFR.

5. *Accident history.* The assessment would not be considered complete with-out an evaluation of the company's existing injury and illness prevention program. This will not only identify those aspects of the program that may be effective, it will also help ascertain those areas where specific improvement may be needed. In this regard, the survey should examine the company's:

Accident injury and illness rates
Accident injury and illness dates
Accident injury and illness types, by location
Worker's compensation costs
Rates of employee turnover and absenteeism
Company policy statements
Documentation on previous safety/health activities
Employee training records
Existing company rules and regulations
Guidelines for proper work practices and procedures
Status of compliance with federal requirements
Other safety- and health-related issues and programs

Once these areas have been properly researched, a site-specific safety and health program can be drafted to address all pertinent issues that must be considered to ensure a safe and healthy workplace.

6.3.5 The Safety Plan

After the rules and regulations have been developed and written into the company safety manual, management must still develop a plan that dictates how the safety program will be implemented. This safety plan describes specific levels of responsibility with regard to workplace safety and health. It establishes accountability through the chain of command for all aspects of the safety program. Beginning with the chief executive and continuing through all levels to first-line supervisors and on to the employees, specific responsibilities for safety program implementation are outlined as company policy. Every single employee in the company should have access to the safety plan to ensure proper dissemination of policy and eliminate any potential for misunderstanding regarding workplace safety and health issues.

The safety plan simply provides clear direction for implementing the safety program rules, regulations, and objectives established in the safety manual. Each document is complementary to the other. Together, they are the foundation of a com-prehensive safety program.

6.3.6 The Benefits of Safety

It is no secret that a company is in business to make a profit. It is the primary objec-tive of any organization to provide quality products and/or services to its identified market at a cost and schedule that are better than any of its competitors. This is the essence of a free enterprise, market economy system. For many years, safety compliance was viewed by industry as a sometimes avoidable obstacle in this process. However, with the right approach, safety can actually be a benefit to companies as they pursue their goal toward perpetual profits. As stated previously, safety should be approached as a team effort, with all elements of the organization working together toward the common goal of providing a safe and healthy workplace. Properly implemented, a comprehensive safety program will pay off in optimized and more efficient production. Typically, there will be fewer work stoppages due to unsafe conditions, injuries, accidents, damaged equipment/tools, and so on. Also, although the initial cost of complying with safety regulations may be considered relatively high, the actual expenditure for the safety effort will reduce over time and therefore effect a similar reduction in long-term implementation costs. Liability insurance, as well as medical and worker's compensation premiums, will also likely be reduced in the long term due to excellent safety performance. Employee turnover may also decline if the workers are convinced that their employer actually cares about safety in the workplace. All of these are examples of the positive impact that safety can have on the bottom line of any business enterprise. Clearly, employers cannot afford to ignore safety as a required aspect of successful business operations.

6.3.7   Summary

People are an organization's most precious resource. Without their collective contributions, success in today's highly competitive market is a hopeless objective. It should follow that the protection and preservation of employee safety and health should be the primary concern of any business endeavor. The federal government certainly maintains this view. In fact, an employer's responsibility to the safety and health of its workers forms the basis of the Occupational Safety and Health Act of 1970. Literally thousands of safety rules and standards are contained in the CFR for Labor (Title 29) which require specific employer actions to ensure employee safety and health. Compliance with these many mandated requirements is rarely achieved without management commitment, which often includes a costly expenditure of resources. But the cost of ensuring safety is, with few exceptions, a great deal less than the cost of ignoring these requirements. To facilitate the implementation of a consistent, comprehensive, and objective safety program, safety rules should be developed and written in a uniform safety manual. The manual must consider general safety requirements, which will typically apply across the organization, as well as very specific safety concerns and issues.

No safety program will succeed unless those elements of the organization responsible for its success are clearly identified. While most companies maintain a safety department within their organizational structure, the responsibility for safety program implementation and enforcement must be a team effort. Safety professionals provide a staff service to line management. They review and research requirements, develop rules and regulations based on those requirements, and recommend implementation strategies to management. They will also conduct assessments to determine program adequacy, and monitor program success through inspections and audits. While the responsibility for actual implementation of the safety program rests with those responsible for daily operations (i.e., the line managers), the safety professional must work together with each functioning element of the organization for the process to be successful. To clarify, these responsibilities should be described in a company safety plan, which establishes clear organizational accountability for safety program implementation.

Strategies for selecting specific rules were also presented and discussed. Without a doubt, every safety program should be different. But for any program to work, there must first be a clear understanding of the reasons for establishing such requirements, a respect for those responsible for the success of the program, and an appreciation for those general and specific safety issues and concerns that each company must identify, understand, and address. Only then will an organization be on its way to achieving a safe and healthy workplace for all its employees.

6.3.8   Bibliography

Boylston, R., *Managing Safety and Health Programs*, Van Nostrand Reinhold, New York, 1990.

Early, M., *The National Electrical Code*, 5th ed., National Fire Protection Association, Quincy, Mass., 1989.

Ferry, T. S., *Modern Accident Investigation and Analysis: An Executive Guide*, John Wiley, New York, 1981.

Larson, M. S. and Hann, S., *Safety and Reliability in System Design*, Ginn Press/Simon & Schuster, Needham Heights, Mass., 1989.

National Safety Council, *Accident Prevention Manual for Industrial Operations*, 9th ed., 2 vols., National Safety Council, Chicago, 1988.

Vincoli, J. W., *Basic Guide to System Safety*, Van Nostrand Reinhold, New York, 1993.

Vincoli, J W., *Basic Guide to Accident Investigation and Loss Control*, Van Nostrand Reinhold, New York, 1994.

6.4   INDUSTRIAL HYGIENE

Paul R. Riedel, certified industrial hygienist, Rockledge, Florida

6.4.0   Introduction to Industrial Hygiene

Of all the resources that a manufacturer has, the most valuable is its people. Without the people, the production stops. Therefore, the primary objective of the manufacturer should be to keep personnel creating, building, assembling, and producing every day, week after week, year after year. The productivity of an employee depends on the state of his or her work environment. If work area conditions are hazardous, or even perceived to be hazardous, morale will suffer and people will not perform effectively. Moreover, if conditions are so bad as to cause worker illness or injury, the consequences may severely impact the company. If the work environment provides a healthful and safe means to allow people to produce, then the employer will achieve profitable results sooner. The successful businesses of the future will be the ones that actively participate in sound industrial hygiene programs that protect their most valuable resource. Another good reason for the company to protect the workforce is the legal requirement. In 1970, the Occupational Safety and Health Administration (OSHA) was created under the U.S. Department of Labor. Many states and municipalities have also promulgated regulations. Failure to comply with legislation can result in fines and criminal penalties and may be very costly to the company.

Hazards are present in many manufacturing processes, some of which are unavoidable. How do you eliminate nuclear hazards in a nuclear power plant, for instance? The answer often is that you cannot eliminate the hazard. The hazard can be controlled, however, so that the potential for injury or exposure is minimized.

This subchapter will discuss the following subjects: What kinds of hazards are present in the work environment? How do such hazards affect personnel? What are the proper ways to protect people from hazards? What programs should the manufacturer implement to ensure employee protection? What regulatory programs are required to be implemented? What is the regulatory agency that enforces laws for occupational health?

6.4.1   History of Occupational Environment

The recording of occupational hazards goes as far back as the fourth century B.C., when Hippocrates described the effects of lead poisoning on miners. The history of occupational hazards describes accidents and deadly diseases due to people's ignorance or indifference to workplace hazards. During the Industrial Revolution, many industrial dangers were created. While public awareness of occupational hazards has increased, work environments have also become more hazardous due to new industrial processes and increased chemical usage. In 1984, more than 2000 people were killed as a result of the release of a deadly chemical at a factory in Bhopal, India. Most catastrophes could have been avoided.

The employee assumes a certain amount of responsibility to ensure that he or she is protected from potential perils at work. However, it is the employer's responsibility to ensure that all employees are protected from occupational hazards that may cause impaired health or injury. These hazards may range from a minor, irritating odor in the office to a potentially life-threatening procedure on the factory floor. Under federal law, severe penalties have been imposed on companies where workers are subjected or exposed to uncontrolled hazards.

6.4.2   Occupational Hazards

Industrial hygiene is the science involved with identifying hazards in the work environment, evaluating the extent of the hazard, and providing recommendations to control such hazards. Simply put, the purpose of a sound industrial hygiene program is to protect the worker. The manufacturing industry involves hazardous work areas and dangerous work processes that can cause serious health effects. Occupational hazards can be divided into four groups: ergonomic hazards, biological hazards, physical hazards, and chemical hazards.

Ergonomics is the science that deals with the interaction of the worker and his or her work environment. Examples of such interaction may be the repetitive use of simple hand tools, tedious assembly-line work, or various activities in a meat-packing plant. An ergonomic hazard can be defined as a process which can cause physiological stress to the human body. For more details on ergonomics see Subchapter 6.1.

Any living organism that can cause disease is a *biological hazard*. These organisms, which include viruses, bacteria, and fungi, can be found anywhere if the right conditions exist. Occupational environments such as in hospitals, slaughterhouses, or farming, in which the source of all kinds of biological agents exist, will have a much higher degree of biological hazards than most manufacturing settings. However, microbial growth may be enhanced in any high-humidity environment. Indoor air-quality problems are often associated with high humidity or improperly designed or maintained air-conditioning systems. Generally, biohazards in the manufacturing plant will be minimized if humidity is controlled and good housekeeping is maintained.

Physical hazards present in the work environment include noise, temperature extremes, ionizing radiation, and nonionizing radiation. These agents at certain levels will severely damage the human body. Noise, which is defined as unwanted sound, will impair hearing to the point of deafness if the noise levels are high enough. One of the dangers of this hazard is that many people do not realize the permanent consequences of working constantly with high noise levels. Many workers think their hearing will return to normal when they leave the noisy work area. In fact, chronic exposure to high noise levels will cause irreversible hearing loss.

Minimizing the hazards of high noise levels can be achieved by engineering controls such as constructing enclosures around the process, isolating the operation from personnel, or utilizing earmuffs or earplugs. A less serious problem than hazardous noise levels is a lower noise range known as speech interference noise. A good example of speech interference noise is a room full of computers. Although it is not damaging to the hearing system, this constant noise may be irritating to personnel and create communication difficulties which may adversely affect productivity. Eliminating speech interference noise levels can be accomplished either by reducing the noise level of the source or by installing acoustical walls in the general area.

Clearly, the danger of working in extreme temperatures can be serious. In the manufacturing world, heat stress is common. Heat stress on the human body is the combined effect of the hot environment and heat generated by the body's metabolism. The heat stress experienced by a worker performing exhaustive work in a hot environment will be significantly more than that of a person doing light work in that same environment because of increased metabolic heat load. Heat stress can cause heat cramps, heat exhaustion, and, most seriously, heat stroke. Ambient temperatures can be controlled by isolating existing hot processes or by providing good mechanical or natural ventilation. To keep the body cool, the metabolic heat load can be lessened by decreasing the workload or implementing scheduled work breaks. Personnel should wear light clothing in hot environments, to allow the body to cool itself. Another very important factor involved with heat stress is humidity. High humidity impairs the human mechanism to cool. Environments with lower humidity will decrease the physical hazards associated with heat stress. Similarly, cold environments can present serious consequences such as frostbite. Reducing the hazard is achieved by increasing the surrounding temperatures or by providing protective clothing for the worker. Certainly, for the sake of productivity, the best solution is to control the temperature, if possible.

Working in slight temperature variations, although not life-threatening, may cause problems. Workers who are simply uncomfortable from being too hot or too cold may be less productive. The American Society of Heating, Refrigerating, and Air Conditioning Engineers (ASHRAE) recommends temperature and humidity ranges for worker comfort in various types of working environments. See Chapter 5 for more information on building design.

The physical hazards of radiation, which are divided into ionizing and nonionizing radiation, may not be as familiar to the reader as other physical hazards. Nonionizing radiation is composed of ultraviolet light, visible light, infrared light, microwaves, and radio frequencies. The sources and uses of nonionizing radiation are becoming more prevalent in industry. The associated hazards for nonionizing radiation are a function of radiation frequency, intensity, duration of exposure, and proximity of the worker to the radiation sources. Overexposure to ultraviolet light can cause severe damage to the skin (i.e., sunburn) or eyes. Besides the sun, sources of ultraviolet radiation in the work environment include electric arc welding, mercury arc lamps, and xenon lights. The infrared energy hazard is heat. Occupational sources include high-temperature processes such as furnace operations or curing ovens. Microwaves and radio-frequency bands are notable in that they penetrate matter, causing a potential hazard to surface organs. The primary effects are thermal in nature.

Ionizing radiation differs from nonionizing radiation in that ionizing radiation has sufficient energy to ionize, or remove electrons from their atomic orbits. The destructiveness of this type of radiation has been seen as a result of nuclear blasts and nuclear accidents, such as in the Chernobyl catastrophe. The danger of ionizing radiation is due not only to the extensive biological damage it can cause, but also to the inability of the senses to detect its presence. Avoiding exposure to radiation can be achieved by reducing the dose, reducing the time of exposure, or increasing the distance from the source.

Typically, the most significant hazards in the work area are *chemical hazards*. Depending on the specific use of the chemical, the hazard is caused by liquid, vapor, gases, dust, fumes, or mists.

What determines the dangers of chemicals? In other words, what is the likelihood that a particular substance will cause damage to the human body? The danger of being exposed to a chemical depends on its properties and the hazard potential associated with its uses.

Significant properties of chemicals affecting safety and health include flammability, reactivity, corrosivity, and toxicity. The first three characteristics are serious safety concerns. The consequential results from flammable materials (e.g., acetone or methyl ethyl ketone) or highly reactive materials (e.g., nitroglycerine or perchloroethylene) are fires and explosion. Corrosive materials such as sulfuric acid, chromic acid, and phosphorus will destroy living tissues on contact.

Toxicity refers to the capability of a substance to cause damage to some biological mechanism. The human body has an amazing system to defend itself from chemicals that have entered the body. Toxic chemicals are either excreted or metabolized in such a way as to render them harmless. However, if the body is overburdened with a chemical, then damage will occur. Brief exposure to a highly toxic chemical or longer exposure to a less toxic chemical may cause severe and irreparable damage.

Some chemical hazards are obvious to the senses. For instance, you can immediately detect the suffocating odor of ammonia in a room, even at low concentrations. However, many hazards exist that you cannot see, smell, or hear. If you enter a room that has depleted oxygen levels, for example, you may not notice anything unusual, although in fact you could be in a deadly atmosphere. Some hazards can cause short-term health effects, while exposure to other hazards can occur for years before the onset of health impairment or disease. Exposure to ammonia or chlorine gas, for instance, can quickly cause burning of the eyes and throat. However, after a short period of time, the symptoms will subside. On the other hand, chronic exposure to low levels of lead fumes or some chlorinated solvents such as trichloroethylene may go unnoticed until irreversible health problems occur long after the initial exposure.

When discussing the dangers of chemicals, it is important to understand not only the physical properties of the materials, but also the hazard potential involved. *Toxicity* refers to how much damage a substance causes once it comes into contact with the human body. The *hazard potential* refers to how easily the material can get to the human body. This potential is determined, in part, by the particular process being used. For example, one process may call for heating the chemical in an open vat with frequent involvement by the worker. On the other hand, a second process may call for the same material to be heated in a closed system, without direct worker involvement. Under the latter scenario the hazard potential is much less than the former scheme. Even if a more toxic substance were used in the second situation than in the first, the hazard potential could be less. The open-vat heating process vapors could easily come into contact with the worker.

Another contributing factor to the hazard potential is the chemical's evaporation rate. Chemicals with high vapor pressures will volatize or evaporate faster, creating vapors in the air and thus increasing the potential for personal exposures. Other important hazard factors include the duration of the process, the general conditions of the room in which the process is located, and the health of the worker involved in the process. Obviously, longer processes increase the chance of worker exposure. An area that is confined or that has limited ventilation would contribute to a more hazardous situation than an area with good mechanical ventilation or good natural ventilation. Finally, the health of the worker is a very significant hazard factor. An individual who is already in poor health will be more susceptible to the external influence of chemicals. Control of these many factors contributing to hazard potential can minimize worker exposure.

Route of Entry

To understand these hazards, one must understand the various routes of entry that the hazardous material can take to inflict bodily damage. If the particular hazard never enters the body, no injury or illness will occur. The four major routes of entry are inhalation, absorption, ingestion, and injection.

The most significant route of entry is by *inhalation*. One reason is due to the rapidity that chemical vapors, mists, dust (including radioactive particles), or fumes can enter the body via the lungs and be absorbed into the bloodstream. Another reason for the importance of inhalation is that the worker can easily become exposed even though the source of the hazard is at a distance. As contaminants infiltrate the air, the hazard can expand throughout an entire building.

Absorption is an important route of entry that is responsible for causing skin diseases, which are the most frequent type of occupational illness. Some chemicals, such as phenol or toluene, are readily absorbed through the skin.

Ingestion occurs if the worker places contaminated hands to his mouth (i.e., eating or smoking). *Injection* of a chemical, usually from the result of an accidental spill, is rare in a manufacturing plant.

Airborne Contaminants

Airborne contaminants are gaseous or particulate. Gaseous contaminants include vapors from the evaporation of a liquid, and gases such as nitrogen, which are in a gaseous state under normal temperature and pressure. Particulate contaminants include dusts, mists, and fumes. Dusts are created from mechanical processes such as grinding or sawing. Mists are suspended liquid droplets that are generated from the condensation of vapors. Fumes are airborne solid particles that have condensed from volatized solids.

Control of Occupational Hazards

Recognizing hazards in the manufacturing industry is the first step in protecting workers, followed by evaluation or measurement of the extent of the hazard. The next step is to control such hazards by following an orderly approach and sound industrial hygiene principles. This approach should evaluate (1) substitution, (2) engineering controls, (3) administrative controls, and (4) PPE.

The best approach, from a health standpoint, is to use a material that is less harmful to the worker. Historically, manufacturers have used chemicals in specific processes that produce the most economical product. That process, however, may have exposed workers unnecessarily to highly toxic chemicals. The implementation of less toxic chemicals with minor process changes can often accomplish the desired results.

When substitution is not possible, engineering controls should be implemented to control the hazard locally. A pneumatic clipper or a punch press, which produces hazardous noise, for example, cannot be substituted. The use of engineering controls to abate the high noise may include isolation of the noise source away from personnel or surrounding the noise source with acoustical materials.

Local exhaust ventilation is another example of engineering controls. The principle is to contain and exhaust contaminates before they reach the worker. Engineering controls also include changing the original process to minimize the hazard. Brush application of chemicals, for example, is much less hazardous than spray application.

Administrative controls are used to protect not only workers involved specifically in hazardous operations, but also personnel in the general area who are not involved directly with the process. These control measures include personnel training in how to avoid exposure by using proper operating procedures. The rotation of employees to minimize exposure is another method.

The last mechanism of protecting the worker is the utilization of PPE such as respirators and protective clothing. PPE should be used when potential exposures exist and engineering controls are not feasible.

Process Design

As in every manufacturing process, the design phase must be complete and accurate before the construction phase begins. With regard to the sound industrial hygiene principles noted above, the manufacturing engineer should design the project or process to ensure that the welfare of the worker is protected and worker exposure is minimized. General ventilation and local exhaust ventilation should be designed to adequately control toxic vapors or particulates. Dangerous processes should be designed to be isolated from nonessential personnel. A less toxic material is the preferred material to be used. Finding an innocuous material that will adequately satisfy the job specification may require technological changes in the process, which could be expensive. However, using the safer product may be prudent. The use of asbestos has cost manufacturers millions of dollars in liability claims and reconstruction fees, in addition to loss of lives.

6.4.3   Implementing an Industrial Hygiene Program

Every manufacturing business with potential hazards, regardless of size, should implement an industrial hygiene program that is proactive. A properly implemented proactive program will prevent exposures and ensure that workers are adequately protected. For everyone, the goal should be to *control the hazard*. This program should, as a minimum, provide the following.

Routine Facility Inspections

Routine inspections, documented by the supervisor, should be performed on a scheduled basis to identify and evaluate hazardous or potentially hazardous operations. Although many work processes change very little, poor habits may develop that potentially spell trouble. Periodic evaluations of the operation from start to finish will uncover undesirable practices. The goal of the inspection should be "How can we do this more safely and efficiently?" not "Who is not working safely?" These inspections should also have full participation of the workers involved in the process, who understand what the real hazards are. Encouraging employees to provide suggestions and solve problems will increase the success of the industrial hygiene program.

Personnel Training

Training sessions should be given to every worker involved in hazardous or potentially hazardous operations. The goal of the training should be to prevent exposures by thoroughly identifying all potential and real hazards involved in the employees' work processes. This training should be performed by an individual who is knowledgeable in operations as well as in industrial hygiene. Sessions can be given in a classroom atmosphere or at the site of the hazardous process. Many companies now are utilizing videotapes to achieve personnel training. An advantage of videotapes is the relatively low cost. Videotapes, however, can be limited due to the one-way communication. Personal interaction between the instructor and the student enhances comprehension. Videotapes can be used in conjunction with interactive training sessions to ensure that the proper level of training is achieved. The employer should never conceal anything regarding occupational hazards from the employee. It is important that the employee has complete trust in the company when it pertains to something that may affect his or her health.

Exposure Monitoring

During hazardous operations, engineering controls and/or PPE may be utilized. Exposure monitoring can determine what airborne contaminants may be present in the worker's "breathing zone" during hazardous operations. For hazardous operations it is important to document that the proper level of protection is actually being utilized by employees.

Contingency Plans

The best way to handle exposure incidents or accidents is to prevent them from occurring. This, unfortunately, is not a perfect world. Episodes will occur during hazardous processes. A contingency plan should be implemented that precisely establishes the proper steps and procedures to take immediately after mishaps, spills, or accidents. Periodic simulated drills should be run to ensure that things go smoothly during actual emergencies. Formalized contingency plans and associated training will greatly reduce the likelihood of catastrophic results.

How to Implement These Programs

What is the best way to ensure that the proper industrial hygiene programs are in place? Who is going to implement and maintain these programs? One way is to keep an industrial hygienist on staff. Another way is to hire an industrial hygiene consultant. Finally, the company may rely strictly on assistance or information obtained directly from OSHA.

There are advantages to retaining an industrial hygienist (IH) on staff. The IH is readily available to provide technical knowledge required in setting up good processes in the first place, and as a point of contact in case of contingencies or problems. The staff IH will be familiar with the company's work processes and knowledgeable about the associated hazards. The IH on staff can build a working relationship with the employees and is able to reinforce the importance of following sound industrial hygiene principles.

The services of a consultant can be on a one-time basis or periodic, depending on the complexities of the company operations. An advantage of using a consultant over having an industrial hygienist on staff is lower cost. The consultant provides the necessary services only when requested. If the company is fairly small and simple, maintaining an IH on staff may not be necessary.

General industrial hygiene information can be obtained by contacting OSHA regional offices. The larger manufacturing companies typically have a number of potential hazards, and find that it is imperative to have an industrial hygienist on staff.

6.4.4   OSHA Programs

The subject of industrial hygiene would be incomplete without a discussion of OSHA. The purpose of OSHA, as stated in the act, is "to provide for the general welfare, to assure so far as possible every working man and woman in the Nation safe and healthful working conditions and to preserve our human resources." This act, which applies to all private employers, has broad powers of enforcement to reduce occupational hazards.

To carry out the mandates under the Occupational Safety and Health Act, OSHA has (1) established standards for hazards or procedures, and (2) prescribed the necessary means to protect employees by referencing such standards. These health standards are called permissible exposure limits (PELs), and are used to identify personal exposures. These PELs are based on research and experimental data from animal studies, human epidemiological studies, and studies of the workplace. If personnel are exposed to airborne concentrations of contaminants exceeding the PEL, a personal exposure is considered to have occurred.

OSHA has established design criteria for industrial operations in addition to health standards. For example, OSHA defines the required ventilation rates for the design of various local exhaust ventilation systems. They have established procedures to enforce these standards to reduce hazards in the workplace. Such procedures include their on-site inspections and company record-keeping requirements.

Not all aspects of the manufacturing industry are covered by specific OSHA standards. For example, no OSHA requirement exists for indoor air quality in offices. Guidelines and parameters that define indoor air quality have been written by scientific and engineering organizations. These guidelines, although not mandated by law, should be adhered to in order to ensure a healthful and safe environment. Where OSHA may not address some specific hazards, the act includes the General Duty Clause, which states, "Each employer shall furnish to each of his employees employment and a place of employment which are free from recognized hazards that are causing or are likely to cause death or serious physical harm to his employees." The employer has the onus of ensuring worker protection from hazards that OSHA has not specifically addressed. OSHA continues to create new standards and revise existing ones. Discussed below are important standards that have been implemented by OSHA. The manufacturing engineer should be familiar with existing applicable OSHA standards and be aware of newly created or revised standards as they appear.

Hazard Communication Standard (29 CFR 1910.1200)

The purpose of the Hazard Communication Standard is to ensure that all employees, in both the manufacturing and nonmanufacturing sectors of industry, are properly informed of the hazards used at work and how to protect themselves from such hazards. The standard establishes that chemicals produced or brought into this country must be evaluated as to their hazards. Manufacturers must provide the pertinent hazard information on the chemicals to their customers, who in turn must ensure that their employees are knowledgeable about the hazards. This means that the employee has the right to know of the hazards involved with the chemical he or she is using. Failure on the part of the employer to implement a hazard communication program can result in serious penalties as well as avoidable exposure incidents.

The basic parts of a hazard communication program are (1) a written hazard communication program, (2) a list of chemicals used in the work area, (3) an MSDS for each chemical used, (4) proper labeling of all chemicals used, and (5) training of all personnel who have a potential exposure to hazardous chemicals.

Respiratory Protection (29 CFR 1910.134)

The Respiratory Protection Standard requires employers to provide respiratory protection to employees when such protection is necessary to protect the employee from chemicals or hazardous atmospheres. The standard mandates the following: (1) written operating procedures on the proper selection and use of respirators based on the hazard involved; (2) training of individuals on the use, limitations, and care of respirators; and (3) routine evaluation of the hazardous operation including processes, workers' duties, and location of the hazardous area.

Noise (29 CFR 1910.95)

The Noise Standard is designed to protect workers from hazardous noise levels. OSHA has established PELs based on a combination of noise levels and duration of exposure. The OSHA PEL decreases as the length of time of exposure increases. For example, an 8-hr. exposure of 90 decibels (dBA) would be equivalent to a 2-hr. exposure at 100 dBA. The standard requires that engineering and administrative controls be implemented to keep exposure limits below these established PELs. The standard also requires that workers who are exposed to the action level of 85 dBA for 8 hr. be enrolled in a hearing conservation program (HCP). The HCP includes exposure monitoring, audiometric testing, hearing protection, employee training, and record keeping.

Confined Space (29 CFR 1910.146)

The Confined Space Standard is designed to protect workers entering confined spaces. A confined space is a tank, vessel, silo, vault, or pit that (1) is large enough for any part of the body to enter, (2) has limited means of egress, and (3) is not designed for continuous occupancy. The standard has identified two types of confined spaces: nonpermit and permit-required. Nonpermit confined spaces do not contain any real or potential health hazards. Permit-required confined spaces, however, have one or more of the following characteristics:

1. A potential for a hazardous or toxic atmosphere
2. A potential for an explosion or flammable atmosphere
3. A potential to have an abnormal oxygen concentration (<19% or >23.5%)
4. A potential hazard of engulfment
5. Poor ventilation

The standard is lengthy and comprehensive, but the basic program elements include:

1. Identifying any confined spaces
2. Identifying permit-required and nonpermit confined spaces
3. Establishing entry control measures
4. Establishing a written confined space program that includes permitting procedures, monitoring procedures to determine if the area is safe to enter, and emergency procedures for rescue
5. Employee training of attendants, entrants, and authorizing persons

Record Keeping

According to OSHA (29 CFR 1910.20), employee medical and exposure records must be kept by the employer. These records include environmental monitoring, MSDS's, and biological monitoring. All employers are required to provide each employee access to his or her medical records. Medical records must be kept for the duration of the employment plus 30 years. According to OSHA (29 CFR 1904), employers with more than ten employees must maintain records (OSHA No. 200 Log) of occupational illnesses and injuries. An incident that occurs on work premises is considered work related. The log, which must be retained for 5 years, tracks the date of the incident, the individual who received the illness or injury and his or her job, the kind of injury or illness, and lost work time. All employers are required to notify OSHA in the event of a fatality or the hospitalization of more than four employees.

OSHA Inspections

OSHA has the authority to inspect companies without advance notice. The employer has the right to refuse the OSHA compliance officer initial entry—but denying access is not recommended. OSHA will take appropriate legal procedures to gain access, and often send a team of inspectors, rather than a single inspector, to proceed with the inspection. It may be best to be cooperative, and have a contented compliance officer inspect the premises, than a team of OSHA inspectors finely examining the premises, thinking you are trying to hide something.

The standards established by OSHA are minimum requirements. Complying with these regulations will satisfy legal obligations. However, in today's competitive manufacturing industry, meeting legal requirements will not be enough. The successful manufacturing engineer will be the one who takes extra steps to stay ahead of the competition. These extra measures include an interactive and comprehensive industrial hygiene program designed to ensure that the employees are working in a safe and healthful environment that enables workers to achieve maximum productivity.

6.4.5 Bibliography

All about OSHA, U.S. Department of Labor, 1985.

ASHRAE Handbook, Fundamentals, I-P ed., American Society of Heating, Refrigerating, and Air Conditioning Engineers, 1989.

Fundamentals of Industrial Hygiene, 3rd ed., National Safety Council, 1988.

The Industrial Environment—Its Evaluation and Control, U.S. Department of Health and Human Services, 1973.

Industrial Ventilation: A Manual of Recommended Practice, 21st ed., American Conference of Governmental Industrial Hygienists, 1992.

Threshold Limit Values for Chemical Substances and Physical Agents, American Conference of Governmental Industrial Hygienists, 1993.

7 Computers and Controllers

Allen E. Plogstedt

with

Marc Plogstedt

7.0 INTRODUCTION TO COMPUTERS AND CONTROLLERS

Businesses today are going through a fundamental shift in the way they rely on information systems to improve their competitive advantage. Reengineering of the total business process makes these new processes more dependent on their information systems. The design of the new information systems could mean the difference between being first in class and being a so-so competitor. Information at the right place and the right time will allow decisions to be made at the lowest possible level in the organization, with the best results.

The mainframe computer system was the only choice of system design available in the 1950s and 1960s. Minicomputers became available in the 1970s, and the personal computer (PC) arrived in 1981. The explosion of PCs and network capabilities in the late 1980s has now provided the tool for businesses to expand the use of computers in an economical way.

7.0.1 Information Systems and the Manufacturing Engineer

The reliance of business today on their information systems will change the role of the manufacturing engineer (ME) dramatically. The ME will be asked to become part of this new order. As part of the new team approach to product development, MEs will need to have skills in information flow just as much as skills in managing the work flow in a manufacturing process. In progressive organizations, the ME may be part of a team to restructure the total manufacturing information flow and the integration of data into a computer-integrated enterprise (CIE). This could affect how the ME generates, controls, and displays process planning to the workforce. Computer-aided process planning (CAPP) is being used by many organizations to speed the generation and distribution of planning to the floor. This can be a paperless process but is more likely to be a "paper-sparse" system. CAPP has shown dramatic process improvements (20 to 50%) in productivity. MEs must be knowledgeable about the interrelationships between ease of training and use of CAPP systems, and the requirements that CAPP puts on the CIE. The technology exists for presenting multimedia CAPP work instructions to a workforce. However, there is a cost for each of the features of that presentation.

Rework and scrap are non-value-added results of poor process control and mis-information. Correct, current information will eliminate most, if not all, of the errors that can creep into a process, either in business or manufacturing areas. To be first in class, a CIE must provide data to and from customers, vendors, and across the total factory. To accomplish your goals of meeting and beating the competition, the ME must look at all the factors that will affect price (and profit). Inventory and work in process must be reduced to an absolute minimum. This can be accomplished by the judicious design of computer interfaces (electronic data interchange, EDI) with ven-dors. With a material-requirements planning (MRP) system, a just-in-time delivery system can and should be used to reduce inventory and work in process.

7.0.2   Definitions of Terms for Information Systems

Application	A software program that carries out useful tasks
Architecture	The manner in which applications, software, and hardware are structured.
Batch	Sequential off-line processes.
Central processing unit (CPU)	The part of the computer in which the fundamental operations of the system are performed
Computer-integrated enter-prise (CIE)	The complete assemblage of computers, network, and applications that support the entire manufacturing enterprise.
Client	The networked machine requesting services.
Data	Recorded information, regardless of form or method of recording.
Data dictionary	A repository of information describing the characteristics of the data elements.
Document image file	A digital-file representation of a human-interpretable document. Examples are raster image files and page-description language files.
Document image standard	A technical standard describing the digital exchange format of a print/display file.
Electronic data interchange (EDI)	A series of standards (X12) for the format and transmission of business-forms data, such as invoices, purchase orders, or

electronic funds transfers. EDI is intended for computer-to-computer interaction.

Government open system interconnection protocol (GOSIP)

A government standard that explicitly identifies the standards that apply to each of the 7-layer OSI model telecommunications protocol stack.

Graphical file

Files containing illustrations, design data, and schematics, in processable vector graphics or nonprocessable raster graphics.

Integrated data file

A digital data file that integrates text, graphic, alphanumeric, and other types of data in a single (compound) file.

Local area network (LAN)

A LAN is geographically the smallest of the network topologies. The LANs have three topologies in general use: the star, bus, and ring.

Mainframe computer

The processor can be configured to receive inputs from all the various devices and networks in your CIE. All of the application manipulations are accomplished by the CPU.

The major features of the mainframe is the very large input/output (I/O) capability built into the processor. The random-access memory (RAM) in the CPU can be partitioned into various configurations to best operate with your applications. In most cases the computer installation requires special facilities, a controlled environment, and an uninterrupted power supply.

Minicomputer

Smaller, cheaper (than the mainframe), stand-alone computer that did not need all of the support systems to operate. In almost all cases these machines were used for a specific stand-alone application or process, with little or no networking.

Motif

A de facto standard graphical user interface that defines the look and feel for systems that utilize it.

Networks	Networks consist of the computers to be connected, the communication media for the physical connection between the computers, and the software for the network communications and management.
Personal computer (PC)	The PC evolved along two lines. Apple introduced the first practical PC in the mid-1970s. The IBM-compatible PC was introduced in 1981.
Protocol	Provides the logical "envelope" into which data is placed for transmission across a network.
Reduced instruction set computing (RISC)	A computer chip architecture that requires all instruction be executable in one clock cycle.
Wide area network (WAN)	The WAN has the capability to allow you to connect your LAN to other LANs within your company or with LANs of vendors and customers. The speed of the WAN will be determined by the communication media used for the long-distance transmission.
X.400	A standard e-mail addressing scheme that is also the e-mail standard identified in GOSIP.
X-Windows	A public-domain, standard software tool kit, developed and distributed by MIT, on which windowing applications can be built.

7.1 MAINFRAME SYSTEMS

The first electronic (electromechanical) computer was designed and built during World War II. In 1945, the Electronic Numerical Integrator, Analyzer, and Computer (ENIAC) went into use, and was used until 1955. Most of the analysis time for the ENIAC was spent on calculation of ballistics tables for artillery. Because of the extensive labor needed to operate and maintain the system, it had limited use in the business world.

The first commercial, programmable, electronic computer, the IBM 701, was introduced in 1953. Most large corporations began to rely on the mainframe computers to process their mission-critical business applications. As the need for more and better business data expanded during the last three decades, the capabilities of, and the dependence of business on, mainframe computers expanded at the same or a higher rate. These business applications are sometimes referred to as "legacy software systems." As the needs for other functions of the organization to use more and more of the capabilities of computing increased, ways and means to integrate these new systems into the legacy systems had to be developed. The concept of this new architecture evolved into the CIE, as discussed in Subchapter 7.5.

7.1.1 System Layout

The general layout of a typical mainframe system is shown in Figure 7.1. This is one of the simplest architectures to implement. The system consists of various modules connected together to form the complete mainframe computer system. The connection to local area networks, wide area networks, and any remote sites is made through the front-end communications processor. This processor can be configured with various cards to receive inputs from all the various devices and networks in the CIE. All of the application manipulations are accomplished by the central processing unit (CPU). The major feature of the mainframe is the very large input/output (I/O) capability built into the processor. A very high-speed bus connects the 64-bit processor, memory, and I/O bus together for the most efficient operation. The random-access memory (RAM) in the CPU can be partitioned into various configurations to best operate with your applications. All of the peripherals are interfaced through controllers. Some of the controllers also contain memory for faster operation. An example of optimizing techniques that will improve the overall performance of the mainframe computer is setting up disk storage units for specific elements of data. The index is to be stored on one disk, the library of subroutines, and the data is stored on individual disk units. This reduces the retrieval time for the data and speeds up the whole process.

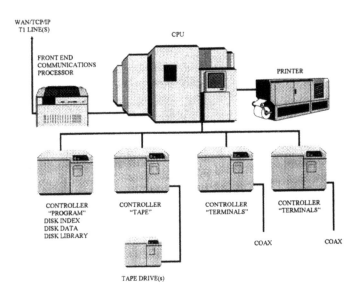

FIGURE 7.1   Mainframe system layout.

The mainframe is connected to each user terminal via coaxial cable. In most cases the computer installation requires special facilities, a controlled environment, and an uninterrupted power source. As the need for additional terminals arises, more coax cables, and controller boxes at the computer, must be added. The information is transmitted from the mainframe to the terminals under IBM's System Network Architecture (SNA), and since most installed mainframes are IBM, this has become the de facto standard. All processing is accomplished at the CPU. The terminals are considered "dumb" terminals. All of the screens are generated by the CPU, and little if any capability exists at the terminal to do any processing. In some advanced terminals for this type of architecture, you can get some graphical capabilities (the regular terminals are character only).

7.2 MINICOMPUTERS

During the first decade of the computer revolution, the mainframe had the computing field to itself. In most cases the programs were built for the business applications of the corporation. The need for productivity improvement on the factory floor caused the manufacturing side of the business to keep the pressure on, and productivity improved at a steady rate (2 to 4% per year). The productivity in the remaining portions of the business (planning, support, and engineering) did not show steady improvement in productivity. The pressure on the other functions of the organization set the stage for the second phase of the computer revolution, the minicomputer. This arrived in the 1960s when DEC introduced computers such as the DEC PDP-8 and later the VAX series. The cost of these computers was so much less than the mainframe that decisions on what and when to buy minicomputers could be handled at a group or department level, not at the corporate level. Minicomputers began to move into many areas of the organization; manufacturing, engineering, and planning represent only a few. In almost all cases these machines were used for a specific, stand-alone application or process, with little or no networking. The productivity of the support functions, even after the introduction of the minicomputer, still did not show the steady improvement that was accomplished in manufacturing. In retrospect, one of the main reasons that the improvements did not materialize was "islands of automation." Each of the organization entities did its own computing, but each could not transfer data to other groups. This failing could not be remedied in a cost-effective way until the introduction of two new technologies, the PC and the network.

7.3 PERSONAL COMPUTERS

The first PC to have any real impact on computing was the ALTAIR, available in kit form from MITS in 1974. The system used an Intel 8080 processor and 256 bytes of memory. The true beginning of the PC was the marriage of Microsoft and IBM to develop the first PC machine in 1981. The Intel 8088 processor was the heart of the machine. The IBM XT was a 16-bit machine with a minimum of 64K of RAM and one disk drive. The IBM AT was introduced in 1982 using the 80286 processor.

The PC industry practically exploded due to the fact that IBM set up a series of PC standards and allowed other manufactures to build PCs to the standards. Standards were so important because they allowed software manufactures to produce software for IBM-type machines that would run on any "clone." Over 100 million PCs have been built since 1981, and the production rate is increasing.

The opportunity of organizations to increase personal productivity was such a powerful lure that workers would do almost anything to get a PC for their own use. This allowed all departments to improve their capability to look at their data and manage their part of the business. The problem was that what was good for the financial department might be bad for the manufacturing department. They had no convenient way of transferring and using each others' data. Thus the original PC revolution did not meet the expectations of management for the total productivity improvements. In some cases the isolated use of the PC made communications between groups even worse than it had been before. At the time of the development of the first IBM PC, the cost of quality peripherals was exorbitant. No networking methods were built into the MS-DOS (Microsoft Disk Operating System) operating system available at the time. To help in this situation, "sneakernet" was used because nothing else existed. In this case data was copied from one PC onto a disk and transported by hand to another machine. A major problem was ensuring that documents were using up-to-date data. Just as important was how documents could be kept from being stolen. Printers could also be shared by more than one PC by the use of a data switch. This was not very fast, however, and did take a large number of hardwires for the few PCs that could be connected to one printer. The real move forward was the development of MS-DOS 3.1 and the development of other networking systems (see Subchapter 7.4.)

7.3.1   Systems and Processors

The PCs evolved along two lines. Apple introduced the first practical PC in the mid-1970s. The IBM-compatible PC was introduced in 1981. The general layout of the PC is shown in Figure 7.2. The performance of the PC depends on the processor type, processor speed, bus type, cache (how much), and graphical interface. Over the years, many benchmark programs have been developed to test the performance of the various subsystems of the PC.

Apple

The Apple PC was designed from the start as a graphical interface. The machine was based on the 6502 processor. The "drag and drop" characteristics of the interface were patterned after the research work by Xerox. The users of this interface did not need much training to become proficient in the use of applications. Users did not have to use any command-line instructions to operate the applications.

Intel

The first Intel microprocessor was designed in the early 1970s. The first unit (4004) had about 1500 transistors on the chip. The first practical microprocessor from Intel

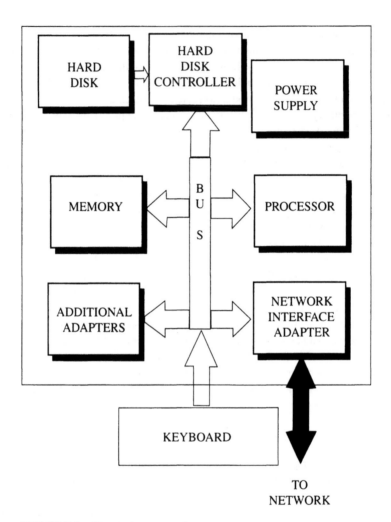

FIGURE 7.2   Personal computer layout.

was the 8080 in 1974. This chip contained about 8000 transistors. The Pentium processor, released in 1993, contains over 3 million transistors on the chip. The evolution of the Intel microprocessor has for many years doubled the processing power
every 18 months. A curve showing the growth in millions of instructions per second
(MIPS) is shown in Figure 7.3. The 8080 could process only about 1 MIPS. The
Pentium can process over 100 MIPS.

When IBM made the decision to release the design specifications of the PC,
a myriad of other vendors began building IBM-compatible PCs. This drove the
expansion of the IBM PC into high gear. In 1993, about 85% of the millions of PCs
used Intel processors. Other microprocessors are appearing in the marketplace and
became a significant source in the late 1990s. In most cases these processors use a

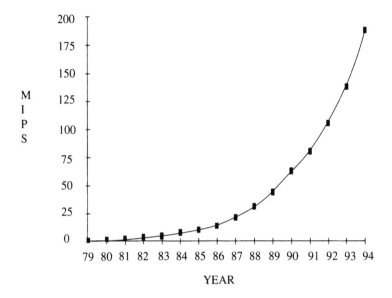

FIGURE 7.3   Microprocessor evolution.

reduced instruction set computers (RISC) structure. The advantage of this type of machine is that it can operate on more that one set of instructions during the processor clock cycle.

7.3.2   Operating Systems

Computer software can be divided into two main categories: system programs, which manage the operation of the computer, and application programs, which solve problems for the user. The operating system is one of the fundamental system programs. The operating system has two main functions: providing the programmer with a "virtual machine" for ease of programming; and acting as a "resource manager" for the efficient operation of the total system. Some of the packaged operating systems also include a graphical user interface (GUI). The GUI provides the user with a simple-to-learn screen system that allows the program to be used with little or no instruction or programming knowledge. Examples are Windows 3.X from Microsoft for DOS, Presentation Manager for OS/2, and X-Windows or Motif for UNIX. Figure 7.4 shows how the various layers of the computer system are interconnected with each other. Without the operating system shielding the programmers from the mundane tasks accomplished by the operating system, the development of new applications would be very difficult and time-consuming.

The development of operating systems has progressed through four phases over the last 50 years. The first phase was from 1945 to 1955. As discussed in Subchapter 7.1, the first computer was developed during World War II and came into use in 1945. The operating system (it did not have a name at the time) for this massive computer was

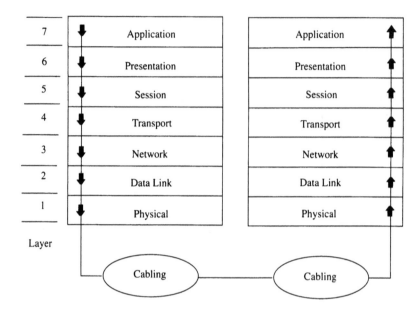

FIGURE 7.4 International Standards Organization (ISO) Open System Interconnect reference model.

accomplished by plug boards. Each user had a plug board with which the operation of the very primitive computer could be controlled. If one of the vacuum tubes failed (which they did regularly), the program would not run correctly, and the user would have to start over after the machine was repaired. Most programs at that time were straightforward numerical calculations. Punched-card input devices were added in the early 1950s. Both programs and data were inputted via punched cards. Imagine what would happen if one of the program cards got out of order.

The invention of the transistor in 1947 and its introduction into computer systems in the mid-1950s changed the picture radically. With the size reduction and the reliability improvements of the transistor, the computer became a viable commercial product. The operating systems of the second phase (1955 to 1965) supported large mainframe university and corporate computer complexes. In this era, programs would be sent to the computer complex for processing. Since most of the time involved in using the computer was not in the processor, but in the I/O portion of the program, the use of batch processing came into vogue. The programs would be placed into a queue, and the processor would work on them in sequence. The input (to tape) was handled by a separate computer designed for that process, not the main processing computer. When the main processor had some free time, the next program would be read off the tape and the processor would carry out the program. After the program was complete, the output would be transferred to an output processor to print or store the output. In this way the main processor would be utilized to it fullest extent and its high cost justified.

The next phase in the development of operating systems occurred from 1965 to 1980. This phase was also driven by the advent of new hardware—the integrated circuit (IC). The IC allowed the mainframe computer to be enhanced to the point where it could handle multiple jobs in the processor at the same time via multiprogramming. When the processor paused for the input or output device to respond, the CPU would run part of another job for greater utilization. The other use of the IC was in the development of the minicomputer, such as the DEC PDP series of computers. These relatively low-cost machines found their way into engineering departments for calculations so the engineers no longer had to wait for the mainframe to be available. This is the time frame in which the UNIX operating system had it beginnings.

The introduction of the PC by IBM in 1981 started a revolution in the use of computers and the beginnings of distributed processing. Apple computers, on the market since 1977, had made inroads into the office environment. Wang had also developed a very good line of machines for the office environment. The problem with both of these computer systems in the long run was the proprietary operating systems and application software. The decision by IBM to make their computer design available to anyone was the key to the incredible expansion in the computer industry. Three major operating systems have been developed for IBM-compatible PCs: MS-DOS, OS/2, and UNIX.

The interface between the operating system and the user application is defined by a set of "extended instructions." The application invokes these extended instructions by the use of system calls. The system calls create, delete, and use various software elements managed by the operating system. In this way the application software, through these system calls, can provide the various functionalities of the computer to the application. While not an integral part of the operating system, important functionality is provided to the system by the editors, compilers, assemblers, and command interpreters.

MS-DOS

When IBM set out to develop the first true PC, the operating system in vogue at the time was CP/M from Digital Research. A new version of CP/M was in the works at the time IBM and Microsoft were developing the PC, but development was behind schedule. IBM had set the date for introduction of the PC, so another operating system had to be found. Tim Paterson of Seattle Computer Products had a CP/M-like operating system that was used for testing memory boards. The product, 86-DOS, was modified by Microsoft and Paterson and became MS-DOS. The first version of MS-DOS occupied 12K of the IBM's 64K memory. The design of the IBM PC set up a separate use of hardware ROM for the BIOS (basic input/output system). The BIOS contained the standard device drivers, thereby allowing MS-DOS to call them to do I/O. MS-DOS 6.0 now requires 2.4 MB of hard disk memory instead of the 12K in Version 1.

To operate MS-DOS, the user must invoke commands to the system via the command line. Through the command line one accesses the file system, system calls, utility programs, and other features. These procedures are similar to those in UNIX but more primitive. MS-DOS is not font case sensitive, as is UNIX. The user of MS-DOS must be familiar with the various command line calls and the format required for the operation of the system.

MS-DOS is not a multitasking system like OS/2 or UNIX. Normally the operating system will support one application at a time. After MS-DOS was in the hands of millions of users in the 1980s, it became obvious that if IBM and Microsoft were going to expand the PC into other uses, it needed an easier, more user-friendly interface. In the late 1980s, Microsoft and IBM were jointly developing the next operating system for the IBM PCs, OS/2. The first version of OS/2 was used on critical process computers, but did not enter the nonindustrial market. In parallel, Microsoft developed Windows, which is a GUI running on MS-DOS. This window environment gave the user some of the functionality of the Apple interface. When Windows 3.1 was introduced in 1991, it became an instant success. Windows supports cooperative multitasking and isolates the user from most of the command-line interface of MS-DOS.

OS/2 2.x

OS/2 2.x is a second-generation replacement of IBM's PC DOS and Microsoft's MS-DOS operating systems. One of the main advantages of OS/2 2.x is that it frees the user/developer from the constraints of the DOS environment. The amount of RAM available to run applications is one of the greatest improvements. OS/2 2.x is not an extension of DOS; rather, it is a completely new operating system that was designed to operate mission-critical business applications. The major advantages of OS/2 2.x are:

> System integrity for applications
> Preemptive multitasking and task scheduling
> More available memory to run DOS programs (~620K)
> Multiple concurrent DOS sessions
> Ability to run concurrently OS/2, DOS, and Windows
> Virtual memory
> Fast, 32-bit architecture
> Fast disk file access
> High-performance file system (HPFS)
> Presentation manager
> National language support
> Object-oriented workplace shell (WPS)
> Ability to run on IBM and IBM-compatible hardware

Presentation Manager is included in the OS/2 operating system. This GUI provides the user with icons and drag-and-drop capability.

UNIX

UNIX is an interactive, time-sharing operating system. It was designed by programmers for programmers. It is not an operating system for the unsophisticated user. When the application shields the user from the operating system via a windowing environment, the system can be used by anyone. The UNIX operating system is generally used in the software development community and in file servers. The major advantage of UNIX over MS-DOS and OS/2 is its capability for multiple concurrent users. The buildup of a UNIX system is shown in Figure 7.3.2.3–1.

Another feature of UNIX not present in MS-DOS is its security system. Logging on to the system requires both your name and a password. This allows your files to be accessed only by you. As in MS-DOS, the interface with UNIX (without a GUI) is through the command line.

The UNIX operating system has been in use since 1974. Thompson and Ritchie of Bell Labs were the originators of UNIX. Thompson also wrote the C programming language for the development of UNIX. The development of UNIX over the years has followed a very crooked road. The original UNIX, licensed by AT&T, has been the basis for all the developments in UNIX. One of the major problems with UNIX is the fact that the machine-dependent kernel must be rewritten for each new machine. The machine-independent kernel is the same for all machines. Many universities used and modified the UNIX operating system. One of the major developers was the University of California at Berkeley, which caused problems for UNIX users and applications developers because their applications would not run on all versions of UNIX. This is a continuing problem. The first real effort to standardize UNIX was by the IEEE Standards Board. The name of this project was the Portable Operating System for UNIX (POSIX). This helped, but it did not solve the problem. Two groups of developers have been set up and are still competing: the open system and the basis protocol for the network that became TCP/IP (see Subchapter 7.9).

7.4   NETWORKS

7.4.0   Introduction to Networks

Networks consist of the computers to be connected, the communication media for the physical connection between the computers, and the software for the network communications and management. Until the development of MS-DOS 3.1, PCs did not have the capability to utilize the full capability of a network. The capabilities added to MS-DOS 3.1 that allowed full use of a network were as follows:

File handles: An identifying number (handle) assigned by the PC operating system so that files can be located anywhere in the network. This leads to a new logical access for the files when they are stored on a file server.

File sharing: Allows programs to specify what kind of access mode they want to a file. This control allows the file to be set up as read-only or read/write.

Byte-level locking: Program can declare exclusive access to a range of bytes in a network file server and the network operating system to enforce that exclusive access.

Unique file names: The network operating system can be requested to set up unique file names so that the users do not ever use the same name.

Redirector support: Acts as the destination for all requests for remote logical drives. The program does not have to know if it is accessing a file on the local drive or on the network.

With these new features, MS-DOS 3.1 gave programmers the capability to write a standard set of programs for network services (interface). This has become the standard interface to the network industry.

7.4.1   Local Area Networks

As the name implies, a local area network (LAN) is geographically the smallest of the network topologies. In general, a LAN is limited to 1 to 2 miles in length and typically 10 to 16 Mbps. On some of the more advanced LANs, 100 Mbps is beginning to be used. Even with the advent of MS-DOS 3.1 and the network interface, PCs were designed to be computers, not communication devices, which presents some limitations in response time and difficulties in expansion. Three LAN topologies are in general use: the star, the bus, and the ring.

LAN Star Topology

The layout for the star topology is shown in Figure 7.5. The terminals are connected by lines branching from a central node. The central node is a single point of failure for the entire network. Expansion is relatively easy, but it may require a large amount of cable to service even a small number of terminals.

LAN Bus Topology

The layout for the bus topology is shown in Figure 7.6. Each of the terminals, or other devices, is connected to the bus (coaxial cable) via connectors. Expansion requires no rewiring, and no single node can bring down the network.

LAN Ring Topology

The ring topology is a point-to-point connection topology that forms a circle; see Figure 7.7. In ring topology, each of the nodes retransmits the signal. Control is distributed, but no single failure in the ring will bring the entire ring down. The ring must be broken to install new nodes.

7.4.2   Wide Area Networks

The wide area network (WAN) has the capability to allow you to connect your LAN to other LANs within your company or with LANs of vendors and customers. The speed of the WAN is determined by the communication media used for the long-distance transmission. At the low end of the speed regime is the modem using regular phone lines. For some applications this analog signal method will suffice. In most cases, however, if you have more than one terminal and wish to transmit large amounts of data in a reasonable time, you need more than just a modem. Most of the data transmitted over long distances is sent over digital carrier signal lines, such as the T1 system, originated by AT&T. There are many new digital data transmission systems available today, and the selection of the proper one can have a great influence on the performance and cost of your commuinication networks.

7.4.3 Network Throughput

One of the most important factors to consider in the design of manufacturing support systems is the amount of data to be transmitted over the network and the response time of the data transfer. The speed of the network will depend on the type of connection medium (copper wire, coax, or fiber optic), the distance to be covered, and the

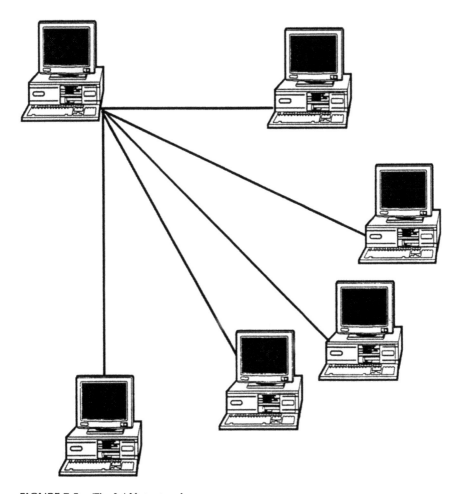

FIGURE 7.5 The LAN star topology.

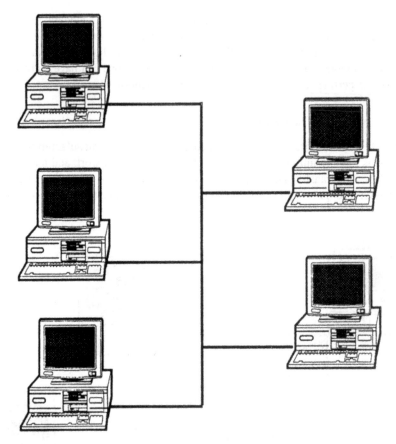

FIGURE 7.6   The LAN bus topology.

network software (Ethernet, Token Ring, Fiber Distributed Data Interface [FDDI], and Apple Talk). Text data does not take as much space as visual aids or photographs. For example, a book of a few hundred pages would be about 300,000 characters (bytes), while a full-color photograph (1024 × 768 pixels) would have 2,359,296 bytes. The transmission time of these two over a network would depend on the raw transmission data rate and the data throughput. The throughput is the limiting factor and is determined by the transport technology and the amount of overhead that must be added to transmit over the network (other data that describes and verifies the data that are sent). Using the Ethernet with a 10-bit data rate, the data rate is 1.25 MB, and with the overhead the effective throughput is 375,000 bytes/sec. Looking at the examples above, the 300,000-byte text file would be transmitted in less than 1 sec., and the 2.4-MB photograph would require about 6 sec. to transmit. Network loading and the advantages of where in the network various data should be stored are

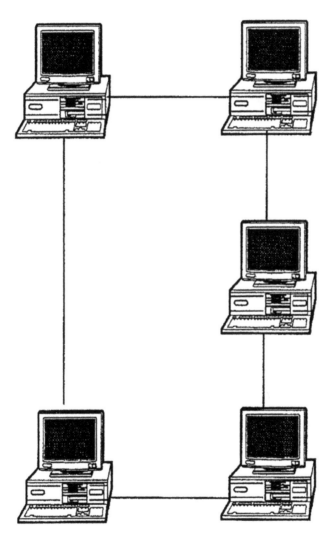

FIGURE 7.7   The LAN ring topology.

discussed in Subchapter 7.7. The following chart shows the data transfer capability of the various network transport products.

Type	In use	Proposed
Ethernet	10 Mbits/sec.	100 Mbits/sec.
Token Ring	4 and 16 Mbits/sec.	
FDDI	100 Mbits/sec.	
Apple Talk	115 Kbits/sec.	

The actual cabling that connects the various components on the network comes in many varieties. The three major types are:

Coaxial cable, where one or more concentric conductors are used
Twin twisted-pair cable, with two pairs of wires, each pair twisted together
Fiber-optic cable, used for FDDI and can be used with Ethernet

Coaxial and twisted-pair cables have been used for some time. The fiber-optic cable is fairly new, but this technology is expanding at a very high rate as all the phone lines are being changed over, and the price of this transmission medium is dropping to the point where the material is very competitive with wire. The cost of installation of fiber optics is still higher than for the other types. The expanded capabilities and the immunity of the fiber optics to electrical interference make it a good choice for many applications.

7.4.4 Network Operating Systems

A network requires an operating system to control and provide the various services to the network users. The operating system should also provide management and maintain functionality for the network operations group. The capability of the network operating systems is expanding at a rate where a new version of the software, with expanded capabilities, is released about once a year. The descriptions provided here will indicate some of the important characteristics of each of the software packages at the time of publication. As each vendor sees the need to incorporate new features, the software packages will become more alike than different.

Banyan VINES

VINES is a UNIX-based operating system from Banyan Systems designed to operate on Intel processors 386 or later. The major feature of this network operating system is the capability to operate multiservers. VINES characteristics are as follows.

Banyan's global naming scheme (GNS), Street Talk, is the premier GNS.
Management of multiple file servers is easy.
It has a built-in mail system.
Large numbers of users (over 100) can be handled on one server.

An unlimited number of files may be open at one time.
Security is provided.
Up to 10 printers are supported per file server.
It is compatible with many multiuser database management software products.
Auto alerts for the administrator on system performance are provided.
It supports Apple Filing Protocol.

Microsoft's LAN Manager

LAN Manager is an OS/2-based application. Basic features of LAN Manager are that it is a mouse-driven point-and-click interface and it is relatively inexpensive. LAN Manager's characteristics include:

Domain server style of global naming scheme
Direct queries to mainframe
OS/2 client-peer service
Autoreconnect when file server comes back online
Administration features
System security

7.4.5   Novell's NetWare

NetWare was designed to be a network operating system from the ground up. It is not designed to run under UNIX or OS/2. This approach gives Novell the advantage of control of the total software package, and therefore it can design in compatibility with a large array of third-party software packages and make the system compatible with all terminal operating systems. Features include:

Hardware independence
Extensive fault tolerance features
Supports OS/2, Named Pipes, and SQL products
Security features
Supports duplicate concurrent file servers for online operational backup
Applied programming interfaces (APIs)
TCP/IP implementation

7.4.6   Network Applications

The network for your computer system will provide various services and functions that were previously not available to your organization. The capability of moving data around in the organization is becoming indispensable. The use of network services for both software and hardware can be a large area for cost reduction. A network version of a software package such as Lotus 1-2-3 for ten concurrent users is more cost-effective than buying ten individual copies of the software. As many users as you have on the network can access the software, as long as no more than ten are using it at the same time. As your usage increases, the probability that a user will have to wait to use the software increases. The network administration can, at relatively small expense, add more users to the network pack when waiting reduces productivity. General office and manufacturing areas may also be served by a network printer. The network operating system will provide the service so that your printing will be stored, directed, queued, and printed on the printer for your area. This is also a good cost-reduction measure. Most services such as printers are not in use 100% of the time for a single user, so group usage makes a lot of sense.

Sharing Files

Through the network you have the capability to share files with anyone on the network who has access to the file under the security system. Access to various files can be controlled by the user profile set up on the network operating system. Once this capability, and the users' profiles, are set up, you can share work across the network. This is especially valuable when various users across the enterprise need to input or review the same document. This capability can also be used to merge files or documents. If only one organization has access to write to a file, and keeps it up to date, the rest of the users can access to read that data so that all of the organization is using the same data. One of the features the network operating system must have is a file naming module, so that no two files ever have the same name. If the file is a database, then all of the users will be inputting and using the most up-to-date online data in real time. This is required for a good MRP system to operate effectively.

Sharing Data

The sharing of data requires that the applications you are using are network compatible and control locking of data records in a file. Data is any information stored by an application (i.e., everything but the executable code). Sharing data gives your organization the opportunity to save various types of information after it has been generated, and then the whole organization can use it. This saves the regeneration of data and prevents the distribution and use of inconsistent data through the organization. In Subchapter 7.5 we discuss the advantages of inputting data into the system only once, with the system providing that data to all the elements of the system that require it.

7.5 COMPUTER-INTEGRATED ENTERPRISE ARCHITECTURE

Once all the elements (i.e., mainframes, minicomputers, PCs) and all the necessary hardware and software parts for the network were in place, the CIE became possible. In the mid-1980s, portions of the total system were available, and some manufactures tried to build computer-integrated manufacturing (CIM) systems. They were not very successful due to the fact that a tie between the planning system and the execution system was not available, especially on a real-time basis. Another problem, which has been the downfall of many good software systems, is the fact that users were not involved in the design, and when the time for implementation came they fought it, so the implementation of the system failed. Management of the enterprise must also embrace new management and organizational concepts to aid in removing walls between functional groups within the factory.

The CIE consists of a combination of various functional units to perform the desired tasks to operate the business. These include finance, purchasing, material control, manufacturing process control, and reporting. To perform these tasks the CIE must have certain capabilities. Processing, input and output terminals, printers, and an interconnecting network among the units make up the minimum capabilities. As technology progressed (see Subchapters 7.2 to 7.4), and the general use of computers became common in the 1950s, options for different architectures appeared. A typical architecture for a CIE system is shown in Figure 7.8. This example assumes that your organization has used computer systems for some time, and you have a mixture of types of hardware and legacy software programs in a multisite system. Medium to small, or new start-up, manufacturers have the luxury of designing a system from the ground up. See Figure 7.9 for an example of a new start-up system. With respect to multisite systems, as in Figure 7.10, the mainframe is at one location while other elements are at remote locations. The data is transmitted from location to location via hard-line (T1 lines, or other digital system) or in some cases by satellite. For the system to be a true CIE, and to be as productive as possible, the user should input data into the system only once. It is then up to the program logic to route and install the data into the correct tables and databases for processing and record keeping. Some of the processing will be handled at the local terminals and servers, while other data will be transmitted to the mainframe for processing. Examples of local processing might be CAPP for manufacturing, while the data from the manufacturing process (i.e., order status, labor charges, time keeping) would be collected locally and transmitted to the mainframe. One thing to remember is that, in a well-designed system, all applications should be available to the users at their local terminals. Timely information at the lowest level of the organization will allow your workers to make decisions and improve productivity. Correct, timely information is one of the most useful productivity improvements you can make.

7.6   PROCESS CONTROLLERS

Marc Plogstedt, CEO, ITEC, Orlando, Florida

7.6.0   Introduction to Process Controllers

Electromechanical and electronic process controls have been used since the 1940s. Early systems used relays for their logical elements. The installation and maintenance of the systems was done by electricians who were not computer experts. The program logic was known as "relay logic." Process controllers today still use relay logic programming. The process controllers became electronic devices in the mid-1970s. Many companies provide a complete line of equipment to meet almost any need.

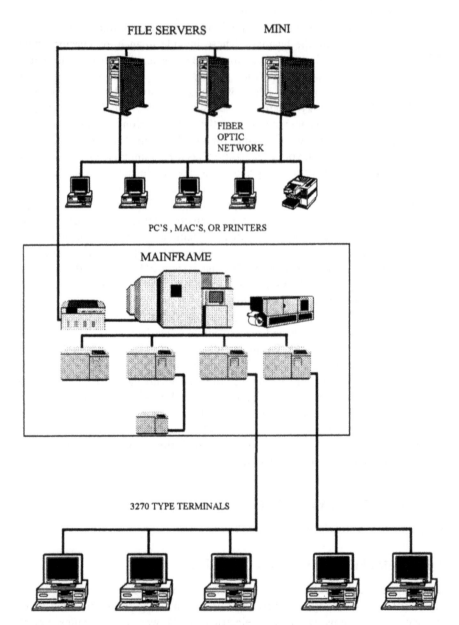

FIGURE 7.8 Computer-integrated enterprise (CIE) system architecture.

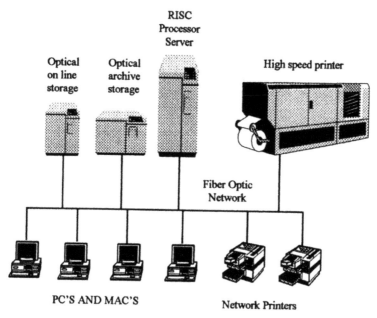

RISC
Processor
Server

Optical   Optical   High speed printer
on line   archive
storage   storage

Fiber Optic
Network

PC'S AND MAC'S   Network Printers

FIGURE 7.9   Start-up CIE system layout.

Controllers have evolved over the years from the simple devices of the early 1970s to the complex multifunctional devices of the 1990s. The first devices were operator-controlled; the operator would activate a simple on-off switch, and the signal from the switch would activate a power relay to operate a motor or other controlled device. The power relay was used to provide the high power required by the controlled device, while the relay power supplied by the on-off switch could be low-power and in most cases low-voltage for simplicity in wiring. As more and more of the operator's functions in the process were taken over by the process controller, the power controller (relay) was replaced by solid-state devices. This provided the system with devices which were more reliable and required less power from the process controller.

With the advent of process controllers that could control many devices anywhere on the factory floor, the positioning of the power controllers becomes more of an issue. If it is a long distance from the process controller to the controlled device, is it better to control all of the devices directly from the process controller, or should the power unit be controlled by a power module at the device—and the signals from the process controller be all that is sent over the long distance from the process controller to the power module? This configuration could simplify and lower the cost of the factory wiring required for the total system.

As more and more of the operator's functions are replaced by the process controller system, the system needs to be set up with a master process controller processor and then the feeding of information to slave or satellite processors positioned at various

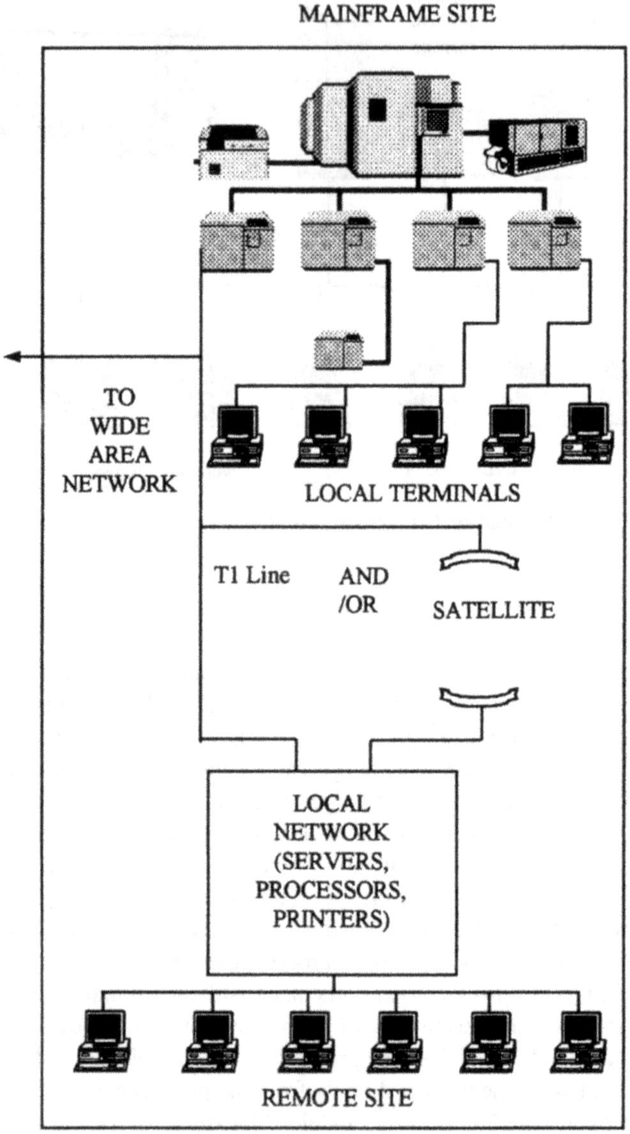

MAINFRAME SITE

TO
WIDE
AREA
NETWORK

LOCAL TERMINALS

T1 Line   AND
               /OR   SATELLITE

LOCAL
NETWORK
(SERVERS,
PROCESSORS,
PRINTERS)

REMOTE SITE

FIGURE 7.10   Multisite CIE system.

points on the factory floor could occur. This configuration could be required if the response of the controlled device is fast enough that the master processor controller cannot service the device at the desired rate.

Once the process to be controlled has been defined, selection of hardware and the system design layout must be made to fit your organization and requirements. If you already have process controllers, your company will probably want to use the same brand for ease of inventory, maintenance, and training. If you do not have any installed base of equipment, then the decision on brand will depend on the availability of your required control elements and the evaluation of cost and performance of the various systems.

In selecting a system, one of the major criteria is not necessarily the installed cost, but rather the reliability of the system and ease of maintenance to keep the process operating. A 1-day shutdown of the factory due to a problem with the control system could be more expensive than the total initial installation cost.

Safety is also very high priority. When a system is designed for a manufacturing process, the safety of the workforce is paramount. When a worker becomes involved with the controlled system, the system must sense any problem and shut down the process so that the worker will not be injured. An unusual example of the use of controllers is found in theme parks. Entertainment parks utilize process controllers to operate many of the rides. Since many people are involved in the controlled process, their safety is of paramount importance. Control of this kind of a process demands accuracy of many control functions at a very high rate. Since these types of systems usually are spread over a large area, multiple controllers with a connecting network are required for ride performance.

Another area for consideration is cost performance. Does your situation allow the setup of the control system in zones (possibly cheaper), or must you distribute the processors over the factory floor (higher performance)?

The following list includes some pertinent functions of process controllers:

An advanced instruction set including relay-type, timer and counter, math, data conversion, diagnostic, shift register, comparison, data transfer, sequence, immediate I/O, program control, and PID control

Use of sequential function chart to implement structured programming Storage of memory files on a disk drive so that you can archive them or download them to other processors

EEPROM backup to protect critical control programs.

Execution of a program file at repeated intervals with a selectable timed interrupt

Program fault routines to respond to system faults

Accessibility of date and time with the real-time clock/calendar

Storage of ladder program files on disk for printing on a serial or parallel printing device

A display of a data file starting with the requested address

Inserting, appending, changing, or deleting instructions or rungs without stopping the machine or process with online programming

Use of search functions to locate items within your program when the processor is in any mode of operation, or when the programming terminal is in off-line mode

Automatic switching from the primary processor to a backup processor when a fault is detected in an acceptable time frame (typically less than 50 msec.) with backup communications

In addition, selection of the type and performance of the specific controller application is dependent on how many I/O elements you need to scan and the maximum time between scans. Table 7.1 shows the specifications of one of the Allen-Bradley controllers series.

7.6.1   System Design and Implementation

Implementation and maintenance of a process control system includes the following phases:

Definition of the system requirements
Development of the total system concept
Selection of the controller hardware
Design of the controller system
Simulation of the controller system
Implementation of the controller system
Debugging and tuning of the controller system
Final acceptance and stress testing of the controller system
Technical and maintenance manuals

For the controller system to meet all of the above objectives, the ME must be the leader of the team that follows the process from beginning to final product. One of the major

TABLE 7.1   Specifications of a Typical Controller Series (Courtesy of Allen-Bradley)

Type	Base memory	I/O	Operating mode	Program scantime	I/O scan
PLC-5/12	6K words	16pt.-256 32pt.-512	Adapter only	8 msec/K words typ.	1 msec local, 7 msec remote
PLC-5/15	14K words	512	Scanner/ adapter	Same as above	Same as above
PLC-5/25	21K words	1,024	Scanner/ adapter	Same as above	Same as above
PLC-5/40	48K words	2,048	Scanner/ adapter	2 msec/K words	0.25 msec local, 7 msec remote
PLC-5/60	64K words	3,072	Scanner/ adapter	2 msec/K words	0.25 msec local, 7 msec remote

problems encountered in the development of the controller system is that the design and development tasks are accomplished by engineers and designers, without the experience and knowledge of the personnel who have used the machines or the elements that will be controlled by the controller system. On small projects, and if the ME has many years of experience in the development of similar systems, he or she may be the only knowledgeable member of the team. More likely, a small group of experienced personnel from several disciplines will be required to develop the system.

As the ME, at the beginning of the project you must determine how much time will be spent on the design and simulation of the final system, and how much time should be spent testing and refining the system with real hardware and machines. If you are designing a large system that will be used at multiple locations, more time should be spent in the design and simulation stages of the process. At some point, however, the real hardware, code, and machines must come into the process for final refinement. The tools provided by the supplier of the controller hardware should be one of the major elements in hardware selection. The job of the design team can be made much easier if the tools are robust and user friendly. Development tools that allow the designer to test new code on the real system, online, is one of the most important elements, and will reduce the total development time for the project. The following suggestions may prove helpful:

Place the high power controllers as close as practical to the controlled device.
Follow rules on cable and raceway installation.
Always plan spares in the cable, and I/O at each rack.

Cable Installation

Conductor Category 1 consists of high-power AC I/O lines and high-power DC I/O lines connected to DC I/O modules rated for high power or high noise rejection. These lines can be routed with machine power up to 600 VAC (feeding up to 100-kp devices). Check local codes. Article 300-3 of the National Electrical Code requires all conductors (AC and/or DC) in the same raceway to be insulated for the highest voltage applied to any of the conductors.

Conductor Category 2 consists of serial communication cables, low-power DC I/O lines connected to proximity switches, photoelectric sensors, encoders, motion-control devices, and analog devices. Low-power AC-DC I/O lines connected to I/O modules rated for low power such as low-power contact-output modules are included. All conductors must be properly shielded, where applicable, and routed in a separate raceway. If conductors must cross power feed lines, they should do so at right angles. Route 1 ft. from 120 VAC, 2 ft. from 240 VAC, and 3 ft. from 480 VAC. Route at least 3 ft. from any electric motors, transformers, rectifiers, generators, arc welders, induction furnaces, or sources of microwave radiation.

Conductor Category 3 interconnects the PLC components within an enclosure. For PLC power cables, provide backplane power to DPLC components. This category includes processor peripheral cables, connecting processors to their communication interface modules. All conductors should be routed external to all raceways, or in a raceway separate from any category 1 or 2 conductors.

The previous information illustrates some of the more important design criteria for process controllers. The specific requirements will be unique to your process. The information will help guide you through the design and evaluation process to select the right hardware and lay out the best system for your application. Keeping in mind that one of the most important requirements for a processor control system is ease of maintenance, you can provide, if desired, a terminal for use in the operations/maintenance center. This terminal can link into any of the processors on the network for system operation or maintenance. The design should also provide connections for portable terminals and annunciators at all racks for maintenance.

System Issues

1. *Environmental.* Hazardous, wet, or corrosive atmospheres will require the controller to be mounted in a sealed NEMA enclosure (see the appendix for a listing of enclosure manufacturers). Depending on the ambient temperature range of the enclosure location, you may need local cooling for the long-term reliability of the controller. One way to reduce the internal temperature of the sealed enclosure is by the use of heat exchangers or a solid-state cooler. The only moving parts in these devices are the external and internal circulation fans. By controlling the polarity of the current through the solid-state device, the unit can be used to remove or add heat to the enclosure.

2. *Lightning protection.* Lightning protection of data lines and network lines may be needed depending on the geographic location of the plant and the exposure of the lines to the elements. If the lines are near the exterior walls of the building or go outside, it is necessary to protect both ends of the data lines with lightning protectors. Small telephone-type protectors will work in most cases.

3. *Electrical noise.* Most factories have a number of rotating machines (motors) and controllers. These devices produce a large amount of electrical noise. If your process is of a critical nature, then noise in the system could be very costly. One of the best ways to eliminate the effects of electrical noise in the process controller system is by the use of fiber optics for transmission of data from rack to rack. This is more expensive in the short term but could be very cost-effective for the smooth, uninterrupted operation of your process. The use of fiber-optic cable can also be useful for a building-to-building network in a lightning-prone environment.

7.7   HUMAN INTERFACE TO THE COMPUTER-INTEGRATED ENTERPRISE

7.7.1   Equipment Selection

The performance of the user, utilizing the terminal interface to the CIE, is very important in providing him or her with timely response and accurate, up-to-date information from the CIE (see also Subchapter 5.1). User requirements will differ between the office environment and the factory environment. Most office users should have access to the integrated databases of the CIE, and will require a terminal (PC) with reasonable performance in the CPU and the graphics card. If the office user is using an analysis package or a CAD package, a coprocessor will probably be required. The performance of the factory (work instructions) units' performance will be determined by how much multimedia processing the units will be required to accomplish, and where that data is stored. A good policy to follow when selecting terminal hardware (PCs) is to test competing PCs with benchmark programs. We found a large variance in the performance of PCs that had the same processors (over 3 to 1). Also, the price of the hardware is not a good indication of the performance. In our case, the lowest-cost unit gave the best performance. When selecting the benchmark software, use the ones that are designed for your operating environment (DOS, Windows, OS/2, or UNIX).

7.7.2 Computer-Aided Process Planning

The generation and electronic display of CAPP work instructions are of great importance to the manufacturing engineer. The effectiveness of the interface will determine in great part the productivity improvements for both the author and the user of the planning. The acceptance level by the users of any new system will depend on their involvement during the design. To provide an effective interface, in a timely, cost-effective manner, you, along with the application developer, should use a rapid development process that includes the users. The planning author and the end user can provide insight into the use of the development software and the interface if they are involved in the development of a new system from the beginning. In most cases, the user interface will be some form of GUI. You may, for your application, use a multimedia GUI. Many development tools are available to design a working model (prototype) with limited capabilities in a very short time (weeks). This prototype can be used to check the look and feel of the interface. As the development progresses and the interface becomes more refined, the user should be brought in from time to time to check the progress. During the rapid development process, you and the application developer must be very careful not to include system errors in the design and then move them into production. These errors are most likely to be in the area of data relationship integrity. Another pitfall to avoid is trying to use too many application developers. Almost all first-class software is developed by a small, dedicated group. All of the members of the group must understand that it is their responsibility to design, code, debug, and document their part of the program.

Providing process planning electronically to the shop floor worker gives you the opportunity to incorporate other features into the process that are not available in a paper system. When your product or process requires that certain employee skills, tools, and/or materials must be certified to build the product, the interface, planning, and a relational database can be used for this control. This will prevent the use of uncertified employees, tools, and materials before they are used in the product, eliminating the need for rework. The control of the process will also provide a complete "as built" configuration of the product.

When you are involved in the design and development of a shop floor terminal (SFT) for the presentation of work instructions for a manufacturing process, the following are major issues that you and the development team should consider.

Is the software package a standard (generic) application, or a package designed for your application?

Are you going to use a keyboard, mouse, or touch screen for the interface (look at the merits and cost of each)?

Will the GUI allow the user only to navigate through the application under the control of the application?

How long would you expect the SFT to provide services with the network down? This will help determine how the work instructions are delivered to the SFT, and how much is stored on the SFT.

What level of education (training) does your workforce have?

How much time are you willing to utilize to train the workforce?

Should all of your displays meet the Swedish MRP I and II specification on low-frequency radiation? If not, why not? It could be a liability issue in the future.

As an example, Figure 7.11 shows the navigation flow diagram of a GUI that will display CAPP (multimedia). The process incorporated into this SFT provides for all of the functions required to operate and manage a work center. The worker logs onto the SFT at the beginning of the shift; the system compares the clock number and PIN for identification. If any problem occurs, the worker must get the supervisor to resolve it prior to start of a job. When he or she logs on for the first job, a list of the jobs available for that work center are shown in priority order. After selecting a job, the system checks to make sure the skill/certifications required for the job are met by that employee. The plan for the job is then presented to the worker on an integrated screen. It contains the text, the visual aids, and the navigation buttons. This SFT utilizes touch screens. Proceeding through the work instructions, as material or product information is required, it is wanded into the SFT. The system checks to make sure that the material is the correct material for this configuration, and/or that the measured value on the required test information is within limits. All this information is stored in the relational database, and will be the "as built" information for this item.

You must have at least a rudimentary flow diagram prior to setting up the prototype GUI. As you work with the prototype and the user community, expand and refine the flow diagram until you have the final requirements defined. Spending a few weeks working with a group of users and the logic developers is a good investment. If this phase of the process is carried out correctly, the final production version of the GUI will require fewer revisions. After following this procedure, we needed to add only two revisions to the GUI during the first year of use. The total development time for the initial version of the GUI was only 3 months. Your application may look

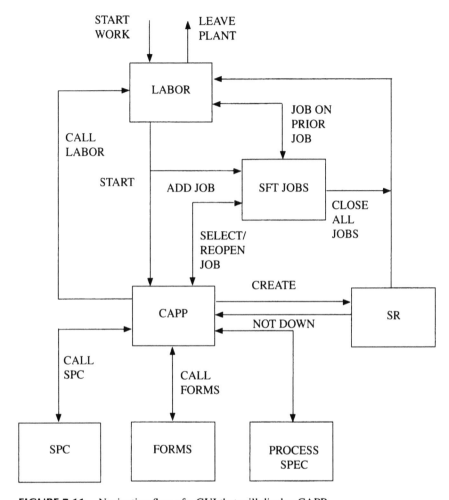

FIGURE 7.11 Navigation flow of a GUI that will display CAPP.

totally different from this example. It is highly dependent on the type of products you build, and the education and turnover of your workers. The workers using this GUI were able to fully utilize the system after less than 1 hr. of training, and were up to full productivity in less than 1 week.

A major consideration in the design of the SFT with a GUI is where and how much data will be stored. With any network system, you must expect the system to be down from time to time. If the production workers are to be provided all the services of the SFT, decisions on where the information is stored must be made. The critical factors are how long the network could be down and how soon new jobs can be expected at the SFT workstation. Another issue to consider, if you are using visual aids to support the work instructions, is the size of the files and when you will transmit them over the network. If your production cycle will allow, the SFTs could be updated at night when the network is not so busy. This can be accomplished automatically by setting up a data transfer log; at the appointed time, the data will be transmitted to the SFTs for local storage. If the various SFTs do not all support the same work content, each of them could be provided with only the jobs accomplished at that work center, to minimize local storage requirements. This functionality was incorporated into the example system described above. The SFTs should have the capability to command a data transfer if a process plan is changed and is required immediately.

7.8   CAD/CAM

7.8.0   Introduction to CAD/CAM

The computer has been used for some time in computer drafting. The capability to produce high-quality drawings by the computer has been utilized in architecture and engineering for many years. This capability has also been used for electrical drawings for wiring. The use of the computer to support computer-aided design/ computer-aided manufacturing (CAD/CAM) of products began in the late 1950s. P.J. Hanratty was one of the pioneers in the field of CAD/CAM. He did a lot of his earliest work for General Motors and McDonnell Douglas. Over the last 30 years, CAD/CAM software products have matured into a marketplace of over $3 billion per year. The software now has the capability to do not only the analysis on the structure using tools such as Nastran, but has at least the beginnings of solid modeling. One of the most exciting newer developments in the world of CAM is the use of simulation to test a process flow prior to building the factory. The optimization of the process and the improvements in process flow can be tuned during the simulation testing along with a graphical demonstration of the process to management. The use of simulation can save large amounts of time and money during development of the process or factory.

From 1984 through 1994, a consortium of industrial, university, and research centers across Europe set up the European Strategic Program for Research and Development in Information Technology (ESPRIT). The output of this research is a set of interface requirements to facilitate the interconnection of the various modules of CAD/CAM systems. Most of the manufacturing software systems in Europe have been designed and built in this environment. Indications are that the European design environment will be more directed to the CAD/CAM systems than the U.S. environment, and many of the new programs are coming from Europe.

The capability and the complexity of the CAD/CAM software packages from the various vendors covers the spectrum from simple to very complex. The price per seat also varies from a few hundred dollars to tens of thousands of dollars per seat. The software package best suited for your application will depend on how you are intending to use it and for what purpose. The major benefits you can realize from an integrated CAD/CAM software package will occur only when your development cycle process changes to utilize the capabilities of the software. If you try to use these software packages without this shift in the processes, you will not see the major benefits. Some of the major suppliers of CAD/CAM software are:

Applicon	Bravo
Autodesk	Autocad Designer
Computervision	CADDS 5
Dassualt Systems	Catia Solutions
EDS Unigraphics	Unigraphics
Hewlett Packard	HP PE Solid Designer
Intergraph	I/EMS
Manufacturing and Consulting Services	Anvil-5000
Mechanical Advantage	Cognition
Parametric Technology	Pro/Engineer
Structural Dynamics Research Corporation	I-Deries

Computer modeling is one of the more dynamic areas of development in the CAD/CAM software world. To utilize this capability, the process in development of a product as mentioned above must change. The process of development of a new product, in general terms, follows.

7.8.1   General Problem-Solving Principles

To resolve all of the complex issues in a new product design, the following process steps of refinement activities must occur over and over again:

Finding a mental solution
Representing the solution
Analyzing the solution

When this process has proceeded to the point of decision, the final solution is compared to the requirements, the solution evaluated against those requirements, and a final decision made on the product. If the design does not meet the requirements, then the process must continue until the requirements are met. The above process will be used during all phases of the design process:

Definition of functions
Definition of physical principles
Preliminary design
Detail design
Prototype (if desired)
Production planning

The results obtained at the end of each phase of the design are the basis for the following phase. The result of the function definition phase is an abstract description of the product. This abstract product description is a function structure that describes how an overall function of the product is realized in terms of a combination of sub-functions. To realize the design, a solution principle must be found for each of the elements in the functional structure.

The definition of the initial shape characteristics during the preliminary design uses the physical principles. A complete design scheme is the result of the pre-liminary design. The documents generated during the detailed design are the basis for the NC programming, process planning, parts lists, and other documentation for building the product. As the total design matures, the design team might return to a previous step in the process due to further definition or other information that has become available during the design phase. Some of the decisions made in previous steps in the process may have to be modified with the changes in the detailed design caused by those changes. (See Chapters 1 and 2 for more on product development and product design.)

7.8.2   Modeling

One of the most powerful tools for the designer in the newer CAD/CAM software packages is the capability to do 2D and 3D modeling. The various packages from the vendors use some form of parametric or variational analysis, or a combination of both for design. The generation of the models depends on the modeling modes utilized. All modeling packages have some overlapping characteristics that allow the designer to show how changing a variable will reconfigure the graphical model. The major types of models are:

Explicit modeling. The geometry of the design is built without any relation-ships between entities. Some software packages combine wire-frame, sur-faces, and solid modeling in a single integrated package. Changes in an explicit model require the designer to redesign for that change, and may entail large changes due to effect on the total requirements.

Variable-driven modeling. The design intent of the product is captured during the product's definition by capturing the relationships that exist within the model and the processes that define the model. By utilizing the intent of the design rather than the explicit geometry of the object, the designer will be able to edit and/or modify the design for design alternatives or revisions.

7.9   EXTENDED ENTERPRISE

For a manufacturing organization to fully utilize the capabilities of automation, it must understand the goals and capabilities of the extended enterprise. First, what do we mean by the extended enterprise? Referring to Figure 7.12, you can see the gen-eral interflow of information among customers, vendors, and your internal operations. Normally, the CIE is your corporation's computer systems. The extended enterprise

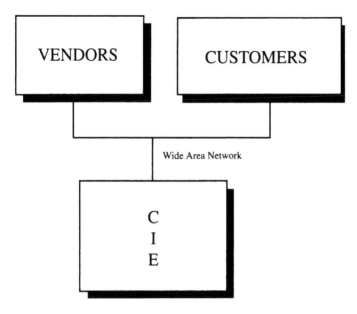

FIGURE 7.12 Extended enterprise information flow.

INTERNET	BANYAN VINES	OSI	TCP / IP					
5 Layer Model		7 Layer Model						
APPLICATION		APPLICATION	KERBOS	XWIN	SMTP	+FTP	TFTP NCS SNMP	
	DOS/WIN/OS2	PRESENTATION						
		SESSION						
TRANSPORT	VINES IPS.SPP	TRANSPORT	TCP				UDP	
INTERNET	VINES IP	NETWORK	IP	ICMP	ARP		RARP	
NETWORK INTERFACE	ETHERNET TOKENRING ARCNET NDIS X25 BSC ASYNC SDLC HDLC	DATALINK	LOGICAL LINK CONTROL MEDIA ACCESS CONNECTION					DRIVER ADAPTER 802.3 ETHERNST 802.4 TOKEN BUSS 802.5 TOKEN RING 802.6 DIX THICKNET DIX+802.3 APPLETALK ARCNET FDDI SDLC BSC HDLC ISDN
HARDWARE	SAME AS LIST AT FAR RIGHT ⟶	PHYSICAL						UNSHIELDED TWISTED PAIR SHIELDED TWISTED PAIR COAX THIN-NET RG58 IBM TYPE 1 IBM TYPE 3 FIBER 625 MICRON FIBER 100/50 MICRON RS232C SATELLTE HYPERCHANNEL RADIO V.35

FIGURE 7.13 Network protocols.

FTAM / ACSE	FTAM / ACSE	FTAM / ACSE	FTAM / ACSE	FTAM / ACSE	X.400	X.400	X.400	X.400	X.400	X.400	X.400
PRESENT*	PRESENT*	PRESENT*	PRESENT*	PRESENT*	SESSION	SESSION	SESSION	SESSION	SESSION	SESSION	SESSION
SESSION	SESSION	SESSION	SESSION	SESSION	TP4	TP4	TP4	TP4	TP4	TPO++	TPO++
TP4	TP4	TP4	TP4	TP4	CLNP	CLNP	CLNP	CLNP	CLNP	X.25	X.25
CLNP	CLNP	CLNP	CLNP	CLNP	LLC	LLC	LLC	X.25	X.25	HDLC	HDLC
LLC	LLC	LLC	X.25	X.25	802.3	802.4	802.5	HDLC	HDLC	RS232C	V.35
802.3	802.4	802.5	HDLC	HDLC				V.35	RS232C		
			V.35	RS232C							

TRANSPORT PLATFORM (ALIAS OSI BRIDGES)

THREE LAYER (ALIAS OSI ROUTER)

PRESENTATION* LAYER ——— ASN.1
** ISO CONNECTION ORIENTED NETWORK SERVER (CONS)USESX.25 PROTOCOL AT THIS LAYER

ISO 8473 CONNECTIONLESS NETWORK PROTOCOL
ISO 9542 END SYSTEM TO INTERMEDIATE SYSTEM ROUTING PROTOCOL
ISO 8208 X.25 PACKET LEVEL PROTOCOL FOR DATA TERMINAL EQUIPMENT
ISO 8802.2 LOCAL AREA NETWORK LOGICAL LINK CONTROL(LLC)
ISO 8202.3 LOCAL AREA NETWORK CSMA/CD-ACCESS METHOD AND PHYSICAL LAYER SPEC
ISO 8802.5 LOCAL AREA TOKEN RING ACCESS METHOD AND PHYSICAL LAYER SPEC

FIGURE 7.14   End system profiles (GOSIP Version 1).

is the shell surrounding the CIE to interconnect your systems to vendors and customers. The actual use of EDI will depend on the type of business you are in. If you have a standard line of products, your customer's order will trigger the shipment of the product to the customer. In a large number of cases when you have standard products, the customer will expect quick delivery of the order. To be able to accomplish quick delivery, you must stock a reasonable number of each product or have a very responsive manufacturing flow that will produce the product in a very short time.

7.9.1 Transmission Protocol

Transmission protocols are a set of standards that allow various computers to communicate with each other and control the transmission of data. The correct protocol must be selected for the type of transmission that will be made when you have set up an EDI procedure for your requirements. Some of the major issues that you and your implementer must resolve are: electronic signoff of POs, transfer of funds, and an audit trail of all business activities. The network protocols are shown in Figure 7.13, and system profiles (GOSIP Version 1) are shown in Figure 7.14.

Index

A

Absorption, chemicals, 341
Abuses of compensation system, 322
Academy of Denmark, 161
Accident history, 334
Accountability, industrial safety,
    328–329, *329*
Accounting systems, 218–219
Accuracy *vs.* cost, machining, 68, 71,
    72–74, 74
Aches, body, 314–316, *316*
Acquisition of equipment, 205–209
Acquisition process, concurrent engineer-
    ing, 12
Adjacencies, building program, 237–239,
    238–239
Advantages and disadvantages
    centralized stores, 197–201
    continuous flow manufacturing, *200*, 203
    decentralized stores, 201–202
    rapid prototyping, 46
Airborne contaminants, 341
Akao (company), 17
Alcoa, 17
Allen-Bradley (company), 17, 374
Alting, Leo, 58
Aluminum alloys, forging, 96
American Society of Heating,
    Refrigeration, and Air Conditioning
    Engineers (ASHRAE), 339
American Society of Mechanical Engineers
    (ASME), 174
Americans with Disabilities Act of 1990, 282
AMP, Inc., 31n
Analysis results, DFA, 41–43, *42–45*
Andrew Linemaster Electrical Discharge
    Machine, 55
Angle of ascent, *307*

Animal life, 252–253
Anthropometry
    dimensions, body, 288, *289–291,* 292
    human elements, 285–298, *288*
    NIOSH lifting equation, 292–293,
        295–296, *297*
    range of motion, 298, *300–302*
    strengths, body, 292, *293–296*
    visual field, 298
Apple, personal computers, 355
Applications
    factory ergonomics, 299–301, *303–307,*
        304–311
    networks, 367–368
    office ergonomics, 312–316, *315–317*
    quality function deployment, 17
Architecture
    computer-integrated enterprise, 368–369,
        370–372
    project design, 255–257
Aria, Ueda, Markus, Monostori, Kals and,
    studies, 158
Aristotle, 221
Arm strength, *294*
Armstrong, Paul, 52
Armstrong Mold Company
    castable urethane, 51
    file formats, 46
    plaster molds, 47
    reaction injection molding, 52
    stereolithography, 46
Army Corps of Engineers, 316
Army Research Institute of Environmental
    Medicine, 292
Ary, Jacobs and Razavieh studies, 150–151
Ascent angle, *307*
ASHRAE, *see* American Society of
    Heating, Refrigeration, and Air
    Conditioning Engineers (ASHRAE)